AF595061

Remote Sensing Applications in Coastal Environment

Remote Sensing Applications in Coastal Environment

Editors

Paweł Terefenko
Jacek Lubczonek
Dominik Paprotny

MDPI • Basel • Beijing • Wuhan • Barcelona • Belgrade • Manchester • Tokyo • Cluj • Tianjin

Editors

Paweł Terefenko
Institute of Marine and
Environmental Sciences
University of Szczecin
Szczecin
Poland

Jacek Lubczonek
Department of Navigation
Maritime University of Szczecin
Szczecin
Poland

Dominik Paprotny
Research Department
Transformation Pathways
Potsdam Institute for Climate
Impact Research
Potsdam
Germany

Editorial Office
MDPI
St. Alban-Anlage 66
4052 Basel, Switzerland

This is a reprint of articles from the Special Issue published online in the open access journal *Remote Sensing* (ISSN 2072-4292) (available at: www.mdpi.com/journal/remotesensing/special_issues/CoastalEnviron_RS).

For citation purposes, cite each article independently as indicated on the article page online and as indicated below:

LastName, A.A.; LastName, B.B.; LastName, C.C. Article Title. *Journal Name* **Year**, *Volume Number*, Page Range.

ISBN 978-3-0365-2613-3 (Hbk)
ISBN 978-3-0365-2612-6 (PDF)

Contents

MDPI

Editorial

Editorial on Special Issue "Remote Sensing Applications in Coastal Environment"

Paweł Terefenko [1,*], Jacek Lubczonek [2] and Dominik Paprotny [3]

1 Institute of Marine and Environmental Sciences, University of Szczecin, 70-383 Szczecin, Poland
2 Institute of Geoinformatics, Department of Navigation, Maritime University of Szczecin, 70-500 Szczecin, Poland; j.lubczonek@am.szczecin.pl
3 Research Department Transformation Pathways, Potsdam Institute for Climate Impact Research, Telegrafenberg, 14473 Potsdam, Germany; paprotny@pik-potsdam.de
* Correspondence: pawel.terefenko@usz.edu.pl; Tel.: +48-914442354

Coastal regions are susceptible to rapid changes as they constitute the boundary between the land and the sea. The resilience of a particular segment of coast depends on many factors, including climate change, sea-level changes, natural and technological hazards, extraction of natural resources, population growth, and tourism [1]. Recent research highlights the strong capabilities for remote sensing applications to monitor, inventory, and analyze the coastal environment [2,3]. This Special Issue contains 12 high-quality and innovative scientific papers that explore, evaluate, and implement the use of remote sensing sensors within both natural and built coastal environments.

Interaction between land subsidence and sea level rise (SLR) increases the hazard in coastal areas, mainly for deltas, which are characterized by flat topography and great social, ecological, and economic value. Coastal areas need continuous monitoring as a support for human interventions aimed at reducing hazards. In Fabris [4], a contribution to the understanding of the future scenarios based on the morphological changes that occurred in the last century on the Po River Delta (PRD, northern Italy) coastal area is provided. Planimetric variations are reconstructed using archival cartographies, multi-temporal high-resolution aerial photogrammetric surveys, and Light Detection and Ranging (LiDAR) datasets. The results, in terms of emerged surface variations, are linked to the available land subsidence rates and to the expected SLR values to obtain projections of changes until 2100.

Citation: Terefenko, P.; Lubczonek, J.; Paprotny, D. Editorial on Special Issue "Remote Sensing Applications in Coastal Environment". *Remote Sens.* **2021**, *13*, 4734. https://doi.org/10.3390/rs13234734

Received: 11 November 2021

[illegible] with regard to jurisdictional claims in published maps and institutional affiliations.

Intermittently closed and open lakes or Lagoons (ICOLLs) are characterized by en[illegible] barriers that form or break down due to the actions of wind, waves, and currents [illegible] with an edge computing [illegible] efficient way to automatically survey the lagoon entrance and estimate the berm profile. Additionally, it estimates the dry notch location and its height, which are critical factors in the management of the lagoon entrances. Data generated by the study provide valuable insights into landscape evolution and berm behavior during natural and mechanical breach events.

Detecting land cover changes requires timely and accurate information, which can be assured by using remotely sensed data and geographic information systems. Bielecka et al. [6] combine these to examine spatiotemporal trends in land cover transitions in the Polish coastal zone of the Baltic Sea, especially urbanization, loss of agricultural land, afforestation, and deforestation. The dynamics of land cover change and its impact were discussed as the major findings of the study.

More detailed spatial and temporal variations in the dune areas of the Pomeranian Bay coast (southern Baltic Sea) were quantified using remote sensing data from the years 1938–2017, supervised classification, and a geographic information system post-classification change detection technique by Giza et al. [7]. The aim of this work was to fill the gap in spatiotemporal analyses of land cover transitions in the Polish coastal zone and, moreover, present a method for assessing indicators of changes in a coastal dune environment that could be an alternative for widely used morphological line indicators. Finally, a novel quantitative approach for coastal areas containing both sea and land surface sections was developed.

Coastal dunes are found at the boundary between continents and seas and represent unique transitional mosaics hosting highly dynamic habitats undergoing substantial seasonal changes. Marzialetti et al. [8] implemented a land cover classification approach specifically designed for coastal landscapes, accounting for the within-year temporal variability of the main components of the coastal mosaic: vegetation, bare surfaces, and water surfaces. Utilizing monthly Sentinel-2 satellite images from 2019, hierarchical clustering, and a Random Forest model, an unsupervised land cover map of coastal dunes in a representative site of the Adriatic coast (central Italy) was produced.

Surface moisture plays a key role in limiting aeolian transport on sandy beaches. However, the existing measurement techniques cannot adequately characterize the spatial and temporal distribution of beach surface moisture. Jin et al. [9] demonstrate mobile terrestrial LiDAR as a promising method to detect beach surface moisture using a phase-based laser scanner mounted on an all-terrain vehicle. Finally, a moisture estimation model was developed that eliminated the effects of the incidence angle and distance. The results show that the MTL is a highly suitable technique to accurately and robustly measure the surface moisture variations on a sandy beach with an ultra-high spatial resolution (centimeter-level) in a short time span.

LiDAR surveys are also widely used for gathering datasets to analyze coastal morphology. Zelaya Wziątek et al. [10] made a study of the volumetric changes in cliff profiles, spatial distribution of erosion, and rate of cliff retreat corresponding to the cliff exposure and rock resistance of the Jasmund National Park chalk cliffs in Rugen, Germany. The study combined multi-temporal LiDAR data analyses with rock sampling, laboratory analyses of chemical and mechanical resistance, and along-shore wave power flux estimation. The rate of retreat for each cliff–beach profile, including the cliff crest, vertical cliff base, and cliff base with talus material, indicates that wave action is the dominant erosive force in areas where the cliff was eroded quickly at equal rates along the cliff profile.

Analyzes of coastal retreat due to strong winter storms have also been carried out by de Sanjosé Blasco et al. [11] for the Cantabrian coast. Different geomatic techniques, such as: orthophotography, photogrammetric flights, LiDAR surveys, Unmanned Aerial Vehicle (UAV) surveys, and terrestrial laser scanner datasets, were used to find volumetric differences in the beach and sea cliff, attributing them to storms. From the results of this investigation, it can be concluded that the retreat of the base of the cliff is insignificant, but this is not the case for the top of the cliff and for the existing beaches in the Cantabrian Sea, where the retreat is evident.

Water areas occupy over 70 percent of the Earth's surface and are constantly subject to research and analysis. Often, hydrographic remote sensors are used for such research, which allow for the collection of information on the shape of the water area bottom and the objects located on it. Information regarding the quality and reliability of the depth data is important, especially during coastal modelling. In-shore areas are liable to continuous transformations, and they must be monitored and analyzed. Presently, bathymetric geodata are usually collected via modern hydrographic systems and comprise very large data point sequences that must then be connected using long and laborious processing sequences, including reduction. As existing bathymetric data reduction methods utilize interpolated values, there is a clear requirement to search for new solutions. Considering the accuracy of bathymetric maps, a new method that preserves real geodata has been presented by Wlodarczyk-Sielicka et al. [12]. This study specifically highlights how to reduce position and depth geodata while maintaining true survey values.

Advances in remote sensing technology have facilitated quick capture and identification of the source and location of oil spills in water bodies, yet the presence of other biogenic elements (lookalikes) with similar visual attributes hinder rapid detection and prompt decision-making for emergency response. To date, different methods have been applied to distinguish oil spills from lookalikes, with limited success. In addition, accurately modeling the trajectory of oil spills remains a challenge. Temitope Yekeen et al. [13] provide further insights on this multi-faceted problem by undertaking a holistic review of past and current approaches to marine oil spill disaster reduction. The scope of previous reviews is extended by covering the inter-related dimensions of detection, discrimination, and trajectory prediction of oil spills for vulnerability assessments.

Coastal upwelling involves an upward movement of deeper, usually colder, water to the surface. Satellite sea surface temperature (SST) observations and simulations with a hydrodynamic model show, however, that the coastal upwelling in the Baltic Sea in winter can bring warmer water to the surface. In a study by Kowalewska-Kalkowska [14], satellite SST data collected by the advanced very high-resolution radiometer (AVHRR) and the moderate-resolution imaging spectroradiometer (MODIS), as well as simulations with a three-dimensional hydrodynamic model of the Baltic Sea, were used to identify upwelling events in the southern Baltic Sea during the 2010–2017 winter seasons.

Urban expansion is one of the most dramatic forms of land transformation in the world and it is one of the greatest challenges in achieving sustainable development in the 21st century. Previous studies have analyzed urbanization patterns in areas with rapid urban expansion, while urban areas with low to moderate expansion have been overlooked, especially in developed countries. In his study, Rifat et al. [15] examined the spatiotemporal dynamics of urban expansion patterns in southern Florida (United States) over the last 25 years (1992–2016) using remote sensing and GIS techniques. The main goal of this paper was to investigate the degree and spatiotemporal patterns of urban expansion at different administrative level in the study area and how spatiotemporal variance in different explanatory factors influence urban expansion.

Funding: This research received no external funding.

Institutional Review Board Statement: Not applicable.

Informed Consent Statement: Not applicable.

Data Availability Statement: Not applicable.

Acknowledgments. The guest editors would like to thank the authors who contributed to this Special Issue and to the reviewers who dedicated their time for providing the authors with valuable and

References

1. Vousdoukas, M.I.; Ranasinghe, R.; Mentaschi, L.; Plomaritis, T.A.; Athanasiou, P.; [illegible] threat of erosion. *Nat. Clim. Chang.* **2020**, *10*, 260–263. [CrossRef]
2. Terefenko, P.; Zelaya Wziatek, D.; Dalyot, S.; Boski, T.; Pinheiro Lima-Filho, F. A high-precision LiDAR-based method for surveying and classifying coastal notches. *ISPRS Int. J. Geo Inf.* **2018**, *7*, 295. [CrossRef]
3. Paprotny, D.; Terefenko, P.; Giza, A.; Czapliński, P.; Vousdoukas, M.I. Future losses of ecosystem services due to coastal erosion in Europe. *Sci. Total Environ.* **2021**, *760*, 144310. [CrossRef]
4. Fabris, M. Monitoring the Coastal Changes of the Po River Delta (Northern Italy) since 1911 Using Archival Cartography, Multi-Temporal Aerial Photogrammetry and LiDAR Data: Implications for Coastline Changes in 2100 A.D. *Remote Sens.* **2021**, *13*, 529. [CrossRef]
5. Arshad, B.; Barthelemy, J.; Perez, P. Autonomous Lidar-Based Monitoring of Coastal Lagoon Entrances. *Remote Sens.* **2021**, *13*, 1320. [CrossRef]
6. Bielecka, E.; Jenerowicz, A.; Pokonieczny, K.; Borkowska, S. Land Cover Changes and Flows in the Polish Baltic Coastal Zone: A Qualitative and Quantitative Approach. *Remote Sens.* **2020**, *12*, 2088. [CrossRef]
7. Giza, A.; Terefenko, P.; Komorowski, T.; Czapliński, P. Determining Long-Term Land Cover Dynamics in the South Baltic Coastal Zone from Historical Aerial Photographs. *Remote Sens.* **2021**, *13*, 1068. [CrossRef]

8. Marzialetti, F.; Di Febbraro, M.; Malavasi, M.; Giulio, S.; Acosta, A.T.R.; Carranza, M.L. Mapping Coastal Dune Landscape through Spectral Rao's Q Temporal Diversity. *Remote Sens.* **2020**, *12*, 2315. [CrossRef]
9. Jin, J.; De Sloover, L.; Verbeurgt, J.; Stal, C.; Deruyter, G.; Montreuil, A.-L.; De Maeyer, P.; De Wulf, A. Measuring Surface Moisture on a Sandy Beach based on Corrected Intensity Data of a Mobile Terrestrial LiDAR. *Remote Sens.* **2020**, *12*, 209. [CrossRef]
10. Zelaya Wziątek, D.; Terefenko, P.; Kurylczyk, A. Multi-Temporal Cliff Erosion Analysis Using Airborne Laser Scanning Surveys. *Remote Sens.* **2019**, *11*, 2666. [CrossRef]
11. de Sanjosé Blasco, J.J.; Serrano-Cañadas, E.; Sánchez-Fernández, M.; Gómez-Lende, M.; Redweik, P. Application of Multiple Geomatic Techniques for Coastline Retreat Analysis: The Case of Gerra Beach (Cantabrian Coast, Spain). *Remote Sens.* **2020**, *12*, 3669. [CrossRef]
12. Wlodarczyk-Sielicka, M.; Stateczny, A.; Lubczonek, J. The Reduction Method of Bathymetric Datasets that Preserves True Geodata. *Remote Sens.* **2019**, *11*, 1610. [CrossRef]
13. Temitope Yekeen, S.; Balogun, A.-L. Advances in Remote Sensing Technology, Machine Learning and Deep Learning for Marine Oil Spill Detection, Prediction and Vulnerability Assessment. *Remote Sens.* **2020**, *12*, 3416. [CrossRef]
14. Kowalewska-Kalkowska, H.; Kowalewski, M. Combining Satellite Imagery and Numerical Modelling to Study the Occurrence of Warm Upwellings in the Southern Baltic Sea in Winter. *Remote Sens.* **2019**, *11*, 2982. [CrossRef]
15. Rifat, S.A.A.; Liu, W. Quantifying Spatiotemporal Patterns and Major Explanatory Factors of Urban Expansion in Miami Metropolitan Area during 1992–2016. *Remote Sens.* **2019**, *11*, 2493. [CrossRef]

MDPI

Article

Monitoring the Coastal Changes of the Po River Delta (Northern Italy) since 1911 Using Archival Cartography, Multi-Temporal Aerial Photogrammetry and LiDAR Data: Implications for Coastline Changes in 2100 A.D.

Massimo Fabris 

Department of Civil, Environmental and Architectural Engineering, University of Padova, Via Marzolo, 9-35131 Padova, Italy; massimo.fabris@unipd.it

Abstract: Interaction between land subsidence and sea level rise (SLR) increases the hazard in coastal areas, mainly for deltas, characterized by flat topography and with great social, ecological, and economic value. Coastal areas need continuous monitoring as a support for human intervention to reduce the hazard. Po River Delta (PRD, northern Italy) in the past was affected by high values of artificial land subsidence: even if at low rates, anthropogenic settlements are currently still in progress and produce an increase of hydraulic risk due to the loss of surface elevation both of ground and levees. Many authors have provided scenarios for the next decades with increased flooding in densely populated areas. In this work, a contribution to the understanding future scenarios based on the morphological changes that occurred in the last century on the PRD coastal area is provided: planimetric variations are reconstructed using two archival cartographies (1911 and 1924), 12 multi-temporal high-resolution aerial photogrammetric surveys (1933, 1944, 1949, 1955, 1962, 1969, 1977, 1983, 1990, 1999, 2008, and 2014), and four LiDAR (light detection and ranging) datasets (acquired in 2006, 2009, 2012, and 2018): obtained results, in terms of emerged surfaces variations, are linked to the available land subsidence rates (provided by leveling, GPS—global positioning system, and SAR—synthetic aperture radar data) and to the expected SLR values, to perform scenarios of the area by 2100: results of this work will be useful to mitigate the hazard by increasing defense systems and preventing the risk of widespread flooding.

Keywords: Po River Delta; archival multi-temporal data; coastline changes; emerged/submerged surfaces; land subsidence; relative sea level rise 2100

Citation: Fabris, M. Monitoring the Coastal Changes of the Po River Delta (Northern Italy) since 1911 Using Archival Cartography, Multi-Temporal Aerial Photogrammetry and LiDAR Data: Implications for Coastline Changes in 2100 A.D. *Remote Sens.* **2021**, *13*, 529. https://doi.org/10.3390/rs13030529

Academic Editor: Paweł Terefenko

Received: 22 January 2021

the effects of this global ... coastal areas, deltas, wetlands, and lagoons, which are becoming increasingly vulnerable to flooding, storm surges, salinization, and permanent inundation [2–5]: half a billion people live, in fact, in delta regions threatened by land subsidence, and concerns for their well-being are increasing [6]. Land subsidence can be origin by natural and/or anthropogenic causes: natural subsidence is due to the compaction of lithological layers of soil and oxidation of peat; anthropogenic subsidence derives from aquifer-system compaction associated with groundwater or hydrocarbon withdrawals, drainage of organic soils, underground mining, natural compaction, sinkholes, and thawing permafrost [7]; the connections and coexistence of these phenomena have a strong negative impact on the territory, and can lead to environmental degradation, damage to buildings, and interruption of services.

The monitoring of these territories, in many cases characterized by high population concentration and low elevation compared to the mean sea level, is crucial to increase and improve the areas from flooding risk [5]: knowing the evolution of the temporal and spatial distribution of the movements is in fact essential to delineate the areas most affected by ground displacements and understand the mechanisms involved [8]. Moreover, the information deriving from the monitoring activities allows us to prevent damage to buildings and infrastructure, plan more sustainable urban development, and hence mitigate the risk.

Several geomatics methodologies can be used to monitor large areas affected by land subsidence: while in the past only geometric leveling had widespread use for the acquisition of high precision data, in recent decades the development of GPS (global positioning system), GNSS (global navigation satellite system), and SAR (synthetic aperture radar) techniques has made it possible to obtain high precision and high resolution data [9,10], which, in many cases, can integrate and/or replace the leveling measurements. Aerial digital photogrammetric [11,12] and LiDAR (with ALS–airborne laser scanning approach) [13] surveys can also be used, but the accuracy in elevation of these methodologies does not allow the high precision analysis necessary in the study of low rates, as typically occurs in the study of land subsidence. In this case, these methodologies can provide useful data in the study of planimetric changes: along the coastal area, with low elevations compared to the mean sea level, low subsidence rates can provide high planimetric changes of the coastline, with submersion of large areas that can be evaluated using photogrammetric and LiDAR data. For more limited areas, unmanned aerial vehicle (UAV) [14,15] and terrestrial laser scanning (TLS) [16–18] systems can also be used successfully.

In this work, archival cartography, digital aerial photogrammetry, and LiDAR (light detection and ranging) geomatic methodologies are used to reconstruct the coastline evolution of the PRD area (Figure 1) from 1911 to 2018, based on two cartographic data (1911 and 1924); 12 photogrammetric surveys performed in 1933, 1944, 1949, 1955, 1962, 1969, 1977, 1983, 1990, 1999, 2008, and 2014; and four LiDAR data acquired in 2006, 2009, 2012, and 2018.

The coastline was extracted from each survey and compared with the previous and subsequent dataset, and the results are provided in terms of emerged/submerged surfaces in the analyzed period. These data are correlated with available land subsidence rates (from leveling-for the past-and from GPS (2012–2017) and SAR data for the period 1992–2017 [19]) and SLR previsions [20]: the expected SLR values in the Po River Delta (PRD) are obtained from [21]: using all these data, a scenario of the study area projected in 2100 is performed in terms of emerged/submerged surfaces outside the levees.

Many authors, using different techniques, have studied land subsidence and coastal changes in PRD in the past [22–24] and recently [19,25,26], but no author has carried out studies on the emerged/submerged surfaces trend linked to the values of land subsidence rates and SLR expected for the next decades in this area.

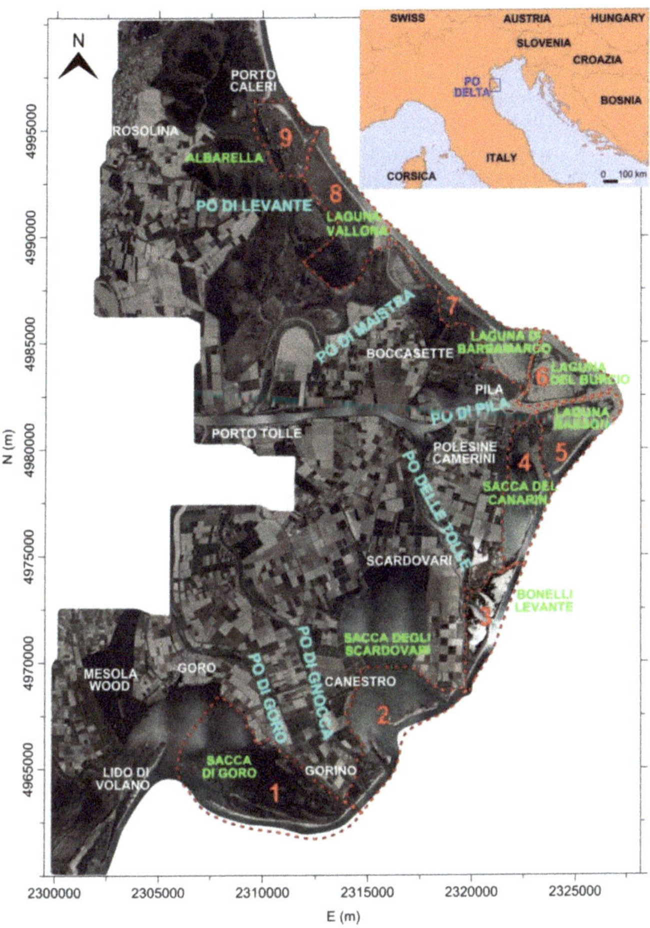

Figure 1. Location of PRD area in northern Italy, the river branches, the main villages, and subdivision of emerged surfaces outside the levees in nine sub-areas (red dashed lines): (**1**) Sacca di Goro; (**2**) Sacca degli Scardovari; (**3**) Bacino Bonelli Levante; (**4**) Sacca del Canarin; (**5**) Laguna Basson; (**6**) Laguna del [illegible] (**8**) Laguna Vallona; (**9**) Isola di Albarella. Coordinates are in the [illegible]

The PRD covers [illegible] of sediments carried by the Po, the largest river in Italy [illegible] 690 km from the Monviso Mont before flowing into the northern Adriatic Sea (Figure 1); the geology of the delta is mainly composed of terrigenous sediments up to 2000 m thick, and it is a complex multi-aquifer freshwater system [27].

Ground deformations in the PRD are mainly due to land subsidence of different origin: tectonic, sediment compaction, and artificial (anthropic). Long-term subsidence is mainly the result of deep tectonics, glacial isostatic adjustments, and geodynamic movements that provide a maximum rate of 2.5 mm/year [28,29]. Artificial subsidence is due to draining of wetlands, land reclamation, and, mostly, pumping of methane water from the medium-depth Quaternary deposits (200–600 m), which was highly intense between 1938 and 1961, when the Italian government suspended such operations [30].

This type of deformation, which may be attributed to large-scale human activities, caused vertical displacements in the analyzed area in the order of 2–3 m during and after the extraction processes (from 1940 to 1980), as documented by many authors: [23,31] indicate a maximum land subsidence rate in the order of 250 mm/year for the period 1951–1957 and 180 mm/year between 1958 and 1962. Later (1962–1967), these rates fell to 33 mm/year, matching the gradual reduction in pumping, and to 37.5 mm/year from 1967 to 1974. These last data clearly show the benefits obtained by halting extraction. Subsequently, the rate decreased still further: recent studies (using geometric leveling, GPS and InSAR data) have shown that land subsidence, albeit reduced, is still ongoing [19,26,32–35] (Figure 2). Nowadays, the effects of the great land subsidence that occurred in the last century are evident: most of the PRD now lies below the mean sea level and is characterized by the lengthening of the deltaic branches, anthropogenic stabilization of the hydrographical network, elevated borders seawards (levees, flood protection structures), and a significant depression in the center [27].

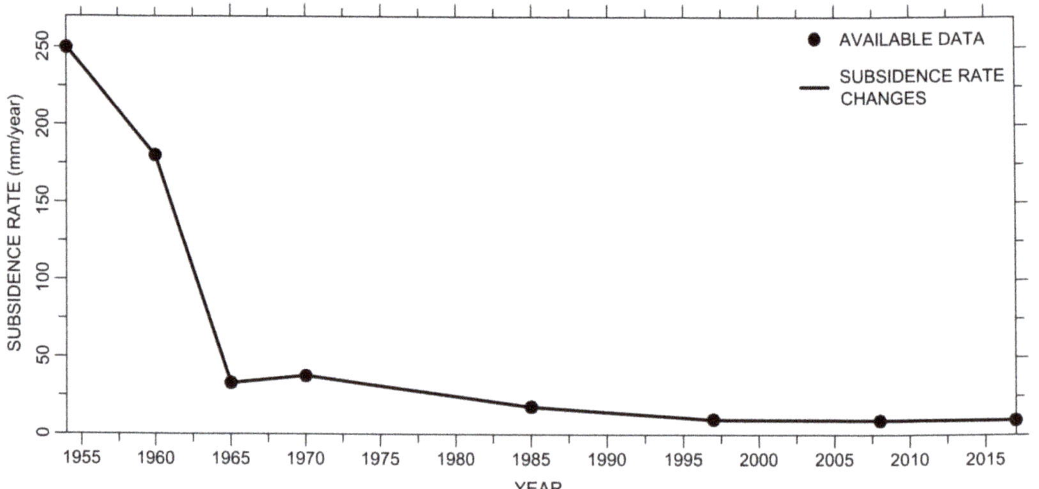

Figure 2. Subsidence rate changes from 1950 to 2017 based on the available data [19,22–26,30–35].

Sediment compaction and anthropogenic land subsidence may lead to serious environmental problems in this area, especially in connection with the relative SLR (RSLR) caused by climate variations worldwide [21,36–43]. Due to great ecological and economic value, high populations, and extensive agriculture, constant monitoring of deformations throughout the complex ecosystem of the PRD is necessary to provide information about displacements in order to implement territorial defense systems against flooding [1–6].

3. Materials and Methods

3.1. Available Datasets

3.1.1. Archival Cartographies

The study of coastline changes was performed using data from two archival cartographies derived from the IGMI (Istituto Geografico Militare Italiano): the representations are overlapped on a map where the complete survey of the eastern portion of PRD was performed in 1911 using classical topographic techniques, while only the easternmost coastal area was updated in 1924. The map, in a local reference system, represents the area from Po di Maistra to Po di Goro, in the north–south direction (Figure 1), and describes the Po river mouths in 1:25,000 scale.

3.1.2. Aerial Photogrammetric Surveys

The aerial photogrammetric surveys of the PRD used in this study were carried out in 1933, 1944, 1949, 1955, 1962, 1969, 1977, 1983, 1990, 1999, 2008, and 2014: in detail, the strips along the coastline were analyzed from Porto Caleri (Rovigo) to the border between the Veneto and Emilia Romagna Regions (Figure 1: a small area in Emilia Romagna). Unfortunately, the coverage of 1933 is incomplete in the northern portion (some photographs on glass plates were lost); the 1944 survey has limited coverage for both Veneto and Emilia Romagna, the 1962 and 1969 surveys are restricted in the southern portion, and the 1999 one is limited to the northern part of the study area: in these cases, the comparisons between the data were reduced in the multi-temporal analysis.

All aerial photos were acquired with analogic cameras, except the 2014 ones: the frames of 1944, 1955, 1977, and 1999 were digitized using the Wehrli Raster Master RM2 photogrammetric scanner at 1000 dpi (for the 1944 ones at 1200 dpi); the 1933, 1949, 1962, and 1969 images were available from IGMI, while the photographs of the 1983, 1990, and 2008 surveys from the Veneto Region were at 800 dpi. The most recent survey of 2014 was performed using a Vexcel UltraCam-Xp digital camera with a pixel size of 6 μm; all these data provide ground sample distance (GSD) between 30 and 80 cm for all images used here. These resolutions are more than adequate to detect the coastline, thanks to the uncertain ground–sea transition, frequent in the coastal areas of the delta [44]. The main characteristics of the aerial photogrammetric surveys are listed in Table 1.

Table 1. Characteristics of the aerial photogrammetric surveys used in this study (GSD: ground sample distance).

Year	Date	N. Strips	N. Images	Scale	Photo Size (cm)	Calibrated Focal Length (mm)	Visibility on Images	Scans (dpi)	GSD (cm)
1933	-	6	106	-	13 × 18	-	very poor	1200	30
1944	-	8	57	1:25,000	24 × 24, 18 × 24	24, 20 inch [1]	poor	1200	55
1949	10–19 July	9	60	1:18,000	30 × 30	200.500 [1]	poor	1200	40
1955	1 June–22 July	8	33	1:35,000	23 × 23	154.170 [1]	poor	1000	80
1962	8–11 July	8	49	1:31,000	23 × 23	153.030	poor	1000	80
1969	16–17 July	6	25	1:29,000	23 × 23	152.720	poor	1000	70
1977	10 October	8	32	1:30,000	23 × 23	152.620	good	1000	70
1983	10 May–5 June	13	78	1:17,000	23 × 23	153.330	good	800	50
1990	5–30 May	13	70	1:20,000	23 × 23	151.770	good	800	60
1999	10 September	5	24	1:34,000	23 × 23	152.900	good	1000	80
2008	21–22 August	14	125	1:16,000	23 × 23	153.875	good	800	50
2014	6–24 May	8	72	1:78,000	11310 × 17310 pi × el	100.500	good	4233	50

[1] No other information about the camera calibration.

The aerial photogrammetric surveys from 1955 to 2014 show stereoscopic coverage of the ground, while those from 1933, 1944, and 1949 partially show the same coverage: [illegible] only possible to generate the photo-plan mosaic of the

LiDAR surveys in PRD coastal area [illegible]
2012, and 2018 to monitor sand islets, which are storm surge barriers and protect the [illegible] from erosive action of sea waves motion. LiDAR data, together with ortho-images acquired simultaneously (with GSD of 20 cm), are available at the Veneto Region (Unità di Progetto per il Sistema Informativo Territoriale e la Cartografia and Unità Organizzativa Genio Civile di Rovigo) and the Local Authority of "Parco Regionale Veneto del Delta del Po".

The surveys were performed from March to September using Optech sensors and georeferencing the 3D acquired points using an integrated GNSS/INS (Inertial Navigation System) system; to increase information in the ground-sea transition area, surveys were carried out during low tide elevation, and with altitude of about 1500 m.

Acquired data cover portions of the eastern PRD coast from the Adige river mouth to the border between Veneto and Emilia Romagna Regions, with a width that only in the last survey covers most of the areas acquired by the aerial photogrammetric surveys (Table 2). Ellipsoidal elevation of the acquired points was converted into orthometric using the geoid model grids provided by the IGMI.

Table 2. Main characteristics of the LiDAR surveys analyzed in this study (ALTM: Airborne Laser Terrain Mapping).

Year	Date	Sensor	Tide (m)	Resolution (Points/m^2)	Average Covered Area (km × km)
2006	26 April	Optech ALTM 3033	−0.37	1.84	0.75 × 57
2009	7 March	Optech ALTM Gemini	−0.32	1.84	0.75 × 57
2012	17 September	Optech ALTM Gemini	−0.30	2.03	1 × 57
2018	14 April	Optech ALTM Galaxy	−0.35	2.38	1–5 × 57

3.2. Methods for the Coastline Changes Evaluation

3.2.1. Aerial Photogrammetric Images Orientation and Coastline Restitution

In a previous paper [44], analyzing the aerial photogrammetric surveys performed in the PRD in 1944, 1955, 1962, 1977, 1999, 2008, and 2014, the same used herein, we described in detail the procedures adopted for (i) external orientation of photogrammetric images using GCPs (ground control points, measured with the GPS technique for the orientation of the most recent surveys, and transferred from the oriented photogrammetric models in order to orientate the oldest images, with the correction of elevation due to the land subsidence occurred in the analyzed periods, based on rates values available in literature); (ii) evaluation of the horizontal co-registration of the extracted photogrammetric models; (iii) coastline restitutions taking into account the tide elevation value referred to the period in which the archival photogrammetric surveys were performed; (iv) emerged surfaces balance and accuracy, with the definition of the nine sub-areas outside to the levees (red dashed lines in Figure 1: sub-areas with borders defined by levees and river banks were chosen, due to the fluvial stability in the multi-temporal analysis; therefore, surface variations depend only on natural and/or anthropic causes, without the influence of the levees; emerged surfaces were measured within each sub-area), using a procedure based on the restitution of the coastline performed by five different operators, to evaluate the accuracy of the computed emerged surfaces and analyze the variations over time of the areas (for more details see [44]).

The same procedures were applied to study the other aerial photogrammetric surveys used here (1933, 1949, 1969, 1983, and 1990).

3.2.2. Co-Registration between Archival Cartographic and Photogrammetric Data

Similarly to the co-registration procedure between the different aerial photogrammetric surveys, the cartography of 1911–1924 was aligned with the photogrammetric data using the coordinates of new points, clearly visible on the 1955 images and map, located in presumably stable areas, manually measured on the photogrammetric model of 1955 (reference model for the orientation of the 1933, 1944, and 1949 surveys) from stereoscopic views and used as GCPs for georeferencing the archival cartography (15 GCPs). Subsequently, restitution of the 1911 and 1924 coastlines was performed in the same reference system of the photogrammetric data (Gauss–Boaga Italian reference system).

The horizontal co-registration between cartographic and photogrammetric data was evaluated by measuring and comparing 2-D coordinates of 18 natural and/or artificial clearly visible points both on the 1911–1924 map and the subsequent photogrammetric model

3.2.3. Emerged Surfaces Computation from LiDAR Data

LiDAR data, acquired using GPS/INS systems, are aligned with the cartographic and photogrammetric surveys from inception. Evaluation of co-registration was carried out measuring homologous natural and/or artificial 3-D points both on LiDAR datasets and the photogrammetric model of 2014 (due to small GSD, Table 1), clearly visible on the two surveys and located in stable areas.

Vertical co-registration between LiDAR surveys can be checked, also comparing the acquired data on stable areas: in this way, involving a large amount of points, more significant statistic parameters can be obtained.

LiDAR data are useful to generate DTMs (digital terrain models) [45,46] of the surveyed areas: from each model, the 0 level contour line can be extracted; this contour line is checked using the ortho-rectified images of the area and manually corrected where the automatic contour level did not follow the coastline visible on the ortho-images, due to outliers that can generate artifacts in the interpolation of the 3-D points, frequent in the ground–sea transition areas. Due to the different coverage of the multi-temporal LiDAR data (Table 2), the comparisons are performed only in common portions, using the boundaries of the more limited survey in the multi-temporal analysis. Finally, the emerged surfaces are computed and compared using the corrected 0 level contour lines for each of the eight sub-areas that subdivide the PRD coastal zone (LiDAR data are limited to the Veneto region: sub-area 1, in the Emilia Romagna region, was not surveyed).

3.2.4. Scenario of Emerged Surfaces by 2100

The DTMs obtained from the LiDAR surveys can be used to perform a prevision of the emerged surfaces outside to the levees in 2100, taking into account SLR and land subsidence values in the analyzed area projected to 2100 [47–49]. These phenomena, which proceed in opposite directions, (i) increase the flooding risks of the PRD due to the decrease in the safety margin between the top of the levees and the mean sea level and (ii) could increase the levees instability due to the reduction of the sandy cords (emerged surfaces) that protect themselves from the erosive action by the sea waves motion.

Projection by 2100 for PRD area was performed by [21] and [50] taking into account several SLR prevision models and considering the contribution of isostasy and tectonics: [50], assuming isostasy and tectonic rates of −1.5 mm/year in the study area, provides RSLR for the year 2100 between 315 mm (lower impact scenario IPCC—Intergovernmental Panel on Climate Change—2007 B1, +180 mm, www.ipcc.ch) and 1535 mm (higher impact scenario [51], +1400 mm); [21], using isostasy rates of −0.21 mm/year and tectonic vertical movements of −0.95 mm/year for the analyzed area, provides RSLR scenario by 2100 between 594 and 999 mm for IPCC 8.5 [52] minimum and maximum, respectively, and maximum of 1395 mm for [51].

4. Results

4.1. Data Processing and 3-D Models Extraction

Following the procedures described in a previous paper [44], natural GCPs, measured in 2008 with GPS technique, were used to generate the photogrammetric model related to [illegible] Exploitation Tool Set) software. [illegible]

1962, and 1969) were processed using GCPs measured on the [illegible] tions until 1955 images (1933, 1944, and 1949 have no stereoscopic coverage) to refer the coordinates at the same time as the flights: in these cases, the photogrammetric models were extracted with residuals of about 0.8–1 m. New points, clearly visible both on the photogrammetric model of 1955 and on the cartography of 1911–1924, were measured with stereoscopic devices and used as GCPs (2-D) for the co-registration of the map in the Gauss–Boaga Italian reference system: the obtained result, with residuals of about 2.4–2.8 m, is shown in Figure 3.

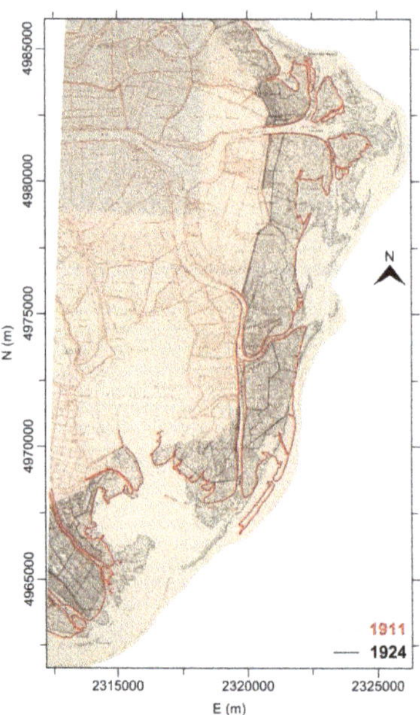

Figure 3. Coverage and alignment of the 1911 (in **red**)–1924 (in **black**) archival cartography (1:25,000 scale) in the Gauss–Boaga Italian reference system.

2-D GCPs were used to generate the photo-plan mosaic of 1933, 1944 (Figure 4, with lack of images on the study area, more for the 1944 ones), and 1949 (Figure 5, with complete planimetric coverage) surveys.

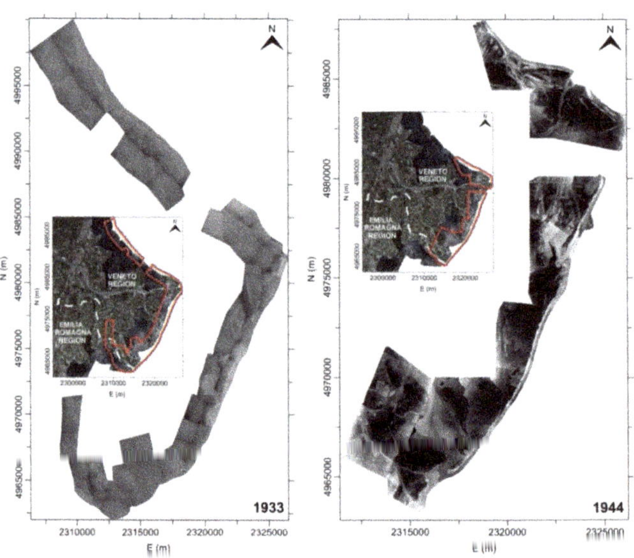

Figure 4. Photo-plan mosaic of the 1933 and 1944 aerial photogrammetric surveys: coverage is restricted to the coastal area; some photographs on glass plates (**1933**) and on film (**1944**) were lost, producing lack of data in the planimetric representation; available aerial images are characterized by poor visibility.

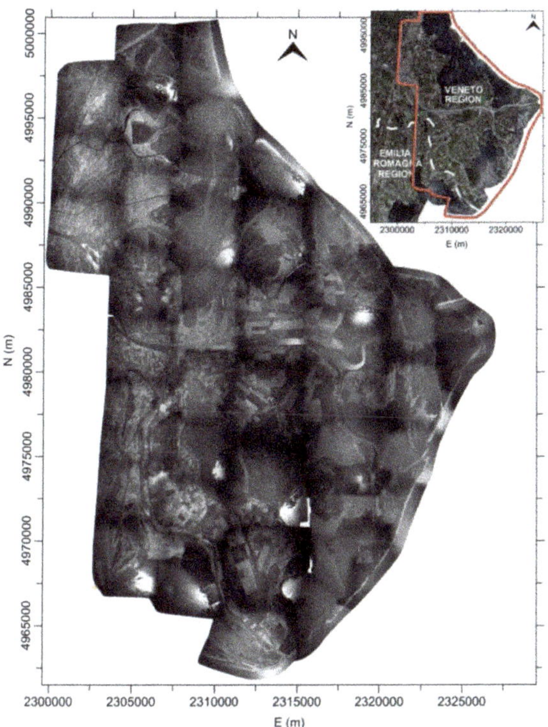

Figure 5. Photo-plan mosaic of the 1949 aerial photogrammetric survey: available images cover most of the PRD area in Veneto region and a portion of the Emilia Romagna region; visibility on aerial photographs increases, although not yet optimal.

Starting from the photogrammetric model, a DEM (Digital Elevation Model) with a grid size of 5 m was extracted automatically for each survey acquired with stereoscopic coverage (from 1955 to 2014) [53]; subsequently, using stereoscopic viewing devices, correction of the DEMs in the areas where automatic correlation failed (mainly in lagoons, due to the sunlight reflected from the water) was performed, adapting the contour lines to the real ground morphology. Thus, an orthophoto with a GSD of 1.5 m was generated automatically (Figures 6–8).

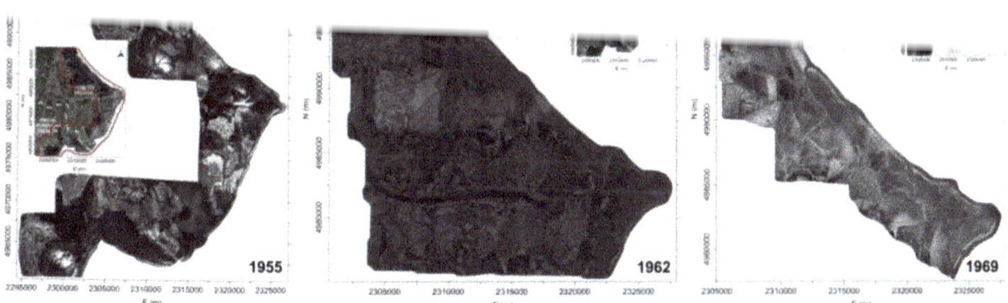

Figure 6. Orthophotos of the **1955**, **1962**, and **1969** aerial photogrammetric surveys: coverage in the PRD coastal area is complete for the 1955 one and restricted in the northern portion both for the **1962** and the **1969** data.

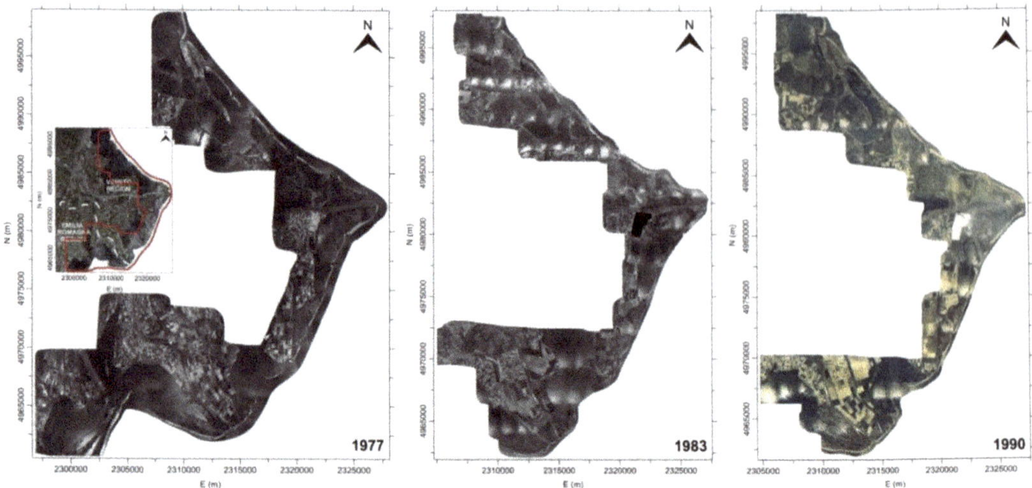

Figure 7. Orthophotos of the **1977**, **1983**, and **1990** aerial photogrammetric surveys: for these data, coverage in the PRD coastal area is complete, together with good visibility on the aerial images.

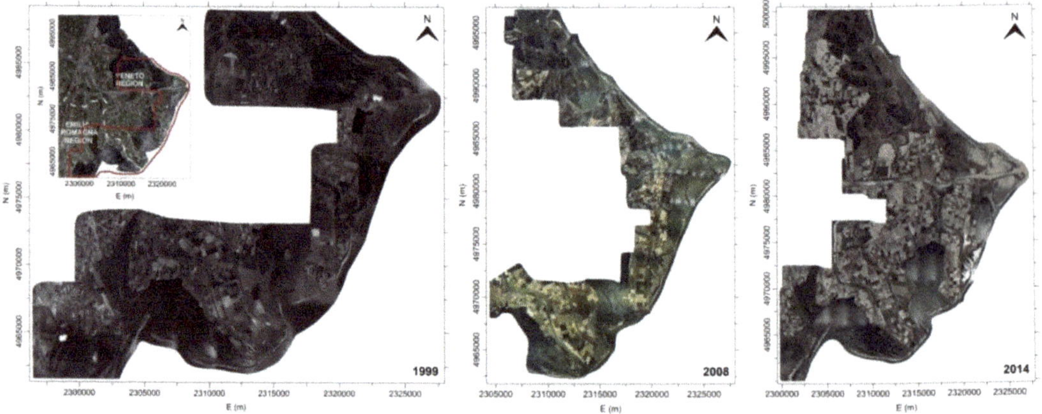

Figure 8. Orthophotos of the **1999**, **2008**, and **2014** aerial photogrammetric surveys: coverage is incomplete only for the 1999 data, and the aerial images are characterized by good/optimal visibility.

LiDAR datasets were used to generate DTMs of the surveyed area: from each acquisition, the last return of the laser beam was used together with procedures for objects/vegetation filtering, and data were gridded with a regular size of 1 m. Finally, four multi-temporal DTMs (2006, 2009, 2012, and 2018) were obtained (Figure 9).

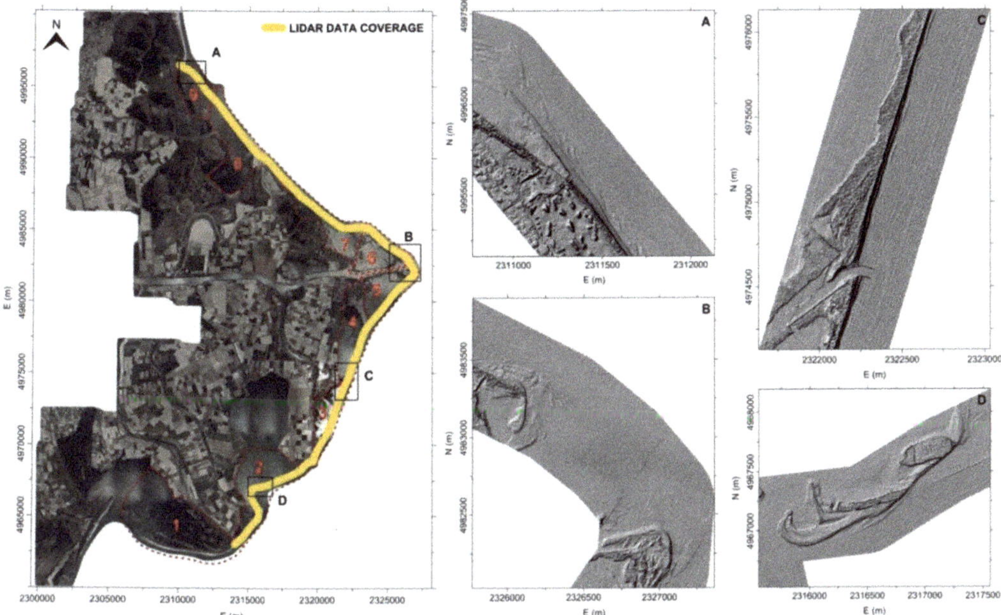

Figure 9. Coverage of the 2006 DTM from LiDAR data (last return laser beam together with procedures for objects/vegetation filtering, gridded with regular size of 1 m) and shaded relief visualization of some details (**A**–**D**) in PRD coastal area, outside the levees.

4.2. Co-Registration Accuracy

Co-registration between photogrammetric models and photo-plans (including the cartography of 1911–1924) was checked, measuring and comparing 2-D coordinates of natural homologous points in the multi-temporal series. The alignment between the cartography and the photogrammetric data should be checked with the first available photo-plan (1933): unfortunately, the lack of a significant number of common points clearly visible in the two subsequent datasets, due to the limited ground coverage both for the 1933 and 1944 surveys, has required the evaluation of the data co-registration using the first aerial photogrammetric survey with significant coverage, that is, the 1949 one. Table 3 lists the results of comparisons.

The standard deviation values of Table 3 show peaks of about 2 m in the comparison [illegible] for proper [illegible] data: it is worth noting that in this case the area was surveyed [illegible] methods for which there is no information about the procedures adopted for tracing the ground–sea line; here, the decision-making component should be added to the co-registration accuracy linked to the restitution choices made by the operators that, at present, cannot be quantified.

The same approach was used for the co-registration check of the LiDAR data: the alignment was verified by identifying and measuring coordinates, in the multi-temporal series of the four surveys, of clearly visible homologous points located in stable areas, like corners of buildings and flat roofs, piers, artifacts, etc. The co-registration between LiDAR and photogrammetric data was checked using the 2014 survey, the most recent and characterized by small pixel size (Table 1). Results of comparison are shown in Table 4.

Table 3. Comparisons between coordinates of homologous natural points, manually measured on archival cartography, photo-plans, and photogrammetric models (for the last, using stereoscopic viewing devices) (E: East; N: North; St. Dev.: Standard Deviation).

Comparison	Measured Points		Average (m)	St. Dev. (m)
1924–1949	18	E N	1.19 −1.25	3.27 2.94
1933–1944	21	E N	0.68 −0.58	1.86 1.75
1944–1949	38	E N	−0.23 −0.58	1.41 1.33
1949–1955	54	E N	0.53 0.32	1.21 1.13
1955–1962	66	E N	−0.34 0.42	0.84 0.75
1962–1969	66	E N	−0.31 0.27	0.78 0.73
1969–1977	66	E N	0.21 −0.08	0.76 0.84
1977–1983	88	E N	0.28 0.19	0.77 0.65
1983–1990	86	E N	−0.25 0.29	0.67 0.71
1990–1999	84	E N	0.19 −0.22	0.63 0.68
1999–2008	84	E N	−0.33 0.27	0.59 0.57
2008–2014	100	E N	0.14 −0.09	0.55 0.49

Table 4. Comparisons between coordinates of homologous natural points, manually measured on LiDAR datasets and the photogrammetric model of 2014 using stereoscopic viewing devices (E: East; N: North; V: Vertical; St. Dev.: Standard Deviation).

Comparison	Measured Points		Average (m)	St Dev. (m)
2006–2009	33	E N V	−0.08 0.11 0.07	0.22 0.20 0.24
2009–2012	31	E N V	0.05 −0.01 0.09	0.18 0.18 0.19
2012–2014	28	E N V	−0.13 0.10 −0.09	0.53 0.57 0.48
2014–2018	42	E N V	0.15 0.11 0.12	0.55 0.49 0.56
2012–2018	31	E N V	−0.06 −0.04 0.05	0.20 0.16 0.18

Table 4 shows standard deviation values greater than 0.5 m only when comparison involves the photogrammetric model of 2014: in the other cases, the values are reduced by more than 50%.

Moreover, the vertical co-registration between LiDAR data was checked, comparing the subsequent surveys on stable areas: due to the variability of the PRD coastal area and the small width that covers non-deformed portions and the data acquired in different periods of the year (which can also introduce changes in the development of vegetation), the availability of stable areas are very limited.

However, this comparison was carried out on about 300,000 points mainly located in the urbanized portions of Albarella island and Barricata village (Table 5).

Table 5. Average and standard deviation of the vertical comparison between subsequent LiDAR surveys performed on PRD in stable areas (portions of Albarella island and Barricata village).

Comparison	Average (m)	Standard Deviation (m)
2006–2009	0.03	0.22
2009–2012	0.06	0.15
2012–2018	0.03	0.16

Table 5 shows that, involving many points that provide a more robust statistic, the vertical co-registration of the LiDAR data is less than 25 cm, and in agreement with the comparison between the coordinates of the manually measured points (Table 4).

4.3. Coastline Restitutions and Changes Detection

Coastlines were drawn taking into account the tide levels (for details on the coastline restitution and analysis of accuracy see [44]: the same procedures are used here): this part of the study was performed only for the photogrammetric surveys of 1983, 1990, 1999, 2008, and 2014, due to the lack of tidal elevation data for the older surveys (no meaning for planimetric data such as photo-plans and cartographies), while the acquisition of LiDAR data was carried out at low tide (Table 2). Coastlines resulting from the 1911, 1924, 1933, 1944, 1949, 1955, 1962, 1969, 1977, 1983, 1990, 1999, and 2008 surveys were overlapped to the orthophoto extracted from the 2014 one (Figure 10).

Figure 10 shows significant changes in the coastline between the analyzed surveys. In particular, the greatest variations occurred at the beginning of the study period and between the 1950s and 1970s of the 20th century (the detail in Figure 10 shows the Bonelli Levante basin, sub-area 3, see Figure 1): while in the first case the changes were due to the land reclamation that took place in the area, which caused a disequilibrium in the coastal area and then in the coastline, in the last case, the great changes occurred during the most intense phase of methane-water pumping in the PRD, with a very high subsidence

this was done to reduce errors [illegible]
produced by the interpolation algorithm in the DTMs extraction. The final corrected contours are assumed to be the real coastline (0 m contour level).

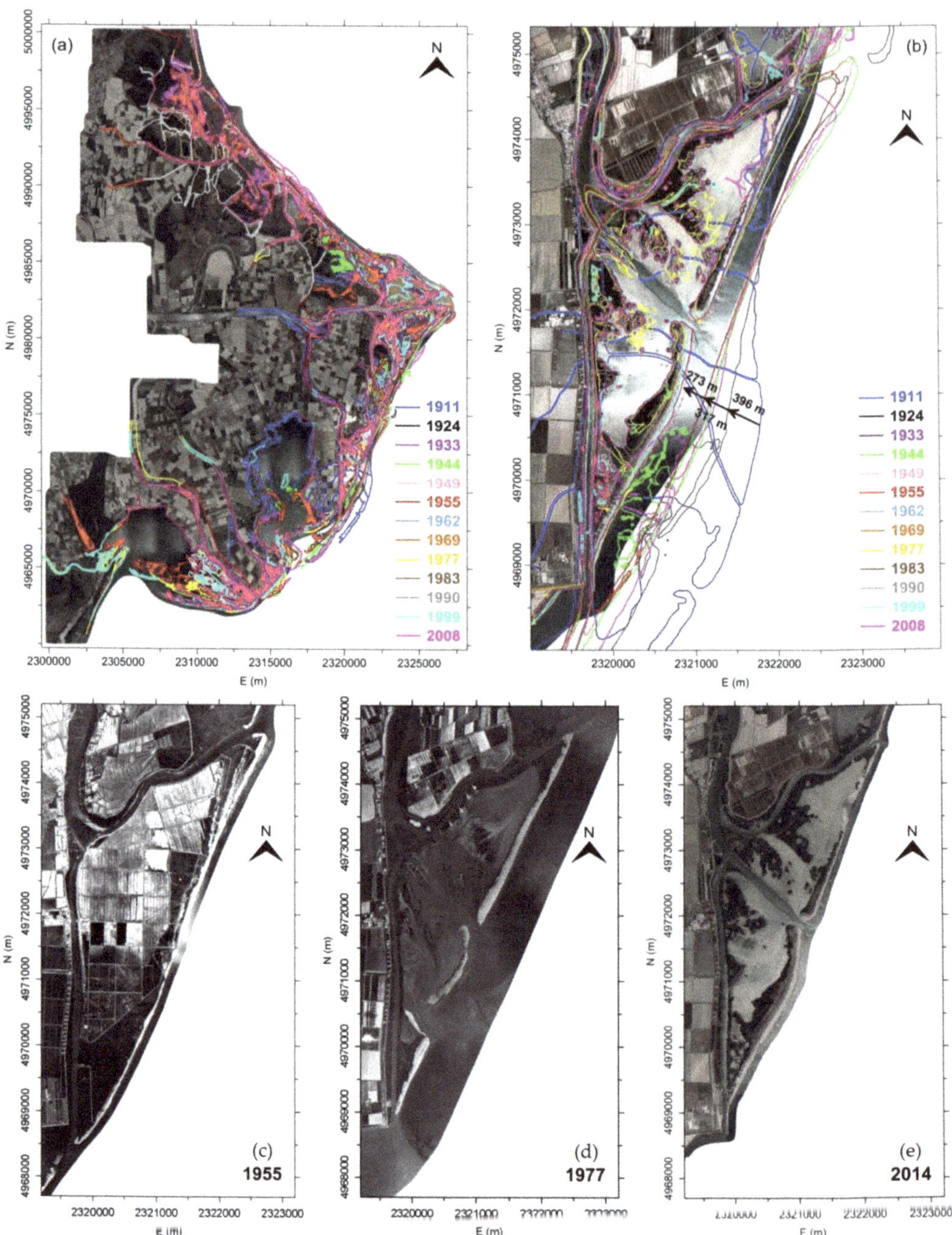

Figure 10. Coastline restitutions from archival cartography (1911–1924), photo-plans (aerial photogrammetric surveys performed in 1933, 1944, and 1949), and photogrammetric models (surveys carried out in 1955, 1962, 1969, 1977, 1983, 1990, 1999, and 2008) using stereoscopic devices were overlapped to the orthophoto of the last aerial photogrammetric survey (performed in 2014), (**a**) a detail related to the sub-area 3 (Bacino Bonelli Levante) is also shown (**b**) together with the orthophotos obtained from 1955 (**c**), 1977 (**d**), and 2014 (**e**) surveys, highlighting large deformations in emerged surfaces in the comparison 1955–1977.

4.4. Sub-Aerial Surface Changes

Surface measurements were performed for the nine sub-areas, analyzing the data in time: Figure 11 shows the result for sub-area 3 of Figure 1 (and the detailed portion of Figure 10, the Bonelli Levante basin) using cartographic and photogrammetric data.

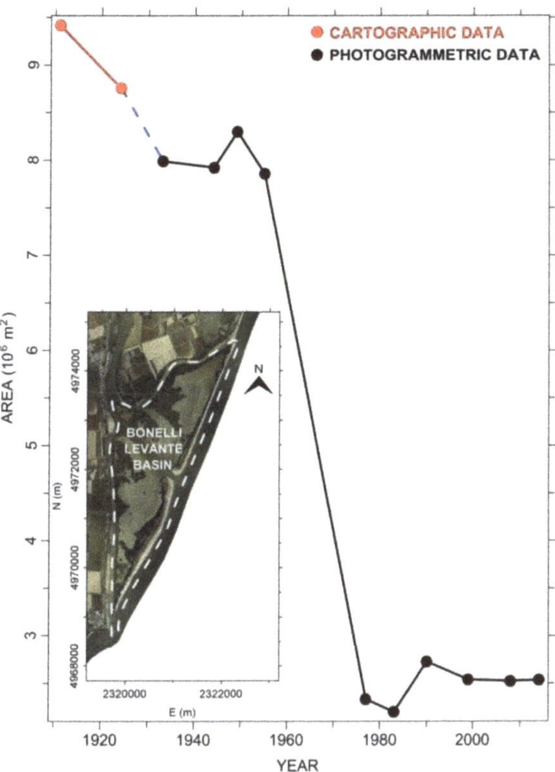

Figure 11. Multi-temporal emerged surfaces relating to sub-area 3 (Bacino Bonelli Levante) from 1911 to 2014 (two cartographic and 12 photogrammetric data): the storm surge event of 10 November 1957 [30], was intercepted in the comparison 1955–1977.

Analysis of Figure 11 shows four phases: after the first period (1911–1933), in which [illegible] period (1933–1955). Subsequently, [illegible] reduced the basin area from 7.86 to 2.17 Mm^2 (millions of m^2) in the [illegible], surveys, the emerged area has reached a new equilibrium phase (Figure 11).

The progression of multi-temporal emerged surfaces measured in the 1911–2014 period (cartographic and photogrammetric data) for the nine sub-areas of Figure 1 is shown in Figure 12.

The trend of emerged surfaces in the analyzed period (1911–2018) requires the integration between photogrammetric and LiDAR data; even if it has been verified by the common reference system, this integration has two problems: (i) the coverage of LiDAR data, which, mostly, involved only small portions of the eight sub-areas relating to the photogrammetric and the cartographic surveys (Table 2) and (ii) the direct comparison of emerged surfaces between photogrammetric and LiDAR data that, in common areas, has provided significant discrepancies. The comparison was made, randomly, between the surfaces obtained from the 2008 (photogrammetry) and 2009 (LiDAR) surveys, and, more completely, between the surfaces of 2014 (photogrammetry) and 2018 (LiDAR, due to the greater coverage of this survey).

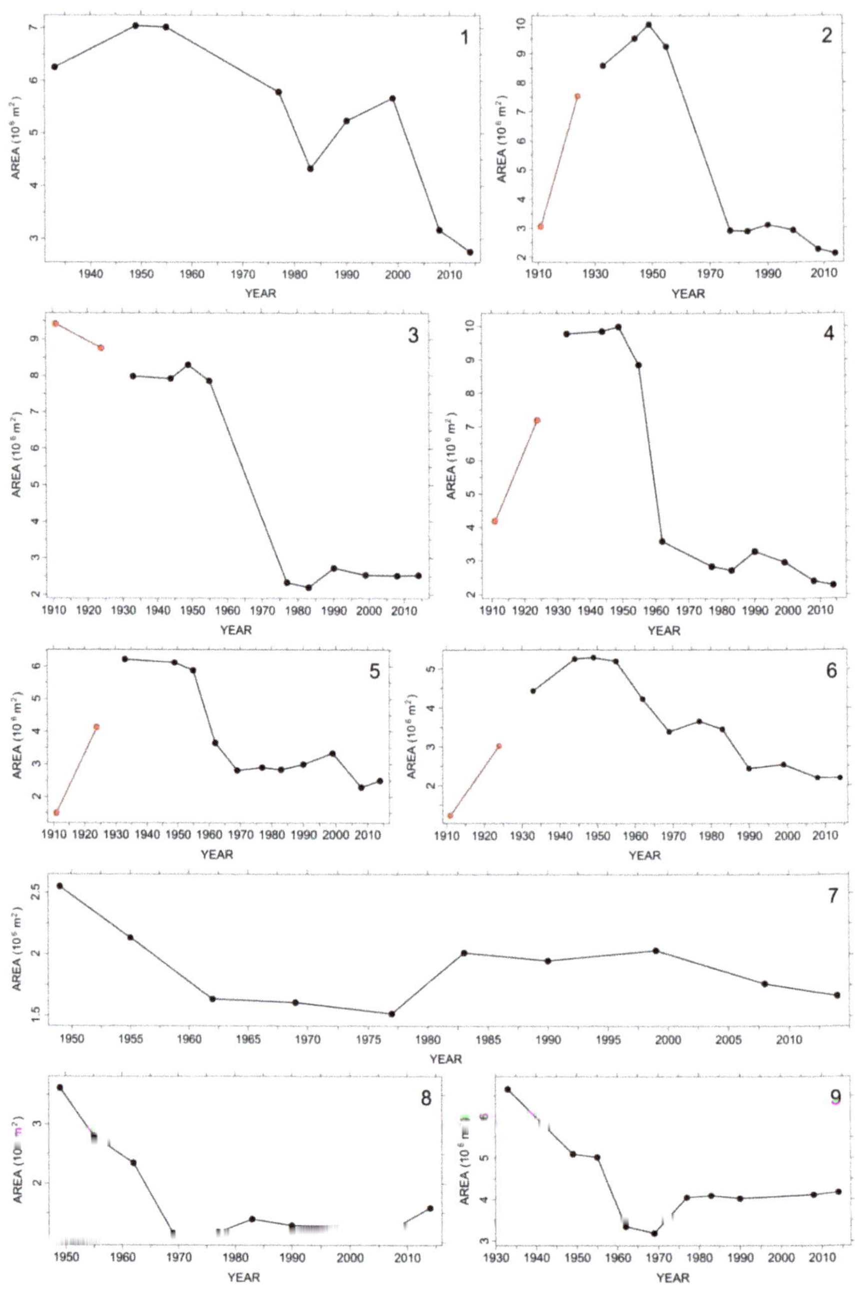

Figure 12. Progression of emerged surfaces for each of the nine sub-areas outside to the levees in PRD coastal area and defined in Figure 1 (**1–9**): due to the different accuracy, the cartographic data are in red (photogrammetric data in black).

Differences range from 2% (sub-area 9) to 25% (sub-area 7), which only partially can be explained by the elapsed time between the surveys (it is worth noting that for sub-area 7 in the 2012–2018 period, using only LiDAR data, the variation of the emerged surface is about 16%). More likely, these large differences may be due to the difficult interpretation of the emerged surfaces performed by the operators in the restitutions of the coastline on the photogrammetric models, as explained in detail in [44]. For these reasons, photogrammetric and LiDAR emerged surfaces are analyzed separately.

From the LiDAR DTMs, the final obtained coastlines together with the common boundaries between the different surveys (common coverage between the multi-temporal data) were used to compute the emerged surfaces of the eight sub-areas of PRD: in Table 6 the differences are shown.

Table 6. Comparison between the emerged surfaces for the eight sub-areas of the PRD computed using the 0 m contour lines level, extracted from LiDAR data, and corrected on the basis of the simultaneous ortho-images (final coastlines) together with the common boundaries between the different surveys.

Comparisons/Area	2	3	4	5	6	7	8	9
2006–2009 (10^6 m^2)	−0.12	−0.02	0.01	−0.08	−0.03	0.00	0.06	0.01
2009–2012 (10^6 m^2)	0.42	0.28	0.17	0.16	0.27	0.29	0.28	0.06
2012–2018 (10^6 m^2)	0.03	0.03	−0.03	0.09	−0.03	0.35	0.21	−0.04

4.5. Prevision of Emerged Surfaces in 2100

The DTM of the latest LiDAR survey (2018), characterized by a large land cover compared to 2006, 2009, and 2012, is used to simulate the emerged surfaces by 2100 for the eight sub-areas analyzed here, based on the RSLR previsions with GIA (glacio(hydro)-isostatic adjustment) and tectonic movements provided by [21]. RSLR projections in 2100 for the IPCC 8.5 scenario minimum (594 mm) and maximum (999 mm) and for the maximum RSLR projection for the investigated area in 2100 based on the [51] model (1395 mm) were used. The contour levels relating to the three cases were extracted from the 2018 DTM (projected to 2100), and the emerged surfaces resulting from the coastlines prevision in 2100 were computed both for each model and sub-area.

Table 7 summarizes the obtained values, and Figure 13 shows the coastlines from the 2018 LiDAR DTM and those obtained using the most pessimistic model of RSLR by 2100 [51] for the most critical sub-areas 2 and 3.

Table 7. Values of emerged surfaces of the eight sub-areas outside the to the levees taking into [illegible] and [51] max

LiDAR DTM (10^6 m^2)	2018	3.60	2.94	1.27	1.49	1.43	[illegible]	[illegible]	[illegible]
IPCC 8.5 min (10^6 m^2)	2100	1.89	1.85	0.85	1.21	0.87	1.50	1.34	4.21
IPCC 8.5 max (10^6 m^2)	2100	1.02	0.92	0.61	1.02	0.59	1.09	1.19	4.12
Rahmstorf max (10^6 m^2)	2100	0.38	0.68	0.38	0.52	0.30	0.70	0.83	4.00

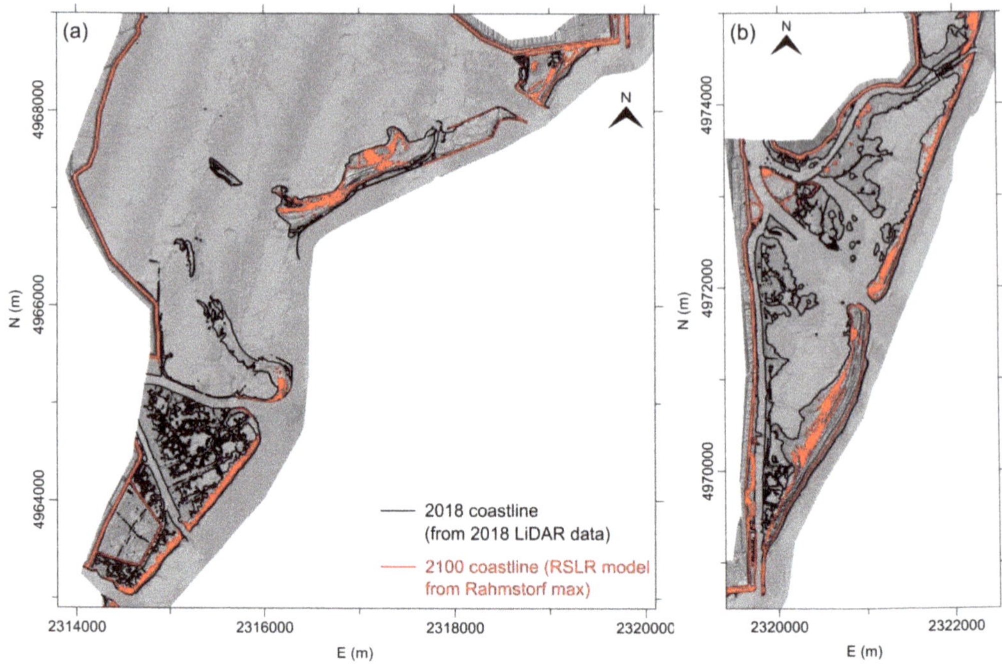

Figure 13. Coastlines representation from the 2018 LiDAR DTM (corrected 0 m contour line level, in black) and with the application of the RSLR projection in 2100 from the [51] max model (1395 mm, in red) for the most critical sub-areas 2 (Sacca degli Scardovari, (**a**)) and 3 (Bacino Bonelli Levante, (**b**)) in the PRD coastal area.

5. Discussion

The changes of the emerged surfaces from 1911 to 2018 in PRD coastal area involved data were characterized by different precision: even if the cartographic products provide co-registration with the photogrammetric models in the order of some meters and the oldest photogrammetric data provide alignments greater than 1 m with each other (Table 3), the great changes of the emerged surfaces that took place up to the 1970s were identified with a rigorous approach that provided reliable computed surfaces. Figure 12 shows large deformations for the nine sub-areas analyzed here: for all the studied portions, the general trend shows an increase in the emerged surfaces (expansion of the delta) in the order of 0.87 Mm^2/year up to 1933 (except sub-area 3); subsequently, a stability phase can be identified in the period 1933–1955 with rates of -0.02 Mm^2/year (except sub-area 9), followed by a rapid decrease in the surfaces from 1955 to 1977 (rates of -1.22 Mm^2/year, less evident in sub-area 7) and a new stabilization up to 2014 (with the exception of sub-area 1, Figure 12).

The different changes of the PRD coastal area are correlated to natural and anthropogenic factors: the increase in the emerged surfaces in the first analyzed period was caused by documented reclamation works that took place in the study area at the beginning of the 20th century [30]; it is worth noting that, given that the cartographic map of 1911–1924 was surveyed with classical topographical methods, during the period of the reclamation activities, with ponds, lakes, and lagoons probably located along the coastline, the description of the coastal area may not be very detailed. On the contrary, the restitution on the photogrammetric models provides a description of emerged surfaces with greater precision and detail. In the period 1955–1977, the strong reduction (in total -26.84 Mm^2) occurred concurrently with the highest measured land subsidence (from 250 mm/year to 37.5 mm/year, as described in Section 2) and intercepted the storm surge event of

November 10, 1957, which contributed to the loss of large areas of emerged surfaces, not only for sub-area 3, but also for portions 2 and 4 (Figure 12: this event has probably also contributed to changes in the other sub-areas analyzed in this study, because a great storm can be more devastating than the anthropogenic activities).

The general stabilization observed from 1977 to 2014 must be related to the accuracy of the aerial photogrammetric surveys (Table 3) combined with coastal works: using high-precision (Tables 4 and 5) and high-resolution (Table 2) LiDAR data, acquired in low tide elevation and together with the ortho-rectified images, the definition of the coastline and then the computation of the emerged surfaces, is improved. In this context, the comparison between the emerged surfaces from LiDAR DTMs provides -0.17 Mm^2 in the first period (2006–2009), +1.93 Mm^2 in the second period (2009–2012), and +0.61 Mm^2 in the 2012–2018 comparison (Table 6). The significant increase observed from 2009 to 2012 could be due to (i) the coastal works documented in the sub-areas [54] or (ii) the seasonal variations linked to the vegetation effects in the ground–sea transition area (mainly reeds that emerges from the water in some seasons that could be interpreted as emerged surfaces: survey of 2012 was carried out in September, while all the others were performed in spring, Table 2).

It is worth noting that the comparison between the 2006–2012 LiDAR data provides variation of the emerged surfaces of +1.76 Mm^2, while the 2008–2014 ones, using the photogrammetric data (without the sub-area 1, not acquired by the LiDAR surveys), provide a difference of +0.25 Mm^2: even if the ground coverage is different (wider for the photogrammetric data), the comparison period is only partially overlapped and the accuracy of the data is hardly comparable, the two techniques provide the same positive trend.

This increase in the emerged surfaces occurred together with land subsidence: [19], using SAR methodology from the A-DInSAR (Advanced-Differential Interferometric Synthetic Aperture Radar) method, processed ERS-1/2 (European Remote-Sensing, data from 10 May 1992, to 13 December 2000), ENVISAT (ENVIronmental SATellite, data from 17 March 2004, to 22 September 2010) and Sentinel-1A (data from 17 November 2014, to 17 May 2017) satellites images, providing land subsidence rates in the coastal area of PRD in the order of 1 cm along the LOS (line of sight) of the satellites. In the same period, using the linear trend, photogrammetric emerged surfaces (sub-areas from 1 to 6 due to the continuity of the data from 1992 to 2014–the 1999 photogrammetric survey does not cover sub-areas 7, 8, and 9, Figure 8) were reduced by 5.7 Mm^2, matching the lowering of the ground with the loss of coastal areas also in the recent time span [44]. This result shows agreement between land subsidence rate and areas submerged by the sea, although they are also strongly influenced by the accuracy of the data, by human activities, and by the SLR, which increasingly exposes the area to the risk of widespread flooding.

However, the comparison between the more precise LiDAR data from 2009 to 2018 shows an opposite trend, with an increase in emerged surfaces occurring together with the land subsidence; this effect can be explained by taking into account the morpho- [illegible] PRD complex [54], the different analyzed period [illegible]

[illegible]

Canarin, 5-Laguna Basson, 6-Laguna del [illegible]

Vallona, Figure 1) to improve the internal circulation of sea water, promoting an adequate change of lagoon water for fish farms aims that needed more efficient environmental hydraulic reorganization [30,55]: these works, necessary for the economic development of the area, have contributed to modified natural trend of emerged surfaces.

Changes in the emerged surfaces in PRD are also linked to the Adriatic Sea eustatism. Although the SLR values for the analyzed area are still limited (were quantified between 1.20 and 2 mm/year [40,41]), and therefore did not contribute significantly to the submersion of the coastal areas, different scenarios for the near future provide increasing values due to climate change. Table 7 shows the emerged surfaces by 2100 in PRD coastal area using data of the RSLR previsions derived from [21] and the 2018 LiDAR DTM projected to

2100: except the sub-area 9 (Albarella Island), all portions could be partially submerged with values from 19% (sub-area 5, IPCC 8.5 min) to 89% (sub-area 2, [51] max) of the current surface. In detail, the IPCC 8.5 min scenario could submerge 29% of the 2018 surface (higher value of 47% for sub-area 2), the IPCC 8.5 max up to 46% (higher value of 72% again for sub-area 2), and [51] max up to 60% of the current emerged surfaces. In this context, sub-area 2, the most important in the PRD coastal area for fish farms activities and, in general, the economic impact for the local population, could suffer the worst consequences (Figure 13).

It is worth noting that results of [21] were obtained by analyzing the CGPS (continuous GPS) observations of the Taglio di Po (TGPO) and Porto Tolle (PTO1) stations. Results obtained by [19] show an increase in land subsidence rates eastward, with values along the coastal area up to double of those related to CGPS stations. This effect, if confirmed in the next decades, could lead to an underestimation of the computed emerged surfaces by 2100.

The great reduction of the emerged surfaces and coastal elements could represent a serious problem in the study area: these portions are essential for the protection of the earthen levees from the erosive action by the sea waves motion; the disappearance of these elements exposes the defense infrastructures to the risk of breaks and/or collapses [56]. In addition, the reduction of the safety margin between the mean sea level and the top of the levees, in the order of 3–4 m at present day, will increase the frequency and extent of flooding and, in general, the overall hydraulic risk of the area.

For all these reasons, risk mitigation strategies in the PRD coastal area must be implemented in the next period, not only with the reinforcement and raising of the earthen levees, but also with activities that must protect and safeguard the emerged surfaces.

6. Conclusions

In this work, archival cartography, multi-temporal aerial photogrammetry, and LiDAR techniques together with available land subsidence rates values, tides elevation data, and RSLR previsions by 2100 were used to study the evolution of the emerged surfaces in PRD coastal area (an area affected by land subsidence, northern Italy), in the 1911–2018 period, and to perform different scenarios by 2100 for the first time.

LiDAR surveys acquired at low tide together with simultaneous ortho-images for the correction of the data in the land–sea transition allowed the extraction of the coastline with high resolution and accuracy in the order of a few tens of centimeters; on the other hand, using the aerial photogrammetric approach, it was possible to extend over time the analyzed area with precision in the restitution of the coastline, which depends on many factors, mainly the relative flight altitude and the resolution and quality of the images (more for archival photos); however, generally, only recent photogrammetric surveys can reach the accuracies of LiDAR data. In any case, the restitution of the coastline should be performed on the photogrammetric models (if stereoscopic acquisitions are available) extracted using a procedure that allows to correct the GCPs elevation according to the archival subsidence rates, to report the reference points to the time in which the aerial photogrammetric survey was carried out, and taking into account the tide elevation value at the time of images acquisition: this procedure allows the increase of the accuracy of the obtained coastline. Archival cartography, which does not permit the 3-D restitution, allows the further extension over time of the study with precision of the data that can deteriorate up to few meters, but is useful in many applications, such as for the study of the coastline modifications in flat areas.

Multi-temporal changes of the PRD coastal area outside the levees were detected by comparing the emerged surfaces of nine homologous sub-areas between the subsequent surveys. Results provided expansion of the Delta up to 1933, high contraction of the emerged surfaces from 1955 to 1977, simultaneously with the higher measured land subsidence rates together with the great storm surge event of 10 November 1957, and a new phase of slight expansion from 2009 to 2018, probably caused by anthropic works which took place in the study area. However, this positive progression of emerged surfaces could collide with the effects of the SLR previsions in near future due to the ongoing climate change: for the investigated area, assuming 594 and 999 mm (IPCC 8.5 min and max respectively) and 1395 mm [51] max of RSLR projection by 2100 with GIA and tectonic movements (www.ipcc.ch, [21,51], using the 2018 LiDAR DTM considered in the projections), the emerged surfaces outside the levees in PRD coastal area could be reduced, except for Albarella island, from 19% to 89% compared to the 2018 surface. These large losses could cause both breaks and/or collapses of the defense infrastructures, exposing the levees to the erosive activity of the sea waves, and the reduction of the safety margin between the mean sea level and the top of the levees (up to 35% of the 2018 value) implies that the investigated coastal area can become highly susceptible to marine inundation: for these reasons, risk mitigation actions must be adopted in the next years.

Finally, the method presented in this work for the PRD coastal area, which includes the integration between different types of data for the evaluation of emerged surfaces and projection scenarios by 2100 based on RSLR previsions, can be applied worldwide in other coastal areas expected to be affected by land subsidence and emerged surface variations also caused by the climate change effects.

Funding: This research was funded by the Department of Civil, Environmental, and Architectural Engineering of the Padova University (Italy) with the Projects "Integration of Satellite and Ground-Based InSAR with Geomatic Methodologies for Detection and Monitoring of Structures Deformations" and "High Resolution Geomatic Methodologies for Monitoring Subsidence and Coastal Changes in the Po Delta area".

Acknowledgments: The Author would like to thank the "Unità di Progetto per il Sistema Informativo Territoriale e la Cartografia" and "U.O. Genio Civile di Rovigo" (Veneto Region) for the supply of aerial photogrammetric surveys of 1983, 1990, and 2008, and the LiDAR surveys of 2006, 2009, and 2012; the "Parco Regionale Veneto del Delta del Po" for supply of the 2018 LiDAR survey; and the students of the Civil and Environmental Engineering Courses of the Padova University for their contribution in data processing.

Conflicts of Interest: The author declare no conflict of interest.

References

1. Ericson, J.P.; Vörösmarty, C.J.; Dingman, S.L.; Ward, L.G.; Meybeck, M. Effective sea-level rise and deltas: Causes of change and human dimension implications. *Glob. Planet. Chang.* **2006**, *50*, 63–82. [CrossRef]
2. Brooks, B.A.; University of Hawaii; Bawden, G.; Manjunath, D.; Werner, C.; Knowles, N.; Foster, J.H.; Dudas, J.; Cayan, D.; Survey, U.G.; et al. Contemporaneous Subsidence and Levee Overtopping Potential, Sacramento-San Joaquin Delta, California. *San Francisco Estuary Watershed Sci.* **2012**, *10*, 10. [CrossRef]
3. Wolstencroft, M.; Shen, Z.; Törnqvist, T.E.; Milne, G.A.; Kulp, M. Understanding subsidence in the Mississippi Delta region due to sediment, ice, and ocean loading: Insights from geophysical modeling. *J. Geophys. Res. Solid Earth* **2014**, *119*, 3838–3856. [CrossRef]
4. Minderhoud, P.S.J.; Erkens, G.; Pham, V.H.; Bui, V.T.; Erban, L.; Kooi, H.; Stouthamer, E. Impacts of 25 years of groundwater [illegible] 2017, 12, 064006. [CrossRef] [PubMed]

Giosan, L.; et al. Sinking deltas due to human activities. [illegible]
7. National Research Council. *Mitigating Losses from Land Subsidence in the United States*; The National Academies Press: Washington, DC, USA, 1991; p. 58. [CrossRef]
8. Yerro, A.; Corominas, J.; Monells, D.; Mallorquí, J.J. Analysis of the evolution of ground movements in a low densely urban area by means of DInSAR technique. *Eng. Geol.* **2014**, *170*, 52–65. [CrossRef]
9. Costantini, M.; Ferretti, A.; Minati, F.; Falco, S.; Trillo, F.; Colombo, D.; Novali, F.; Malvarosa, F.; Mammone, C.; Vecchioli, F.; et al. Analysis of surface deformations over the whole Italian territory by interferometric processing of ERS, Envisat and COSMO-SkyMed radar data. *Remote. Sens. Environ.* **2017**, *202*, 250–275. [CrossRef]
10. Hung, W.-C.; Hwang, C.; Chen, Y.-A.; Zhang, L.; Chen, K.-H.; Wei, S.-H.; Huang, D.-R.; Lin, S.-H. Land Subsidence in Chiayi, Taiwan, from Compaction Well, Leveling and ALOS/PALSAR: Aquaculture-Induced Relative Sea Level Rise. *Remote. Sens.* **2017**, *10*, 40. [CrossRef]

11. Baldi, P.; Fabris, M.; Marsella, M.; Monticelli, R. Monitoring the morphological evolution of the Sciara del Fuoco during the 2002–2003 Stromboli eruption using multi-temporal photogrammetry. *ISPRS J. Photogramm. Remote. Sens.* **2005**, *59*, 199–211. [CrossRef]
12. Baldi, P.; Cenni, N.; Fabris, M.; Zanutta, A. Kinematics of a landslide derived from archival photogrammetry and GPS data. *Geomorphology* **2008**, *102*, 435–444. [CrossRef]
13. Fabris, M.; Baldi, P.; Anzidei, M.; Pesci, A.; Bortoluzzi, G.; Aliani, S. High resolution topographic model of Panarea Island by fusion of photogrammetric, lidar and bathymetric digital terrain models. *Photogramm. Rec.* **2010**, *25*, 382–401. [CrossRef]
14. Mancini, F.; Dubbini, M.; Gattelli, M.; Stecchi, F.; Fabbri, S.; Gabbianelli, G. Using Unmanned Aerial Vehicles (UAV) for High-Resolution Reconstruction of Topography: The Structure from Motion Approach on Coastal Environments. *Remote. Sens.* **2013**, *5*, 6880–6898. [CrossRef]
15. Huang, C.; Zhang, H.; Zhao, J. High-Efficiency Determination of Coastline by Combination of Tidal Level and Coastal Zone DEM from UAV Tilt Photogrammetry. *Remote. Sens.* **2020**, *12*, 2189. [CrossRef]
16. Pesci, A.; Fabris, M.; Conforti, D.; Loddo, F.; Baldi, P.; Anzidei, M. Integration of ground-based laser scanner and aerial digital photogrammetry for topographic modelling of Vesuvio volcano. *J. Volcanol. Geotherm. Res.* **2007**, *162*, 123–138. [CrossRef]
17. Pesci, A.; Teza, G.; Casula, G.; Fabris, M.; Bonforte, A. Remote Sensing and Geodetic Measurements for Volcanic Slope Monitoring: Surface Variations Measured at Northern Flank of La Fossa Cone (Vulcano Island, Italy). *Remote. Sens.* **2013**, *5*, 2238–2256. [CrossRef]
18. Jaud, M.; Bertin, S.; Beauverger, M.; Augereau, E.; Delacourt, C. RTK GNSS-Assisted Terrestrial SfM Photogrammetry without GCP: Application to Coastal Morphodynamics Monitoring. *Remote. Sens.* **2020**, *12*, 1889. [CrossRef]
19. Fiaschi, S.; Fabris, M.; Floris, M.; Achilli, V. Estimation of land subsidence in deltaic areas through differential SAR interferometry: The Po River Delta case study (Northeast Italy). *Int. J. Remote Sens.* **2018**, *39*, 8724–8745. [CrossRef]
20. Anzidei, M.; Bosman, A.; Carluccio, R.; Casalbore, D.; Caracciolo, F.D.; Esposito, A.; Nicolosi, I.; Pietrantonio, G.; Vecchio, A.; Carmisciano, C.; et al. Flooding scenarios due to land subsidence and sea-level rise: A case study for Lipari Island (Italy). *Terra Nova* **2017**, *29*, 44–51. [CrossRef]
21. Antonioli, F.; Anzidei, M.; Amorosi, A.; Presti, V.L.; Mastronuzzi, G.; Deiana, G.; De Falco, G.; Fontana, A.; Fontolan, G.; Lisco, S.; et al. Sea-level rise and potential drowning of the Italian coastal plains: Flooding risk scenarios for 2100. *Quat. Sci. Rev.* **2017**, *158*, 29–43. [CrossRef]
22. Zambon, M. Abbassamento del Suolo per Estrazioni di Acqua e Gas: Deduzioni ed Indirizzi Logicamente Conseguenti per la Sistemazione del Delta del Fiume Po. In Proceedings of the 23nd National Conference on "Bonifiche", Rome, Italy, 20 May 1967; pp. 345–370.
23. Caputo, M.; Pieri, L.; Unguendoli, M. Geometric investigation of the subsidence in the Po Delta. *Boll. Geofis. Teor. Appl.* **1970**, *47*, 187–207.
24. Caputo, M.; Folloni, G.; Gubellini, A.; Pieri, L.; Unguendoli, M. Survey and geometric analysis of the phenomena of subsidence in the region of Venice and its hinterland. *Riv. Ital. Geofis.* **1972**, *21*, 19–26.
25. Bock, Y.; Wdowinski, S.; Ferretti, A.; Novali, F.; Fumagalli, A. Recent subsidence of the Venice Lagoon from continuous GPS and interferometric synthetic aperture radar. *Geochem. Geophys. Geosystems* **2012**, *13*, 03023. [CrossRef]
26. Tosi, L.; Da Lio, C.; Strozzi, T.; Teatini, P. Combining L- and X-Band SAR Interferometry to Assess Ground Displacements in Heterogeneous Coastal Environments: The Po River Delta and Venice Lagoon, Italy. *Remote. Sens.* **2016**, *8*, 308. [CrossRef]
27. Simeoni, U.; Corbau, C. A review of the Delta Po evolution (Italy) related to climatic changes and human impacts. *Geomorphology* **2009**, *107*, 64–71. [CrossRef]
28. Carminati, E.; Martinelli, G. Subsidence rates in the Po Plain, northern Italy: The relative impact of natural and anthropogenic causation. *Eng. Geol.* **2002**, *66*, 241–255. [CrossRef]
29. Carminati, E.; Martinelli, G.; Severi, P. Influence of glacial cycles and tectonics on natural subsidence in the Po Plain (Northern Italy): Insights from14C ages. *Geochem. Geophys. Geosystems* **2003**, *4*, 1–14. [CrossRef]
30. Colombo, C.; Tosini, L. *Sessant'anni di Bonifica Nel Delta del Po*, 1st ed.; Papergraf S.p.a. Publisher: Padova, Italy, 2009; pp. 3–217. ISBN 978-88-87264-70-8.
31. Borgia, G.; Brighenti, G.; Vitali, D. La Coltivazione dei Pozzi Metaniferi del Bacino Polesano e Ferrarese. Esame Critico della Vicenda. *Ina Georisorse Territ.* **1982**, *425*, 13–23.
32. Bitelli, G.; Bonsignore, F.; Unguendoli, M. Levelling and GPS networks to monitor ground subsidence in the Southern Po Valley. *J. Geodyn.* **2000**, *30*, 355–369. [CrossRef]
33. Baldi, P.; Casula, G.; Cenni, N.; Loddo, F.; Pesci, A. GPS based monitoring of land subsidence in the Po Plain (Northern Italy). *Earth Planet. Sci. Lett.* **2009**, *288*, 204–212. [CrossRef]
34. Tosi, L.; Teatini, P.; Strozzi, T.; Carbognin, L.; Brancolini, G.; Rizzetto, F. Ground surface dynamics in the northern Adriatic coastland over the last two decades. *RENDICONTI Lince-* **2010**, *21*, 115–129. [CrossRef]
35. Cenni, N.; Viti, M.; Baldi, P.; Mantovani, E.; Bacchetti, M.; Vannucchi, A. [illegible] vertical movements in Central and Northern Italy from GPS data. Possible role of natural and anthropogenic causes. *J. Geodyn.* **2013**, *71*, 74–85. [CrossRef]
36. Gornitz, V. Sea-level rise: A review of recent past and near-future trends. *Earth Surf. Process. Landforms* **1995**, *20*, 7–20. [CrossRef]
37. Carbognin, L.; Tosi, L. Interaction between Climate Changes, Eustacy and Land Subsidence in the North Adriatic Region, Italy. *Mar. Ecol.* **2002**, *23*, 38–50. [CrossRef]

38. Carbognin, L.; Teatini, P.; Tosi, L. Eustacy and land subsidence in the Venice Lagoon at the beginning of the new millennium. *J. Mar. Syst.* **2004**, *51*, 345–353. [CrossRef]
39. Carbognin, L.; Teatini, P.; Tomasin, A.; Tosi, L. Global change and relative sea level rise at Venice: What impact in term of flooding. *Clim. Dyn.* **2009**, *35*, 1039–1047. [CrossRef]
40. Carbognin, L.; Teatini, P.; Tosi, L.; Strozzi, T.; Tomasin, A. Present relative sea level rise in the northern Adriatic coastal area. In *Coastal and Marine Spatial Planning. Marine Research at CNR-Theme 3*; Brugnoli, E., Cavarretta, G., Mazzola, S., Trincardi, F., Ravaioli, M., Santoleri, R., Eds.; CNR-DTA Publisher: Rome, Italy, 2011; pp. 1123–1138.
41. Vilibić, I.; Šepić, J.; Pasarić, M.; Orlić, M. The Adriatic Sea: A Long-Standing Laboratory for Sea Level Studies. *Pure Appl. Geophys. PAGEOPH* **2017**, *174*, 3765–3811. [CrossRef]
42. Antonioli, F.; De Falco, G.; Presti, V.L.; Moretti, L.; Scardino, G.; Anzidei, M.; Bonaldo, D.; Carniel, S.; Leoni, G.; Furlani, S.; et al. Relative Sea-Level Rise and Potential Submersion Risk for 2100 on 16 Coastal Plains of the Mediterranean Sea. *Water* **2020**, *12*, 2173. [CrossRef]
43. Elias, P.; Benekos, G.; Perrou, T.; Parcharidis, I. Spatio-Temporal Assessment of Land Deformation as a Factor Contributing to Relative Sea Level Rise in Coastal Urban and Natural Protected Areas Using Multi-Source Earth Observation Data. *Remote. Sens.* **2020**, *12*, 2296. [CrossRef]
44. Fabris, M. Coastline evolution of the Po River Delta (Italy) by archival multi-temporal digital photogrammetry. *Geomat. Nat. Hazards Risk* **2019**, *10*, 1007–1027. [CrossRef]
45. Baldi, P.; Coltelli, M.; Fabris, M.; Marsella, M.; Tommasi, P. High precision photogrammetry for monitoring the evolution of the NW flank of Stromboli volcano during and after the 2002–2003 eruption. *Bull. Volcanol.* **2007**, *70*, 703–715. [CrossRef]
46. Marsella, M.A.; Baldi, P.; Coltelli, M.; Fabris, M. The morphological evolution of the Sciara del Fuoco since 1868: Reconstructing the effusive activity at Stromboli volcano. *Bull. Volcanol.* **2011**, *74*, 231–248. [CrossRef]
47. Vecchio, A.; Anzidei, M.; Serpelloni, E.; Florindo, F. Natural Variability and Vertical Land Motion Contributions in the Mediterranean Sea-Level Records over the Last Two Centuries and Projections for 2100. *Water* **2019**, *11*, 1480. [CrossRef]
48. Anzidei, M.; Doumaz, F.; Vecchio, A.; Serpelloni, E.; Pizzimenti, L.; Civico, R.; Greco, M.; Martino, G.; Enei, F. Sea Level Rise Scenario for 2100 A.D. in the Heritage Site of Pyrgi (Santa Severa, Italy). *J. Mar. Sci. Eng.* **2020**, *8*, 64. [CrossRef]
49. Scardino, G.; Sabatier, F.; Scicchitano, G.; Piscitelli, A.; Milella, M.; Vecchio, A.; Anzidei, M.; Mastronuzzi, G. Sea-Level Rise and Shoreline Changes Along an Open Sandy Coast: Case Study of Gulf of Taranto, Italy. *Water* **2020**, *12*, 1414. [CrossRef]
50. Lambeck, K.; Antonioli, F.; Anzidei, M.; Ferranti, L.; Leoni, G.; Scicchitano, G.; Silenzi, S. Sea level change along the Italian coast during the Holocene and projections for the future. *Quat. Int.* **2011**, *232*, 250–257. [CrossRef]
51. Rahmstorf, S. A Semi-Empirical Approach to Projecting Future Sea-Level Rise. *Sci.* **2007**, *315*, 368–370. [CrossRef]
52. IPCC. Climate Change 2013: The Physical Science Basis. In *Contribution of Working Group I to the Fifth Assessment Report of the Intergovernmental Panel on Climate Change*; Stocker, T.F., Qin, D., Plattner, G.-K., Tignor, M., Allen, S.K., Boschung, J., Nauels, A., Xia, Y., Bex, V., Midgley, P.M., Eds.; Cambridge University Press: Cambridge, UK; New York, NY, USA, 2013; p. 1535.
53. Fabris, M.; Pesci, A. Automated DEM extraction in digital aerial photogrammetry: Precisions and validation for mass movement monitoring. *Ann. Geophys.* **2005**, *48*, 57–72. [CrossRef]
54. Bosman, A.; Romagnoli, C.; Madricardo, F.; Correggiari, A.; Remia, A.; Zubalich, R.; Fogarin, S.; Kruss, A.; Trincardi, F. Short-term evolution of Po della Pila delta lobe from time lapse high-resolution multibeam bathymetry (2013–2016). *Estuar. Coast. Shelf Sci.* **2020**, *233*, 106533. [CrossRef]
55. *Consorzio di Bonifica Delta del Po. Le Lagune del Delta del Po*; Consorzio di Bonifica del Delta del Po: Rovigo, Italy, 2013; pp. 4–83.
56. Özer, I.E.; Rikkert, S.J.H.; Van Leijen, F.J.; Jonkman, S.N.; Hanssen, R.F. Sub-seasonal Levee Deformation Observed Using Satellite Radar Interferometry to Enhance Flood Protection. *Sci. Rep.* **2019**, *9*, 2646. [CrossRef]

Article

Autonomous Lidar-Based Monitoring of Coastal Lagoon Entrances

Bilal Arshad *, Johan Barthelemy and Pascal Perez

SMART Infrastructure Facility, University of Wollongong, Wollongong, NSW 2522, Australia; johan@uow.edu.au (J.B.); pascal@uow.edu.au (P.P.)
* Correspondence: bilal@wizedynamics.com.au; Tel.: +61-2-4239-2329

Abstract: Intermittently closed and open lakes or Lagoons (ICOLLs) are characterised by entrance barriers that form or break down due to the action of wind, waves and currents until the ocean-lagoon exchange becomes discontinuous. Entrance closure raises a variety of management issues that are regulated by monitoring. In this paper, those issues are investigated, and an automated sensor solution is proposed. Based upon a static Lidar paired with an edge computing device. This solar-powered remote sensing device provides an efficient way to automatically survey the lagoon entrance and estimate the berm profile. Additionally, it estimates the dry notch location and its height, critical factors in the management of the lagoon entrances. Generated data provide valuable insights into landscape evolution and berm behaviour during natural and mechanical breach events.

Keywords: coastal monitoring; estuaries; IoT; lidar; remote sensing

Citation: Arshad, B.; Barthelemy, J.; Perez, P. Autonomous Lidar-Based Monitoring of Coastal Lagoon Entrances. *Remote Sens.* **2021**, *13*, 1320. https://doi.org/10.3390/rs13071320

Academic Editor: Paweł Terefenko

Received: 6 February 2021
Accepted: 29 March 2021
Published: 30 March 2021

[illegible] iations.

1. Introduction

Within ecosystem coastal areas provide critical services, such as ecotourism, climate regulation, storm and wave protection, and recreational regions; yet they only comprise 5% of the earth's land surface [1]. The impact of global temperature change, rising sea-levels, and rapid urbanization is increasingly affecting coasts. Thus, there is a demand for long-term coastal monitoring and effective management initiatives, including coastal lagoons. Coastal lagoons are common features (nearly 13%) of the coastal systems [1]. They are regions where water, atmosphere, and land interact in a complex environment, that is constantly been changed by humans and natural influence [1]. Along the coastline of Australia, approximately 61 out of 135 estuarine systems have intermittently open and closed lagoon entrances. Of these, 44 are artificially opened when the berm height exceeds a pre-defined threshold. The rest are kept open by breakwaters and training walls, primarily to provide boating access but also to improve water quality and maximize flushing [2]. [illegible] between being closed and open to the ocean are [illegible] (ICOLLs) [3] [illegible] river or tidal dominated systems. [illegible] consequences during flash flood events or water contamination by hazardous substances. Thus, understanding the dynamics of such process is crucial to managing lagoons and estuaries, especially in densely populated coastal areas.

Due to the intermittent nature of rainfall in south-eastern Australia, the open/closed cycles of ICOLLs are not seasonal. The frequency and timing of the entrance opening are dependent on the factors such as catchment size, water levels in creeks or rivers and height of the sand berm. Lagoon entrances are usually characterized by a sand berm, formed from sediments deposited by tides, winds, and waves from the ocean. This natural process increases the vertical growth of the berm and the peak height is determined by the wave runup [4]. The berm prevents water flow from the lagoon to the ocean and vice versa, which can cause the lagoon to overflow and inundate low-lying residential areas, due to the water build-up behind the berm [5]. Therefore, monitoring berm height is crucial as this is one of the important factors for authorities to consider when planning to breach the entrance.

The decision to mechanically breach an entrance (lowering or removal of the berm) entails several considerations., such as the negative impacts on local flora and fauna. Henceforth, artificial breach is carefully considered by the local authorities, in order to strike a balance between the social, economic and ecological risks. As the timing of the intervention is a key to success, regular monitoring of lagoon entrances and sand berms is necessary.

Usually, the effective management of lagoon entrances is based on a pre-determined tolerated threshold for the maximum berm height (This is known as "maintain a dry notch"). A dry notch is the minimum height of the berm that is not affected by normal hydraulic beach processes (wave run-up and tides). A well-designed dry notch allows water behind the berm to naturally open the lagoon during a flood event. However, in many instances, there is a need to excavate the dry notch in order to provide such an outlet and avoid upstream inundation (Figure 1). Therefore, it is important to regularly monitor the berm through observation height markers or/and regular in situ surveys to maintain the presence of a dry notch [6].

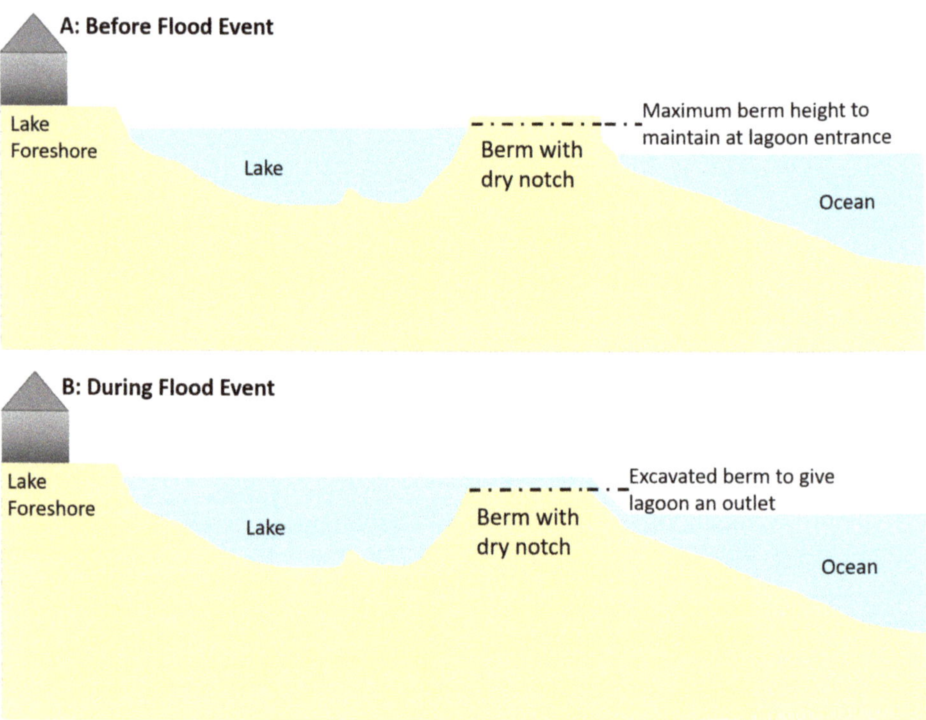

Figure 1. Importance of berm and dry notch management for lagoon entrances.

This study seeks to address the need for active and on-demand monitoring of lagoons via automating the task of in situ surveying. As the systematic review by Arshad et al. [7] highlighted that lagoon monitoring is still conducted manually and it requires further investigation. In this paper, various remote sensing technologies [8] have been explored to automate the management of the lagoon entrances. It is hypothesized that lagoon entrance management can be automated by using a remote sensing station consisting of a continuous scanning light detection and ranging (Lidar) sensor paired with an edge computing device to construct a topographic profile of the lagoon entrance. This study further contributes towards estimating the height and location of a dry notch, thus assisting local authorities to manage lagoon entrances effectively. This paper is organized as follows: Section 2 reviews recent advancements in coastal monitoring and management. Methodology including study site, monitoring setup and data processing are presented in Section 3. Section 4 presents discussion and lastly, Section 5 presents conclusion and future direction for the research.

2. Recent Advancements in Coastal Monitoring and Management

Remote sensing technologies such as computer vision and internet of things (IoT) sensors available to develop an early warning system have previously been covered in a systematic review [7,9]. This section will look at the recent advancements in coastal management from a monitoring perspective, including ground-based camera systems, satellite, unmanned aerial vehicles (UAVs) and Lidar-based technologies.

2.1. Camera-Based Remote Sensing

Frequent in situ surveying of estuaries is challenging and costly therefore an alternate approach is required which is free from manual work and thus serves the purpose of continuous monitoring with no human input. Since 1996 in Australia, ARGUS coastal imaging system has been commonly used to monitor the coastal environment [10]. A typical coastal imaging system consists of four or five elevated cameras to provide a 180-degree view of the shoreline. This automated coastal monitoring station takes images and patches them together to cover the whole coastal area. Image/video analysis is then performed to observe shoreline behaviour [11], sandbar behaviour [12], nearshore morphology [13], and contributes to other coastal-related research areas [10]. Such monitoring has an advantage over traditional surveys as it captures both spatial and temporal data that can be utilized to understand coastline response to storms. A study conducted at Narrabeen–Collaroy in Australia by Harley et al. [14] utilized approximately 5 years of both time-exposure and variance image data to develop an empirical relationship between change in beach width and wave energy of storms. This work was extended by Beuzen et al. [15] and utilized approximately 10 years of coastline data to develop a Bayesian Network to predict coastline response to storms.

Video-based remote sensing data provide flexibility and a long-term source of data, for which estuary behaviour can be explored under the influence of changing environmental conditions. Another example of video-based remote monitoring is surf cameras [16]. Surfcams are used around the world to provide surfers with information about conditions of the beach, so they can plan their trip accordingly. Mole et al. [16] in their work identified [illegible] networks for capturing both shoreline change and the real-time [illegible] of coastline [illegible] The study used surfcams [illegible] Their study compared an estimate of shoreline elevation between surfcams and [illegible] kinematic global navigation satellite system (RTK-GNSS) surveys. After calibrating the surfcams, standard deviations (SDs) error of 1 to 4 m (horizontal) was observed when compared to RTK-GNSS survey. When compared to Argus-derived shoreline dataset an error of 2 m (horizontal) was observed. Recently Umberto et al. [18] presented a methodological approach to exploit surfcams for coastal morphodynamic studies. This study offered a procedure to geo-rectify the online streamed images by utilizing ground control points and estimating the unknown camera parameters to generate accurate rectified planar images for quantitative analysis of coastal behaviour. Furthermore, Sanchez-Garcia et al. [19] presented a method of projecting terrestrial images into geo-referenced planes to minimise the error of existing low-elevation surfcams. As surf cameras continue to grow in number, such resources will become increasingly valuable in expanding coverage of coastal monitoring.

Following from surfcams, recent advancements in camera technology and availability of such technology in smartphones indicates that there are unexploited opportunities for coastline monitoring [20] using crowd-sourced photos. For instance, citizen science launched the CoastSnap program [21], where several known photo points have been installed, creating monitoring locations. These crowd-sourced images are provided by the general public visiting beaches via widely used digital media platforms or through instructions provided on photo points to share photos with researchers and the local community [22]. The first photo point station was commissioned at Narrabeen–Collaroy embayment and provided the opportunity to generate a shoreline dataset for coastal management and research. The photo point station consists of a basic camera cradle that controls and stabilizes the viewpoint of the smartphone camera. The general public can then place their phone in the bracket and take pictures of the view and upload them to their preferred database. This, however, brings some challenges such as the wide range of smartphone, low-resolution images, manual adjustment of smartphones, and uncertainty in image capture times. Similar to surfcams, ground points were measured at the time of installation for every CoastSnap, and such information is used to geo-reference and rectifies each image submitted by the community participants. However, coastal monitoring via both surfcams and ARGUS monitoring station relies on known camera parameters, whereas for CoastSnap intrinsic and extrinsic parameters (i.e., azimuth pitch, focal length, and roll) of the smartphone cameras are computed on the fly and computed numerically via surveyed ground control points in the photo [23]. This crowd-sourced method opens new opportunities for emerging countries, where coastal research is limited due to lack of resources, but social media and smartphone usage are high.

2.2. *Satellite-Derived Remote Sensing Approach*

High-resolution satellite imagery, i.e., Landsat 1–5 multispectral scanner (MSS) (60 m), Landsat 4–5 thematic mapper (TM) (30 m), Landsat 7 enhanced thematic mapper (ETM+) (30 m), Landsat 8 operational land imager (OLI) (30 m) and Sentinel-2 (10 m/20 m/60 m) are becoming freely available, hence there are new opportunities to remotely monitor coastal areas more frequently. Such data offer mapping of estuaries immediately after or during events, such as floods due to change in sea level, or via human disturbances. Additionally, there is potential for monitoring the sea level and intertidal zones of coastal lagoons. For example, Salameh et al. [1] demonstrated the ability of altimetry to retrieve the landscape of the intertidal zone and the sea surface height. Furthermore, a study conducted by Liu et al. [24] explored the super-resolution technique that utilized 29 years of Landsat imagery data to derive monthly, seasonal and annual trends to demonstrate coastline variability. More recently, machine learning is playing an important role in mapping satellite imagery data. For example, Park et al. [25] utilized a support vector machine (SVM) learning classifier on high-resolution imagery data (acquired from PlanetScope satellite) to automatically map the coastal area. Moreover, Vos et al. [26] explored a machine learning approach to extend the approach used by Liu et al. to detect the coastline in the satellite imagery data. The authors then compared their enhanced coastline method to five coastline datasets. They reported a cross-shore root mean square error (RMSE) value of 8.2 m; at the Narabeen-Collaroy between in situ (Emery method [27] with RTK-GPS data) and satellite-derived data. The authors showed that by utilizing a machine learning approach they were able to capture storm-scale variations and shoreline behaviour. Satellite monitoring offers a long-term source of the coastal dataset and such monitoring is vital in defining monthly and seasonal variability. However, it is a delayed response and therefore, not suited for real-time monitoring of the site.

2.3. Unmanned Aerial Vehicle Remote Sensing Approach

Unmanned Aerial Vehicles (UAVs) have been increasingly used in different fields of geoscience such as beach dune evolution [28], rocky cliff erosion [29], gully erosion [30], tidal inlet evolution [31] and coastal monitoring [32]. UAV extends the use of aerial photogrammetry for coastal surveying by utilizing autonomous flight capabilities, advancements in Lidar technologies and state of the art computer vision technologies. Ian et al. [33] suggested that coastal monitoring can be carried out accurately by using photogrammetry techniques. Structure from Motion (SfM) [28] is a photogrammetric approach which produces a 3D-point cloud data by stitching together a series of 2-dimensional (2D) overlapping images. However, this generated point cloud data are dependent on ground control points. Ian et al. [33] compared the survey accuracy conducted by UAV (using SfM technique) and RTK-GPS ATV at Narrabeen—Collaroy in Australia. The authors mentioned that the difference between these two methods has a mean difference of 0.026 m and a standard deviation of 0.068 m. Furthermore, authors reported from 1:1 comparison that both survey data are highly correlated to each other (i.e., linear slope = 0.996 and R^2 = 0.998). In addition, authors mentioned that in complex environment the elevation variability of UAV approach is comparable to expected vertical accuracy of RTK-GPS. However, UAV does not provide a permanent monitoring solution and such monitoring is dependent on the presence of a trained UAV operator.

2.4. Lidar-Based Remote Sensing Approach

Lidar sensors operate on the principle of time of flight (TOF) by computing a distance between the target and sensor [34]. The basic functioning of Lidar sensor is shown in Figure 2, and the distance calculation between the target object and the lidar sensor is shown in Equation (1). Hydrodynamic, beach profile, and morphodynamic features can be obtained by using high-frequency laser pulses. In recent years, due to advancements in Lidar technology, Lidar data have been extensively used to obtain subaerial beach profiles, transformation in the surf zone, and near-shore breaking waves [35,36]. The accuracy of laser scanners is comparable to in situ RTK-GNSS surveying methods. For instance, Philip et al. [37] investigated post-storm effects on the beach face and studied different modes of berms via utilizing Lidar. The authors reported a root-mean-square error (RMSE) of 4 cm when compared with 11 monthly RTK-GPS ground control in situ surveys. The authors also provided useful insights into tide-by-tide building of the berm and beach face. Moreover, Brodie et al. [38] reported a RMSE in between 3 cm to 7 cm when compared with the Lidar data and pressure-based measurements for the range of hydrodynamic surf zone properties. The approach of using static Lidar for remote sensing applications provides an efficient way to acquire data with centimetre level accuracy.

$$\text{Distance between the target object and Lidar sensor} = \frac{\text{Pulse travel time}}{2} \times \text{speed of light} \tag{1}$$

... is based on the pulse travel time and speed

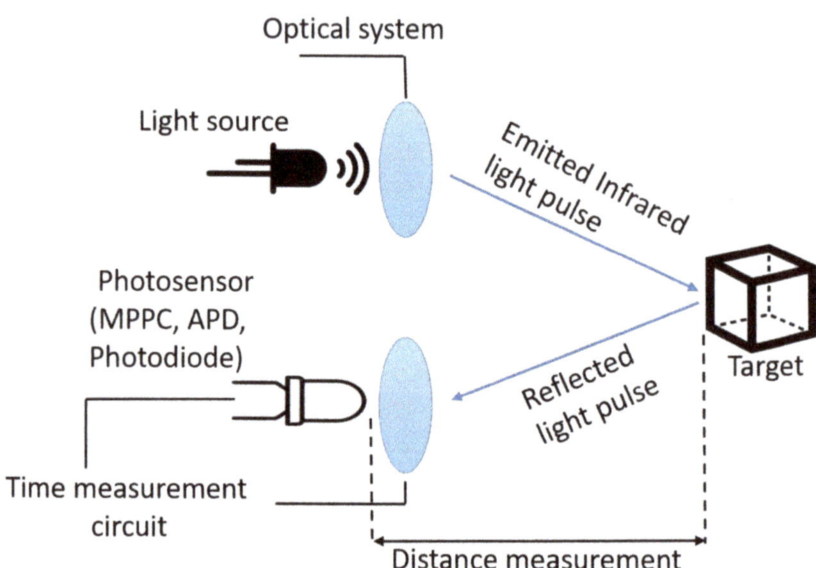

Figure 2. Lidar and time of flight (TOF) principle.

3. Methodology

This section provides details for the procedure involved regarding study site, and the monitoring setup. This section also lists pre-processing steps undertaken to minimize the noise in the 3D-point cloud data acquired from the Lidar. Moreover, lagoon entrance extraction from the subaerial beach profile, and parametric information estimation of the berm and dry notch is also discussed in this section.

3.1. Study Site

Fairy Lagoon is located at the southern end of Fairy Meadow beach, within the northern part of the Wollongong, New South Wales, Australia (refer to Figure 3). The behaviour of Fairy Lagoon immensely impacts on water quality, water levels and the ecology of the coastal area. Fairy Lagoon can be characterized as follows:

- A popular recreational and tourist area; open access to tourists and the general public due to urbanization around the lagoon entrance.
- Intermittently open and close to the ocean due to formation of a sand bar. After breaching the entrance, it stays open for a few weeks or months, and sand is deposited due to the longshore drift and wave action.
- Entrance is subject to periodic flooding; receives a large quantity of water due to runoff from an urbanized area, and a recent flood study has indicated that main component of the flood risk is associated with the elevated water level in the lagoon concerning the ocean.

The greater Fairy Creek catchment (also includes Cabbage Tree Creek) has an area of about 20.76 km^2, which further extends from Illawarra escarpment to the coast and includes Fairy Meadow, North Wollongong, Balgownie, and Mount Ousley residential areas. Rapid urbanization around catchment has caused a negative impact on the conditions of Fairy Lagoon. During an extended period of heavy rainfull and closed entrance, the lagoon would naturally breakout, and that can result in flooding of neighbouring urban development. To manage this risk effectively, real-time monitoring is required.

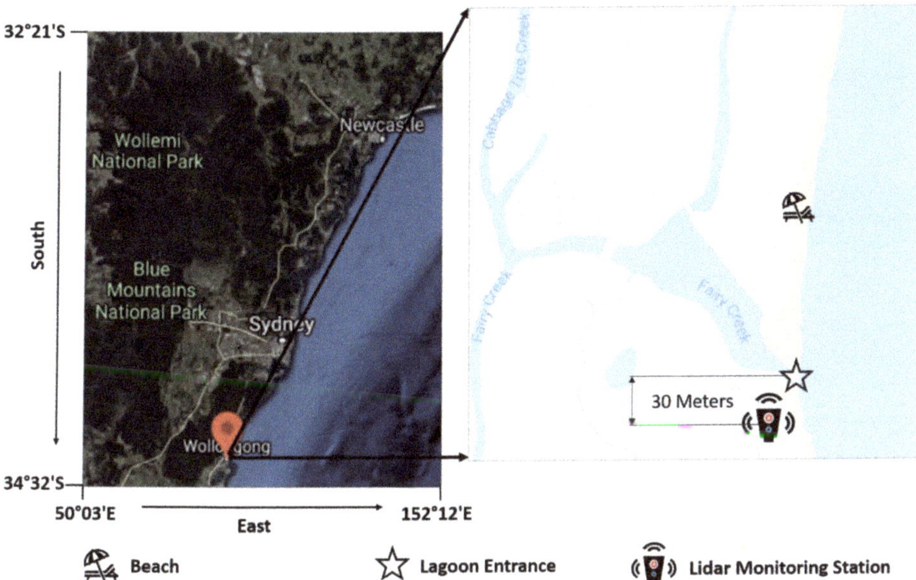

Figure 3. Fairy Lagoon Entrance situated at the Southern end of Fairy Meadow beach, within the northern part of the Wollongong, New South Wales, Australia (Google Maps, 2020).

3.2. Monitoring Setup

The static monitoring station developed for this study can be seen in Figure 4. The monitoring station is equipped with a Lidar (Cepton Vista—P60), along with an edge computer (Nvidia Jetson Nano), 4G universal serial bus (USB) modem (Huawei e83272) and an inertial measurement unit (IMU) sensor (Phidget Precision 3/3/3). This remote sensing station is solar-powered, making it fully autonomous.

Figure 4. (**A**) Remote monitoring station; (**B**) Field of View from Lidar Sensor.

The monitoring station is deployed at the Fairy Lagoon entrance and located approximately 7.8 m above the mean sea level (MSL). The Lidar is used to extract the 3D-point cloud data of the lagoon entrance. The Field of View from Lidar to the entrance is about 60° × 22° (H × V) with an angular resolution of approximately 0.25° both horizontally and vertically. The Lidar provides information for each ground point: x, y, z coordinates and intensity of the reflected light pulse. The intensity corresponds to the strength of the reflected pulse and reflectance of a target object. For instance, in coastal environment reflectance of dry sand is about 35 to 45 percent whereas, for wet sand, it varies from 20 to 30 percent [39]. The Lidar used for this research can detect an object with the minimum reflectivity coefficient of 30 percent at 200 m. Additionally, as that object moves closer to the Lidar sensor, that object can be detected with even less reflectivity score. However, as that object moves further away from the Lidar sensor, that object needs to be more reflective to be able to get detected by the Lidar sensor. The reason for using a long-range Lidar sensor for this application is due to the lagoon entrance being far from the nearest infrastructure, where the monitoring station is mounted.

For monitoring the entrance, NVIDIA's Jetson Nano platform acquires Lidar data via Ethernet. The Jetson unit is an edge computing [40] device meaning that it locally performs all computations to extract the topographic profile from raw data. The IMU sensor is connected to the edge computer via USB and outputs the orientation of the Lidar sensor in x, y and z directions. The 4G provides the IoT connectivity for the monitoring setup and automatic data transmission to an external database via MQ Telemetry Transport (MQTT) [41]. The block diagram of a Lidar-based monitoring setup is shown in Figure 5. As part of this monitoring setup, an IoT-based [42] water level sensor is also deployed in the field to obtain water level data of the lagoon. This water level sensor is located next to the static Lidar station. The sensor provides the water level data in the local units of Australian Height Datum (AHD) (m, equivalent to Mean Sea Level (MSL). Moreover, these data also provide the minimum elevation point, for which the subaerial beach profile needs to be calibrated.

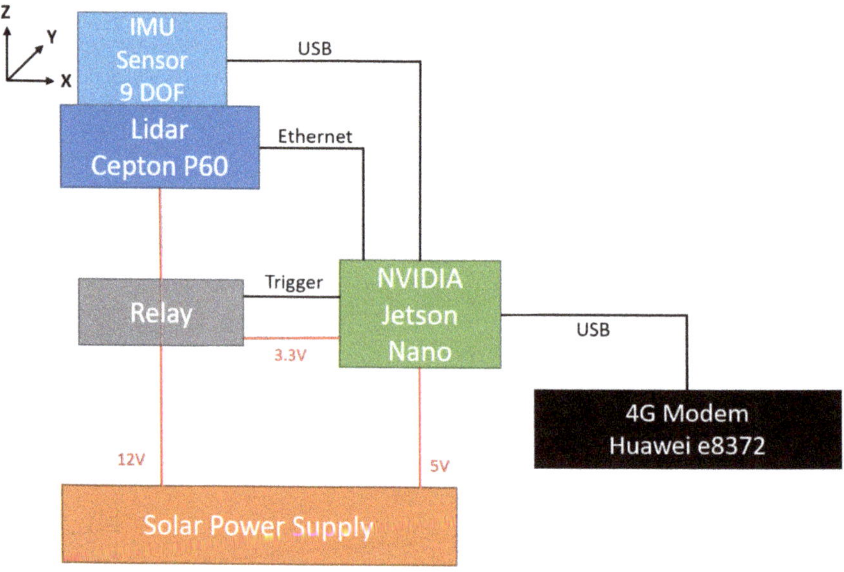

Figure 5. Block Diagram of Remote Monitoring Station.

3.3. Subaerial Lagoon Entrance Data Extraction and Preprocessing

The beach profile is extracted and analysed with a frequency of about every 3 h. The data acquired from the Lidar contain noise such as people, birds, and high-intensity reflections from objects being too close to the Lidar sensor (as shown in Figure 6A). To alleviate such noise, several pre-processing steps are required to obtain the subaerial lagoon profile from the raw 3D-point cloud data. The pre-processing steps are listed as below.

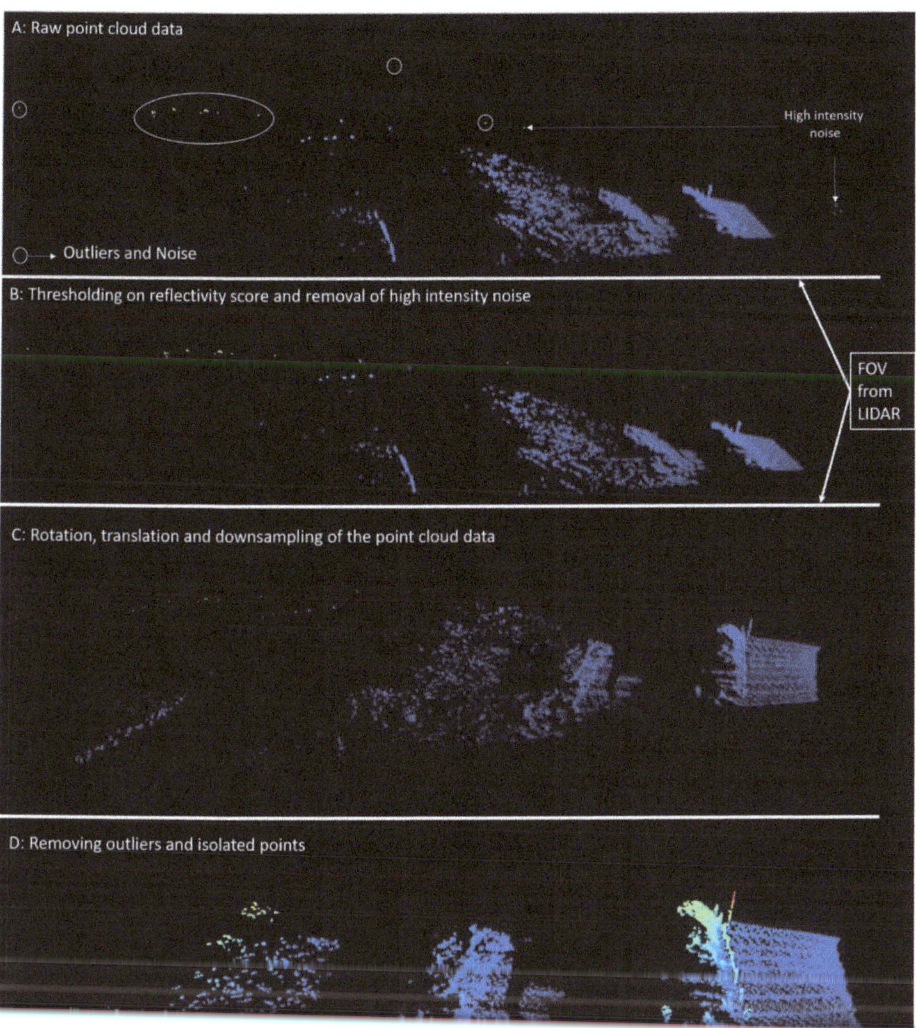

Figure 6. (**A**) Raw point cloud data acquired from the LiDAR; (**B**) Removal of high-intensity noise via thresholding on reflectivity values; (**C**) Output of rotation, translation and down-sampling of the point cloud data; (**D**) Removing outliers and isolated points from resultant point cloud data.

3.3.1. Thresholding on Reflectivity Score

The main source of noise in the data is due to the high-intensity reflections from objects being too close to the Lidar sensor, which causes the saturation of the returned signal. Saturated and high-intensity points are then filtered out, based on their reflectivity score. The only points with a reflectivity coefficient value between 5 to 95 percent are stored for further analysis as shown in Figure 6B.

3.3.2. Rotation of the Point Cloud Data

Following the thresholding, the 3D-point cloud data are then rotated about the z-axis to represent the coastal elevations in metres. The IMU sensor automates the task of rotation and calibration of the Lidar data. It provides the rotation matrix information, including axis and the angle by which the 3D-point cloud data need to be rotated and this information is also used to counter the vibrations of the Lidar sensor during data acquisition.

3.3.3. Translation of Z-Axis

The z-axis (elevation) is then translated to find the correct elevations concerning for the sea level. The mean sea level (MSL) is used as a minimum elevation reference point to translate the elevation data along the z-axis.

3.3.4. Down-Sampling of Point Cloud Data

Following the rotation and translation, the next pre-processing step is down-sampling of the point cloud data. As the acquired point cloud data are highly dense; voxel down-sampling [43] is used to uniformly reduce the density of the point cloud data. The algorithm works by bucketing the points into voxels and then extract one point by averaging all the points in that voxel. The size of the voxel is determined by the user and usually depends on the number of points and application. In this case, after data analysis, a voxel size of 0.8 cm was selected. This voxel size was selected due to the high-density point cloud data. The reason for choosing a larger voxel size is to reduce the number of points for further processing. The down-sampled point cloud data are shown in Figure 6C.

3.3.5. Removing Outliers

The objects and people are then removed via thresholding on the elevation data. To remove birds and other small objects, a radius outlier removal technique [44] is employed to remove isolated points from the subaerial lagoon profile. This filter removes the points that have few neighbouring points around them in each sphere. The parameter which defines the minimum number of points in a given sphere is tuned based on the point cloud density and noise in the acquired data. In this application, the sphere radius is set to 3 m and the minimum number of points is 70. The final 3D-point cloud data are shown in Figure 6D.

It should be noted that, the radius outlier removal technique worked best when compared to other filters such as the statistical outlier removal [45] method due to the continuity of the natural landscape and interesting points being clustered together.

3.3.6. Interpolation and Visualization

The interpolation [46] is then performed on spatially unstructured data to uniformly distribute it on the plane. The elevation data are then projected on a plane to form a contour plot [47]. This step is only performed for visualization purposes. The contour plot shown in Figure 7 represents the subaerial beach profile and provides an information about the elevation of the sand berm.

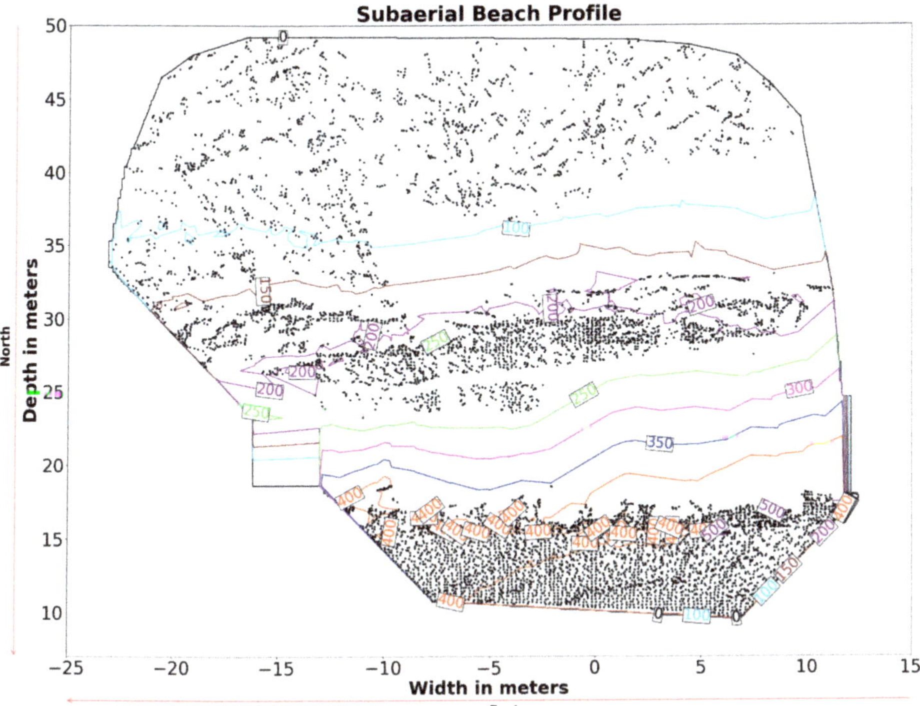

Figure 7. Subaerial Beach Profile; where x-axis represents the width and y-axis represents the depth concern to sensing station. The elevation data are also shown on the contour plot.

3.4. Extracting the Berm and Dry Notch Heights

Following the subaerial beach profile generation is the estimation of the dry notch location and its elevation. The first step to find the location of the dry notch is the identification of the lagoon entrance in the data (obtained after removing outliers in Section 3.3.5).

As mentioned earlier, the 3D-point cloud data represent the width (x-axis), depth (y-axis) and elevation (z-axis) of the entrance, and the surrounding area. The Lagoon entrance can be extracted by finding the continuous stream of points along the x-axis of the Lidar data divided in a grid of squared windows of size $2 \times 3\ m^2$. For a closed entrance, there [illegible] Figure 8B). The pseudo-code for extracting the lagoon [illegible] Figure 8C.

A: Pseudocode for finding lagoon Entrance

```
Input:
- Lidar data where x-axis represents the width and y-axis the depth of the view from Lidar
- threshold1 = min depth required for the lagoon entrance
- threshold2 = min width required for the lagoon entrance

## first step is to extract the region of interest from subaerial beach profile (non-interpolated data) via thresholding
## and grid the data as shown in this Figure-B
entrance = [ ]
width (x-axis), depth (y-axis) = region after 30 meters (subaerial beach profile)
total number of x steps = width / single x step
total number of y steps = depth / single y step

for current_depth_step in total number of y steps:
    for current_width_step in total number of x steps:
        count points in each grid
    end
    sum of points for given depth region = $\sum_{0}^{n} count\ points\ in\ each\ grid$
    number of non zero grids = nonzero elements in [count points in each grid]
    if (sum of points for given depth region > threshold1) and (number of non zero grids > threshold2):
        entrance.append(current_depth_step)
    else:
        current_depth_step is water or non-continuous subaerial surface
    end
end
## extracted Lagoon entrance is shown in this Figure-C
```

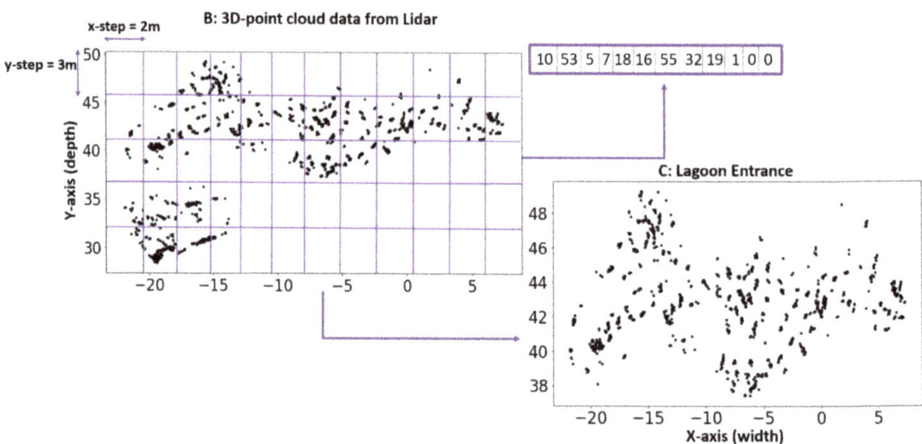

Figure 8. (**A**) Pseudocode to extract the Lagoon Entrance; (**B**) Gridding of the point cloud data and number of points in each grid is also shown for the given y-step; (**C**) Lagoon entrance extracted by following steps shown in (**A**).

3.5. Berm and Dry Notch

The dry notch height can be determined after the lagoon entrance has been identified. Estimating the dry notch height requires first determining the berm height. The berm height is typically classified as the highest point between the two highest elevations points, whereas the dry notch is normally defined as the lowest elevation point between the two highest crest points.

For this process, firstly, elevation data (z-axis) are extracted from the 3D-point cloud data which belongs to the entrance. The acquired elevation data are still dense and noisy. To alleviate such noise, data are then down-sampled to single decimal point resolution, i.e., decreasing resolution of points taken for analysis (as shown in Figure 9B). After down-sampling, maximum elevation points are filtered out in each window. Crests (peaks) and troughs (minimum dips) are then calculated based on the neighbouring elevation points. The dry notch is then located, and its elevation can be estimated (as shown in Figure 9B) and the pseudo-code of this process is shown in Figure 9.

A: Pseudo code for finding berm and dry notch height

```
Input:
- lagoon entrance
- min_region = Minimal region require along the width for the lagoon entrance

## decrease the resolution of points
find unique values along the depth = unique elements in [x-axis]
decrease the x-axis resolution to single decimal point = single decimal resolution [find unique values along the depth ]
## create a small x-step and find the average along the y-axis
x-step = width of lagoon entrance / min_region
take an average for x-step along the y-axis = take an average for x-step in [y-axis]

## choose a length of step and find maximum elevations
find maximum elevation in chosen step = find maximum points in [take an average for x-step along the y-axis ]

## find the crest and troughs in the data
crest = find all crests [find maximum elevation in chosen step]
trough = find all troughs [find maximum elevation in chosen step]
berm height = maximum elevation in [crest] (shown in Figure-B: Berm and dry notch)

## once crest and troughs are found dry notch would be the minimum dip between two maximum crests
first maximum crest, second maximum crest = two maximum values in [crest]
index1, index2 = indexes [first maximum crest, second maximum crest]
dry notch height = minimum elevation in between [index1, index2] (shown in Figure-B: Berm and dry notch)
```

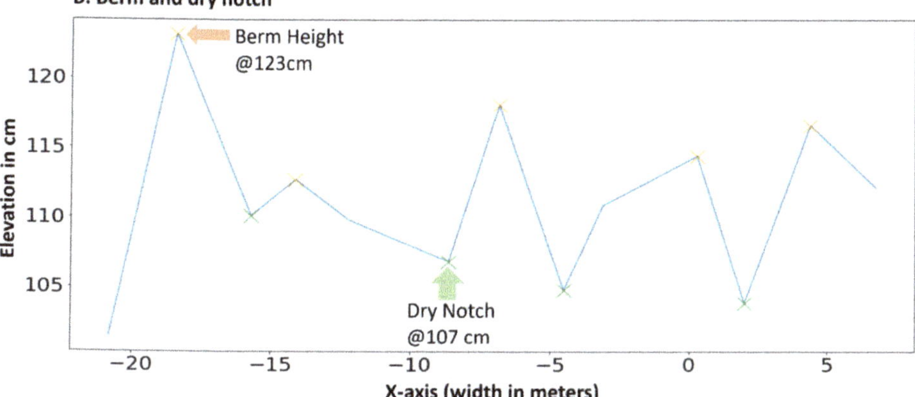

Figure 9. (**A**) Pseudo code for estimating the dry notch height from lagoon entrance; (**B**) Down-sampled elevation data versus width. Berm and dry notch estimation from the elevation data by following steps shown in (**A**).

4. Results

[illegible] the proposed approach with the in situ survey conducted on the

4.1. Accuracy Comparison [illegible]

For comparison purposes, the processed Lidar 3D-point cloud data are compared with an in situ survey. Root mean square error (RMSE) is used to calculate the error between Lidar data and the ground truth. RMSE of 12.4 cm was reported during such comparison. The overlapping points were estimated by using a minimum distance threshold of 4 cm between both Lidar and survey data. This error can be minimized by using a lower overlapping distance threshold such as for 1 cm difference the RMSE was reported to be 6.2 cm. A visual comparison of the ground truth and the Lidar data can be seen in Figure 10. Moreover, the empirical cumulative distribution function shows that 92.3% of the squared error difference is less than 2.7 between the actual and processed Lidar point as shown in Figure 11.

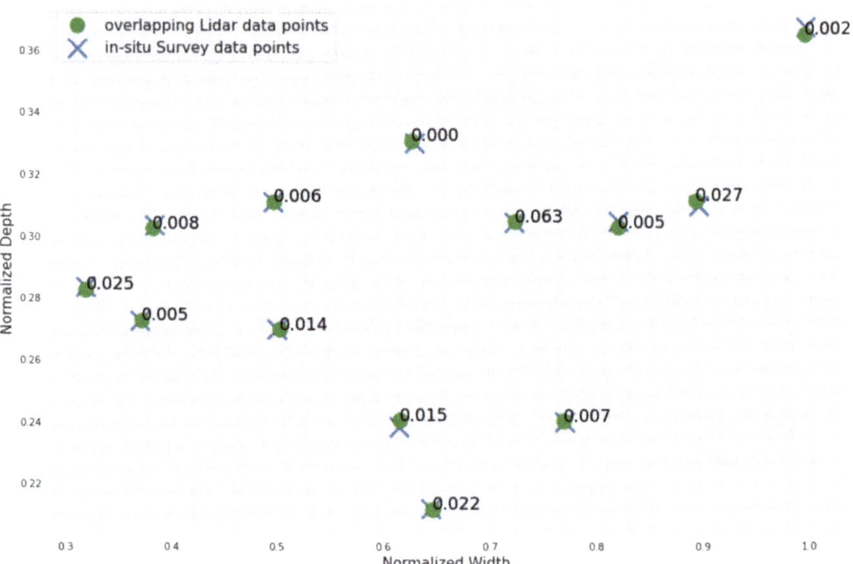

Figure 10. Difference between Lidar and in situ survey data in meters.

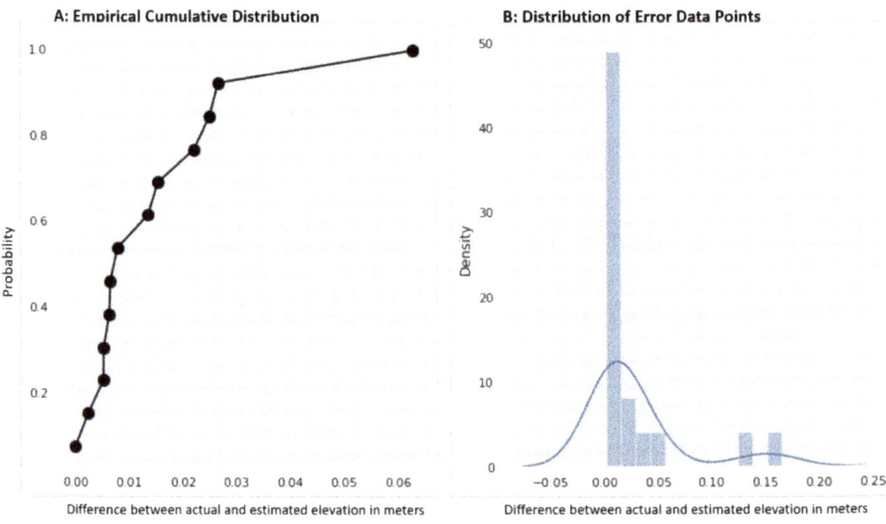

Figure 11. (**A**) Empirical Cumulative Distribution of the error; (**B**) Distribution of the resultant error.

4.2. Berm and Dry Notch Height Comparison between Lidar and In Situ Survey

Berm and dry notch height are then estimated via the algorithm detailed in Figure [illegible]. These parameters are then compared with the ground truth values acquired from the in situ survey data. The results are shown in Table 1 and the visual comparison can be seen in Figure 12. The difference of 4.5 cm was then found between actual berm height and the estimated height from the point cloud data, whereas for dry notch the difference of 1.6 cm was obtained between ground truth and the Lidar data. These small deviations demonstrate the proposed approach's potential for lagoon monitoring applications.

Table 1. Comparison of difference between Lidar 3D-point cloud data and ground truth.

	Berm Height (m)	Dry Notch Height (m)
Lidar 3D-point cloud data	0.685	0.267
Ground Truth from survey data	0.730	0.283
Difference between Lidar data and ground truth	0.045	0.016

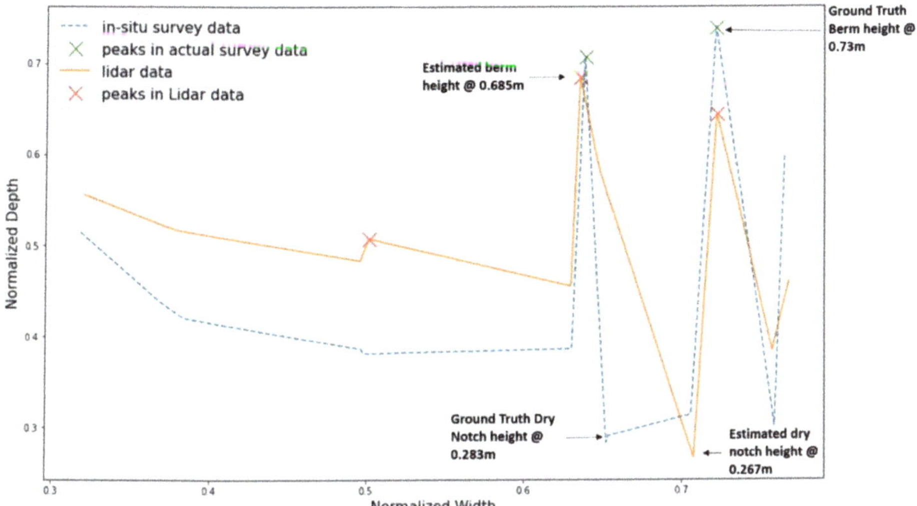

Figure 12. Sand berm height comparison between ground truth and Lidar point cloud data.

5. Discussion

In this paper, a remote sensing-based approach for regular monitoring of estuaries is presented to counter the flood threat that they pose to the local communities. In addition, the literature is examined to gain a better understanding of the emerging technologies for [illegible] early warning and real-time monitoring system, [illegible] remote sensing [illegible] scientist [illegible] control points for calibration and achieving [illegible] vision algorithms are major shortcomings. In addition, both UAV and satellite [illegible] coastal remote sensing approaches come with the limitations of continuous and real-time monitoring, respectively. Static Lidar has been used in literature for coastal applications such as understanding the behaviour of near-shore breaking waves [35], cross-shore 2d scanning of the dune and for general coastal behaviour applications [10]. However, to the best of authors' knowledge, Lidar technology has not been used as early warning system for managing lagoon entrances.

An initial objective of the project was to estimate the berm height and it was hypothesized that Lidar can be used to extract the topographic information of the lagoon entrance. The Lidar-based remote-sensor architecture described in this paper is an alternative to an in situ survey and works on the functionality of extracting the elevation data along with the estimation of parameters related to berm and dry notch. The most important aspect of this work is that proposed system aims to alert the operator about the conditions of the entrance based on the calculated information. Another interesting finding was that the breaching of an entrance is also dependent on the water level in the lagoon and ocean level. To counter this, an IoT-based water level sensor is also deployed in the lagoon to monitor the water level. This information along with the topographic profile of the entrance is also transmitted to a remote database for facilitating local authorities in decision making process.

The elevation data acquired from static monitoring station are compared with the in situ survey data in Section 4.1. Comparison reported that 92.3% of the squared error difference is less than 2.7 cm^2 between point cloud data and actual ground truth. For the presented comparison, overlapping ground truth points and 3D point cloud data are acquired by using a window size of 1 cm, due to unavailability of exactly overlapping data points. This error can be reduced if the ground truth points exactly align with the Lidar 3D point cloud data. Moreover, Section 4.2 shows that the error difference for berm and dry notch height is 4.5 cm and 1.6 cm, respectively. From the presented investigations, and encouraging results, it is highlighted that Lidar based approach performance was admissible and has the potential for lagoon monitoring applications.

This research work is unique in-essence of providing a near-real-time monitoring application of the lagoon entrance. The presented results are notable in terms of the scalability of the proposed approach and autonomous efficient site monitoring. This paper further contributes towards the development of a generic framework and algorithm that can be implemented to other flood prone lagoon entrances. Furthermore, the proposed solution has the potential to be embedded in an early warning system for providing entrance parametric information (i.e., berm height, dry notch height) to the human operator and recommending an optimal time to breach the entrance. Knowing when to dredge the berm is crucial for effective flood management at lagoon entrances, therefore, the proposed solution will help protecting local communities residing next to flood-prone areas.

6. Conclusions and Future Work

The effective sand management of closed and open lakes or lagoons (ICOLLs) is critical to mitigate the impact of future flood events in their neighbourhood. In particular, if the height of the sand berm of a lagoon entrance is above a threshold determined by flood engineers, then the local authorities must mechanically break the entrance to allow the water to flow during extreme weather events. Currently, the height is determined via infrequent and time-consuming manual surveys. In addition, when there is a high risk of flood and due to unavailability of data, local authorities may rush towards opening the entrance while it is unnecessary. This sub-optimal policy increases the number opening and closing of the entrance which negatively impact on the local biodiversity. Hence, determining the optimal opening schedule to lower the number of human interventions to open the entrance will lower the environmental impact.

The monitoring station presented in this paper addresses the issue of autonomous and on-demand monitoring of the entrance while providing actionable information to local authorities. The autonomous monitoring station relies on a Lidar facing the coast to obtain an accurate topographic profile of the lagoon entrance. The Lidar is paired with an edge computer running a novel algorithm to provide an efficient and flexible way to continuously monitor the dynamics of berm formations and extract the heights of the berm and its dry notch from the 3D-point cloud data generated by the Lidar. Thanks to the inertial measurement unit the solution is self-calibrating. In addition, being solar-powered and using 4G connectivity for data transmission, the device is fully autonomous and can be easily deployed to other estuaries.

Moreover, the acquired data from the Lidar are stored in a database and therefore accessible for future research. These include the development of models to better understand coastal erosion, as well as the berm behaviour and evolution during the opening and closing phases of the entrance. Finally, the device will be paired with other sensors such as water levels, disdrometers, and automated rain-gauges as well as data sources about the weather forecast to design complete early flood warning system for the city of Wollongong (NSW, Australia).

Author Contributions: Conceptualization, B.A. and J.B.; Formal analysis, B.A. and J.B.; Funding acquisition, P.P.; Investigation, B.A. and J.B.; Methodology, B.A. and J.B.; Software, B.A. and J.B.; Supervision, J.B. and P.P.; Writing—original draft, B.A.; Writing—review and editing, B.A., J.B. and P.P. All authors have read and agreed to the published version of the manuscript.

Funding: This work received funding from the Australian Government under the Smart Cities and Suburbs Program Round 2 Grant number SCS69244.

Data Availability Statement: No new data were created or analyzed in this study. Data sharing is not applicable to this article.

Acknowledgments: This work was supported by Wollongong City Council in partnership with Shellharbour, Kiama and Shoalhaven Councils, Lendlease and the University of Wollongong's SMART Infrastructure Facility. The authors also gratefully acknowledge the support of NVIDIA Corporation with the donation of the Titan V used for this research.

Conflicts of Interest: The authors declare no conflict of interest.

References

1. Salameh, E.; Frappart, F.; Marieu, V.; Spodar, A.; Parisot, J.-P.; Hanquiez, V.; Turki, I.; Laignel, B. Monitoring sea level and topography of coastal lagoons using satellite radar altimetry: The example of the Arcachon Bay in the Bay of Biscay. *Remote Sens.* **2018**, *10*, 297. [CrossRef]
2. Roy, P.; Williams, R.; Jones, A.; Yassini, I.; Gibbs, P.; Coates, B.; West, R.; Scanes, P.; Hudson, J.; Nichol, S. Structure and function of south-east Australian estuaries. *Estuar. Coast. Shelf Sci.* **2001**, *53*, 351–384. [CrossRef]
3. Hanslow, D.; Davis, G.; You, B.; Zastawny, J. Berm height at coastal lagoon entrances in NSW. In Proceedings of the Proc. 10th ann. NSW coast. conf., Yamba, NSW, Australia, 3–5 November 2000.
4. Weir, F.M.; Hughes, M.G.; Baldock, T.E. Beach face and berm morphodynamics fronting a coastal lagoon. *Geomorphology* **2006**, *82*, 331–346. [CrossRef]
5. Booysen, Z.; Theron, A.K. Methods for predicting berm height at Temporarily Open/Closed Estuaries. *Estuar. Coast. Shelf Sci.* **2020**, *245*, 106906. [CrossRef]
6. Gordon, A.D. Coastal acclimatisation to intermittently open river entrances (IORE). In Proceedings of the Coasts and Ports 2013: 21st Australasian Coastal and Ocean Engineering Conference and the 14th Australasian Port and Harbour Conference, Sydney, NSW, Australia, 11–13 September 2013; p. 313.
7. Arshad, B.; Ogie, R.; Barthelemy, J.; Pradhan, B.; Verstaevel, N.; Perez, P. Computer Vision and IoT-Based Sensors in Flood Monitoring and Mapping: A Systematic Review. *Sensors* **2019**, *19*, 5012. [CrossRef] [PubMed]
8. Barthelemy, J.; Amirghasemi, M.; Arshad, B.; Fay, C.; Forehead, H.; Hutchison, N.; Iqbal, U.; Li, Y.; Qian, Y.; Perez, P. Problem-Driven and Technology-Enabled Solutions for Safer Communities: The case of stormwater management in the Illawarra-Shoalhaven region (NSW, Australia). *Handb. Smart Cities* **2020**, 1–28. [CrossRef]
9. Iqbal, U.; Perez, P.; Li, W.; Barthelemy, J. How Computer Vision can Facilitate Flood Management: A Systematic Review. *Int. J.*

Observations and modelling at an open-coast beach. [illegible]

12. Splinter, K.D.; Holman, R.A.; Plant, N.G. A behavior-oriented dynamic model for sandbar migration and 2DH evolution. *J. Geophys. Res. Ocean.* **2011**, *116*. [CrossRef]
13. Lippmann, T.C.; Holman, R.A. Quantification of sand bar morphology: A video technique based on wave dissipation. *J. Geophys. Res. Ocean.* **1989**, *94*, 995–1011. [CrossRef]
14. Harley, M.D.; Turner, I.L.; Short, A.D.; Ranasinghe, R. An empirical model of beach response to storms-SE Australia. In Proceedings of the Coasts Ports: In a Dynamic Environment, Wellington, New Zealand, 16–18 September 2009; pp. 600–606.
15. Beuzen, T.; Splinter, K.; Marshall, L.; Turner, I.; Harley, M.; Palmsten, M. Bayesian Networks in coastal engineering: Distinguishing descriptive and predictive applications. *Coast. Eng.* **2018**, *135*, 16–30. [CrossRef]

16. Mole, M.A.; Mortlock, T.R.; Turner, I.L.; Goodwin, I.D.; Splinter, K.D.; Short, A.D. Capitalizing on the surfcam phenomenon: A pilot study in regional-scale shoreline and inshore wave monitoring utilizing existing camera infrastructure. *J. Coast. Res.* **2013**, *65*, 1433–1438. [CrossRef]
17. Bracs, M.A.; Turner, I.L.; Splinter, K.D.; Short, A.D.; Lane, C.; Davidson, M.A.; Goodwin, I.D.; Pritchard, T.; Cameron, D. Evaluation of opportunistic shoreline monitoring capability utilizing existing "surfcam" infrastructure. *J. Coast. Res.* **2016**, *32*, 542–554. [CrossRef]
18. Andriolo, U.; Sánchez-García, E.; Taborda, R. Operational use of surfcam online streaming images for coastal morphodynamic studies. *Remote Sens.* **2019**, *11*, 78. [CrossRef]
19. Sánchez-García, E.; Balaguer-Beser, A.; Pardo-Pascual, J.E. C-Pro: A coastal projector monitoring system using terrestrial photogrammetry with a geometric horizon constraint. *Isprs J. Photogramm. Remote Sens.* **2017**, *128*, 255–273. [CrossRef]
20. Jaud, M.; Kervot, M.; Delacourt, C.; Bertin, S. Potential of smartphone SfM photogrammetry to measure coastal morphodynamics. *Remote Sens.* **2019**, *11*, 2242. [CrossRef]
21. Roger, E.; Tegart, P.; Dowsett, R.; Kinsela, M.A.; Harley, M.D.; Ortac, G. Maximising the potential for citizen science in New South Wales. *Aust. Zool.* **2020**, *40*, 449–461. [CrossRef]
22. Hart, J.; Blenkinsopp, C. Using Citizen Science to Collect Coastal Monitoring Data. *J. Coast. Res.* **2020**, *95*, 824–828. [CrossRef]
23. Harley, M.D.; Kinsela, M.A.; Sanchez-Garcia, E.; Vos, K. Shoreline change mapping using crowd-sourced smartphone images. *Coast. Eng.* **2019**, *150*, 175–189. [CrossRef]
24. Liu, Q.; Trinder, J.C.; Turner, I.L. Automatic super-resolution shoreline change monitoring using Landsat archival data: A case study at Narrabeen–Collaroy Beach, Australia. *J. Appl. Remote Sens.* **2017**, *11*, 016036. [CrossRef]
25. Park, S.J.; Achmad, A.R.; Syifa, M.; Lee, C.-W. Machine learning application for coastal area change detection in gangwon province, South Korea using high-resolution satellite imagery. *J. Coast. Res.* **2019**, *90*, 228–235. [CrossRef]
26. Vos, K.; Harley, M.D.; Splinter, K.D.; Simmons, J.A.; Turner, I.L. Sub-annual to multi-decadal shoreline variability from publicly available satellite imagery. *Coast. Eng.* **2019**, *150*, 160–174. [CrossRef]
27. Krause, G. The "Emery-method" revisited—performance of an inexpensive method of measuring beach profiles and modifications. *J. Coast. Res.* **2004**, *20*, 340–346. [CrossRef]
28. Mancini, F.; Dubbini, M.; Gattelli, M.; Stecchi, F.; Fabbri, S.; Gabbianelli, G. Using unmanned aerial vehicles (UAV) for high-resolution reconstruction of topography: The structure from motion approach on coastal environments. *Remote Sens.* **2013**, *5*, 6880–6898. [CrossRef]
29. Mancini, F.; Castagnetti, C.; Rossi, P.; Dubbini, M.; Fazio, N.L.; Perrotti, M.; Lollino, P. An integrated procedure to assess the stability of coastal rocky cliffs: From UAV close-range photogrammetry to geomechanical finite element modeling. *Remote Sens.* **2017**, *9*, 1235. [CrossRef]
30. Marzolff, I.; Poesen, J. The potential of 3D gully monitoring with GIS using high-resolution aerial photography and a digital photogrammetry system. *Geomorphology* **2009**, *111*, 48–60. [CrossRef]
31. Long, N.; Millescamps, B.; Guillot, B.; Pouget, F.; Bertin, X. Monitoring the topography of a dynamic tidal inlet using UAV imagery. *Remote Sens.* **2016**, *8*, 387. [CrossRef]
32. Laporte-Fauret, Q.; Marieu, V.; Castelle, B.; Michalet, R.; Bujan, S.; Rosebery, D. Low-Cost UAV for high-resolution and large-scale coastal dune change monitoring using photogrammetry. *J. Mar. Sci. Eng.* **2019**, *7*, 63. [CrossRef]
33. Turner, I.L.; Harley, M.D.; Drummond, C.D. UAVs for coastal surveying. *Coast. Eng.* **2016**, *114*, 19–24. [CrossRef]
34. Liu, J.; Sun, Q.; Fan, Z.; Jia, Y. TOF lidar development in autonomous vehicle. In Proceedings of the 2018 IEEE 3rd Optoelectronics Global Conference (OGC), Shenzhen, China, 4–7 September 2018; pp. 185–190.
35. Brodie, K.L.; Slocum, R.K.; McNinch, J.E. New insights into the physical drivers of wave runup from a continuously operating terrestrial laser scanner. In *2012 Oceans*; IEEE: Hampton Roads, VA, USA, 2012.
36. Martins, K.; Blenkinsopp, C.E.; Power, H.E.; Bruder, B.; Puleo, J.A.; Bergsma, E.W. High-resolution monitoring of wave transformation in the surf zone using a LiDAR scanner array. *Coast. Eng.* **2017**, *128*, 37–43. [CrossRef]
37. Phillips, M.; Blenkinsopp, C.; Splinter, K.; Harley, M.; Turner, I. Modes of berm and beachface recovery following storm reset: Observations using a continuously scanning lidar. *J. Geophys. Res. Earth Surf.* **2019**, *124*, 720–736. [CrossRef]
38. Brodie, K.L.; Raubenheimer, B.; Elgar, S.; Slocum, R.K.; McNinch, J.E. Lidar and pressure measurements of inner-surfzone waves and setup. *J. Atmos. Ocean. Technol.* **2015**, *32*, 1945–1959. [CrossRef]
39. Allaby, M. *A Change in the Weather*; Facts on File: New York, NY, USA, 2004; p. 200.
40. Verstaevel, N.; Barthélemy, J.; Forehead, H.; Arshad, B.; Perez, P. Assessing the effects of mobility on air quality: The Liverpool Smart Pedestrian project. In Proceedings of the Transportation Research Procedia, Mumbai, India, 26–30 May 2019; pp. 2197–2206. [CrossRef]
41. Standard, O. MQTT Version 5.0. *Retrieved June* **2019**, *22*, 2020.
42. Forehead, H.; Barthelemy, J.; Arshad, B.; Verstaevel, N.; Price, O.; Perez, P. Traffic exhaust to wildfires: PM2. 5 measurements with fixed and portable, low-cost LoRaWAN-connected sensors. *PLoS ONE* **2020**, *15*, e0231778. [CrossRef]
43. Miknis, M.; Davies, R.; Plassmann, P.; Ware, A. Near real-time point cloud processing using the PCL. In Proceedings of the 2015 International Conference on Systems, Signals and Image Processing (IWSSIP), London, UK, 10–12 September 2015; pp. 153–156.
44. Zhou, Q.-Y.; Park, J.; Koltun, V. Open3D: A modern library for 3D data processing. *arXiv preprint* **2018**, arXiv:09847.

45. Balta, H.; Velagic, J.; Bosschaerts, W.; De Cubber, G.; Siciliano, B. Fast statistical outlier removal based method for large 3D point clouds of outdoor environments. *Ifac-Pap* **2018**, *51*, 348–353. [CrossRef]
46. Ashraf, I.; Hur, S.; Park, Y. An investigation of interpolation techniques to generate 2D intensity image from LIDAR data. *IEEE Access* **2017**, *5*, 8250–8260. [CrossRef]
47. Nelli, F. *Python Data Analytics: With Pandas, Numpy, and Matplotlib*; Apress: New York, NY, USA, 2018.

Article

Land Cover Changes and Flows in the Polish Baltic Coastal Zone: A Qualitative and Quantitative Approach

Elzbieta Bielecka, Agnieszka Jenerowicz, Krzysztof Pokonieczny * and Sylwia Borkowska

Faculty of Civil Engineering and Geodesy, Military University of Technology, 00-908 Warsaw, Poland; elzbieta.bielecka@wat.edu.pl (E.B.); agnieszka.jenerowicz@wat.edu.pl (A.J.); sylwia.borkowska@wat.edu.pl (S.B.)

* Correspondence: krzysztof.pokonieczny@wat.edu.pl

Received: 26 May 2020; Accepted: 27 June 2020; Published: 29 June 2020

Abstract: Detecting land cover changes requires timely and accurate information, which can be assured by using remotely sensed data and Geographic Information System(GIS). This paper examines spatiotemporal trends in land cover changes in the Polish Baltic coastal zone, especially the urbanisation, loss of agricultural land, afforestation, and deforestation. The dynamics of land cover change and its impact were discussed as the major findings. The analysis revealed that land cover changes on the Polish Baltic coast have been consistent throughout the 1990–2018 period, and in the consecutive inventories of land cover, they have changed faster. As shown in the research, the area of agricultural land was subject to significant change, i.e., about 40% of the initial 8% of the land area in heterogeneous agriculture was either developed or abandoned at about equal rates. Next, the steady growth of the forest and semi-natural area also changed the land cover. The enlargement of the artificial surface was the third observed trend of land cover changes. However, the pace of land cover changes on the Baltic coast is slightly slower than in the rest of Poland and the European average. The region is very diverse both in terms of land cover, types of land transformation, and the pace of change. Hence, the Polish national authorities classified the Baltic coast as an area of strategic intervention requiring additional action to achieve territorial cohesion and the goals of sustainable development.

Keywords: Baltic coast; Poland; CORINE Land Cover; land cover flow; urbanisation; afforestation; deforestation; spatial analysis; SDGs

---

changes, in particular their quantitative and qualitative ...
them, have been reported in literature since the 1960s. It was noted that the change in land cover modifies albedo and heat fluxes [1], the exchange of energy in the surface–atmosphere interaction [2], carbon sequestration [3], and evapotranspiration [1], which finally affect the climate. Later, a much broader scope of the impact of land cover changes on ecosystem goods and services was identified, e.g., the vulnerability of places and people to climatic, economic, or sociopolitical perturbations, the ability of biological systems to support human needs [4], biotic diversity worldwide [5], and soil degradation [6]. Many of these changes remain a serious challenge today. As stated by Lambin et al. [7], many thought land changes mainly consisted in the conversion of arable land and forest to urban

and industrial areas (urban sprawl), the transformation of forests to agricultural uses (deforestation), as well as the devastation of natural vegetation by overgrazing, which leads to desert conditions (desertification). These conversions were assumed to be irreversible and spatially homogeneous and to progress linearly [7]. However, as noticed by [7,8], not all impacts of LC changes are negative, as some are associated with growths in food and fiber production, in the efficiency of resource use, wealth, and well-being.

The coastal zone has been the main goal of the development of human society for centuries, which resulted in a land cover transformation. Urbanisation and industrialisation, tourism, fishing and other agricultural activities very often lead to environmental degradation, loss of biodiversity, and simplification of the landscape [9]. Halting the loss of biodiversity is an issue of both local and global concern, and it is one of the main strategic goals of the environmental policy of the European Union [10] as well as the United Nations Sustainable Development Goals (SDGs) defined in the "2030 Agenda" [11]. The most important of them are those explicitly related to coastal regions, i.e., Sustainable Cities and Communities—Goal 11, Climate Action—Goal 13, and Life below Water—Goal 14, as well as Goal 15—Protect, restore and promote sustainable use of terrestrial ecosystems, sustainably manage forests, combat desertification, and halt and reverse land degradation and halt biodiversity loss [12,13]. Satellite-derived information on land cover (use), as well as geographical distribution of population at the fine scale, provide an essential contribution to establishing measures to achieve the above objectives [14].

The European Union (EU) coastal regions are densely populated, being home to more than 41% of the total European population. However, population density varies depending on the geographical characteristic of the region, e.g., in Cyprus and Denmark 100% of people live in coastal regions, while in Romania only 5% live in coastal regions, and in Poland, only 12% do [15].

The coastal landscapes of the Baltic Sea are varied and show considerable regional differences. Based on a profound analysis of meteorological parameters of the Baltic Sea coastal region, e.g., air temperature (including the heat and cold waves), precipitation (rainfall), changes of the sea level (including storm surges and erosion of the coast), and sea monitoring data (e.g., salinity, acidity), the Institute of Meteorology and Water Management [16], identified warming of the troposphere and hydrosphere and change in rainfall intensity. The direct consequences of climate change comprise an increase of air temperature in all seasons of the year and changes of precipitation. Besides, such extreme phenomena as heat waves and longer dry and wet periods are also expected to occur more frequently. Climate change affects both ecosystems and the economy, but the effects vary depending on the geographical location and development. However, in the Baltic Sea region, despite climate change, transformation processes in the economy and agriculture that have been going on since the 1990s still have substantial and multiple impacts on ecosystems and human beings [17,18].

Land cover in the Polish Baltic coastal region is considered to be highly sensitive to observed climate change, since it is directly linked to weather and the environmental conditions [16,19]. However, the changes in land cover have not yet been assessed. Our study attempts to fill this gap by providing the first comprehensive quantitative and qualitative analysis of land cover changes and their nature in the Polish Baltic coastal zone in the last three decades, namely 1990–2018, the dates of the first and fifth CORINE Land Cover (CLC) data releases. Particularly, the trajectory of land cover changes in 1990–2018 was analysed with certain emphasis on urbanisation, agriculture intensification, and extensification, as well as afforestation and deforestation. Moreover, the study derives environmentally related indicators that could help monitor coastal management plans, and Spatial Development Goals 2030, especially goals 11 and 15. The study contributes to both research and policy fields. The introduction of CLC-based indicators makes it possible to determine the size and direction of land cover changes, and thus assess whether their impact on sustainable development is positive or negative. Considering the comprehensive analysis of land cover flows in the last three decades derives an effective footprint for establishing further sustainable development rules in the

Polish Baltic Coast Region. Moreover, the research helps raise awareness to follow the spatial planning policy, particularly during rapid economic growth and spatial development.

The remainder of this paper is structured as follows: Section 2 describes the study area, data, and methods; Section 3 describes the results; Section 4 discusses the obtained results along with other achievements; finally, Section 5 concludes the research.

2. Materials and Methods

2.1. Area Description

The study covers the Polish Baltic coastal zone as defined by Eurostat [20], namely the NUTS3 administrative division having a border with a coastline or with more than half of their population living less than 50 km from the sea. It comprises the eight regions: Elblaski, Gdanski, Trojmiejski, Slupski, Koszalinski, Starogardzki, Szczecinski, and the city of Szczecin, as shown in Figure 1. They are further divided into 37 counties (NUTS4). Socioeconomic transformations and the incessant development of tourism, recreation, and associated services, as well as intense exploitation of marine resources have recently increased the role of the Polish coastal regions.

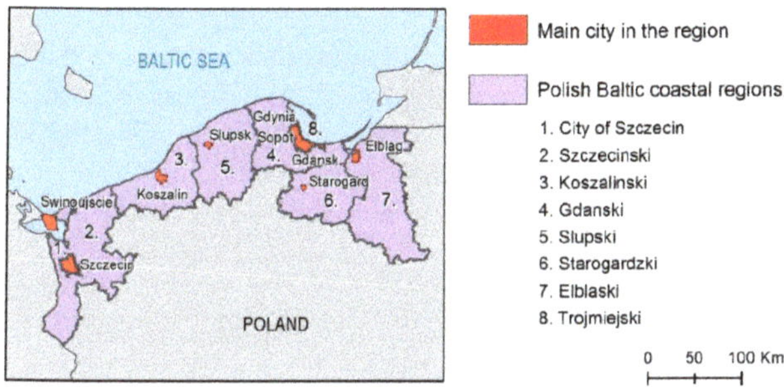

Figure 1. The Polish Baltic coastal zone.

Approximately 3.9 million people live in the study area, which encompasses 33,482 km^2 (see Table 1). With a mean population density of about 117 people per km^2 (the country average is 123 people per km^2), the study area is predominantly rural. Trojmiejski (comprises three cities: Gdansk, Gdynia, and Sopot) and the City of Szczecin regions are of significant industrial, economic, and cultural importance. The Baltic coastal zone is characterised by a 20.7% share of high nature value areas, which attract tourism development and improve the residents' quality of life, but on the other hand, also restrict investment opportunities. The highest proportion of legally protected areas (NATURA2000, [illegible] reserves, national parks, landscape parks), up to 30%, is found in the Szczecin (city), Gdanski,

Coastal Region (NUTS3) Names	Number of Counties			
Elblaski	7(1)	526,321	72	
Gdanski	5(0)	590,198	137	4299.30
Trojmiejski	3(3)	748,986	1805	415.05
Slupski	4(1)	335,402	64	5240.16
Koszalinski	5(1)	357,478	82	4363.27
Starogardzki	5(0)	432,488	106	4095.79
Szczecinski	7(1)	513,412	69	7444.81
City of Szczecin	1(1)	402,465	1339	300.50
Total	37(8)	3,906,750		33,482.34

[1] Based on statistical data from the year 2018 [21]. [2] The number of city counties (cities with poviat rights) is shown in brackets.

The Polish Baltic coast is 468 km long, without internal lagoonal coasts, including 428 km from Russia to the German border and 72 km of both sides of the Hel peninsula [22]. Dunes and sandy beaches dominate most of the Polish open coast (358 km), while cliffs are approximately 100 km long [22,23]. Two major gulfs: Pomerania Gulf in the west, and the Gulf of Gdansk in the east, as well as two large lagoons, the Szczecin Lagoon and the Vistula Lagoon, limit the Polish Baltic coast. About 26% of the coast is protected by groynes (98 km) and light and heavy revetments (41 km), which mostly protect the environment and densely populated areas [23].

2.2. Data Used

2.2.1. CORINE Land Cover

CORINE Land Cover (CLC) is the European land cover database containing information on the physical and biological cover of the Earth's surface including artificial surfaces, agricultural areas, forests, (semi-)natural areas, wetlands, and water bodies [24]. The information capture is generally based on computer-assisted photo-interpretation of satellite images (see Table 2), according to agreed, standardised, and hierarchical nomenclature. The temporal span for satellite acquisition data, as well as the sensor used for capturing the images, are the primary sources of uncertainty, significantly affecting the interpretation of land cover [25]. The first (highest) level comprises five land cover types, namely (1) artificial surfaces, (2) agricultural areas, (3) forest and semi-natural areas, (4) wetland and (5) water bodies, which are further divided into 15 land cover classes (level 2), and 44 classes at level 3. Until now, five CLC inventories are available for the reference years: 1990, 2000, 2006, 2012, 2018, as well as four LC changes for the periods 1990–2000; 2000–2006; 2006–2012; and 2012–2018 (see Figure A1). In Poland, the datasets are derived free of charge from the Chief Inspectorate of Environmental Protection [26].

Table 2. Overview of satellite data used for land cover interpretation in Poland.

Reference Year	Satellite	Temporal Extend (Poland) [1]
1990	Landsat 4/5 TM	1986–1995
2000	Landsat 7 ETM+	1990–2001
2006	SPOT 4/5, IRS-P6	2005–2006
2012	RapidEye, IRS-P6	2011–2012
2018	Sentinel 2, Landsat-8	2017

[1] Based on National Reports on CORINE Land Cover inventories (available at Chieef Environmental Inspectorate-Glowny Inspektroat Ochrony Srodowiska (GIOS) web site http://clc.gios.gov.pl/).

Our study considers the second level of CLC nomenclature, and only those classes that are mapped in Poland, as presented in Table 3.

Table 3. CORINE Land Cover nomenclature (level 1 and level 2).

CLC Level 1		CLC Level 2	
Code	Name	Code	Name
1.	Artificial surfaces	1.1	Urban fabric
		1.2	Industrial, commercial and transport units
		1.3	Mine, dump and construction sites
		1.4	Artificial, non-agricultural vegetated areas
2.	Agricultural areas	2.1	Arable land
		2.2	Permanent crops
		2.3	Pastures
		2.4	Heterogeneous agricultural areas
3.	Forests and semi-natural areas	3.1	Forest
		3.2	Scrub and/or herbaceous vegetation associations
		3.3	Open spaces with little or no vegetation
4.	Wetlands	4.1	Inland wetlands
		4.2	*Coastal wetlands* [1]
	Water bodies	5.1	Continental waters
		5.2	Marine waters

[1] *Italics* show the land cover class that does not occur in Poland.

2.2.2. Satellite and Aerial Data

Prior to the analysis of changes in land cover and land flow, CLC change data were checked based on aerial orthophotomaps, high and very high-resolution mosaics elaborated within the frame of Copernicus land monitoring services as well as optical satellites (medium and high resolutions) available from open map services, as specified in Table 4 and shown in Figure A2.

Table 4. Satellite and aerial imageries used for CLC change verification.

Products	Time Horizon	Repository	Comments
Aerial orthophotomapas	1990s, since 2000 every 2 years	Geoportal maintained by the Head Office of Geodesy and Cartography [27]	For the 1990s. data available only for Gdansk, Sopot, Szczecin; 0.25 resolution
Very High-Resolution Image Mosaic 2012	2012	Copernicus land monitoring services [28]	True Colour (2.5m) of pan-sharpened: SPOT-5, 6 and FORMOSAT-2
The HR Mosaic for 2018	2018	Copernicus land monitoring services [28]	Sentinel-2 (10 m) true and false colour compositions
Sentinel-2	2018	European Space Agency (ESA) [29]	Sentinel-2 (10 m, 20 m, 60 m); false colour compositions
IKONOS, QuickBird, GeoEye, WorldView, SPOT, Pleiades	Since 2000 [1]	Google Earth	True-colour compositions; Very High-resolution images 2.5–5 m resolution
Landsat	1990 [1], 2000, 2006, 2012, 2018	USGS Earth Explorer [30]	For 1990 partial coverage; Pansharpened images (15 m); True and false colour compositions

[1] Depending on the period and place being analysed, different Very High Resolution (VHR) satellite images are available.

2.2.3. Administrative Boundaries

The boundaries of NUTS3, small regions for specific diagnoses, were obtained from the Eurostat GISCO (Geographical Information and maps) [31,32]. Each NUTS unit is attributed a NUTS_CODE (unique code of the NUTS region as defined and published by Eurostat; NUTS_LABEL (name of the NUTS region as defined and published by Eurostat), and TAA (the type of the administrative area, i.e., land area or inland water). Further statistical and administrative division of Poland (counties) of the year 2018 was derived from the National Register of Boundaries (PRG) maintained by the Head Office of Geodesy and Cartography.

2.3. Methods Applied

2.3.1. Verification of Land Cover Changes

Data checking mainly involved verification of the CLC change code correctness. For the analyses, the correctness of the results of automatic change detection between classes resulting from the comparison of CLC databases from two relevant periods (e.g., 2000 and 2006). The analysis consisted mainly of visual analysis of remote sensing satellite imagery and verification, whether the change in the CLC database corresponds to the situation from multispectral satellite images. The analyses used data from open-source imagery data, i.e., GoogleEarth, USGS, and ESA Copernicus, that were not used [illegible] About 0% of land cover change polygons, randomly

[illegible]

imagery data (Table 4). Prior to visual analysis, pan-sharpening of Landsat7 and Landsat 8 images was conducted. The nearest-neighbour diffusion-based pan-sharpening algorithm [33] was used to enhance the visual quality and improve the spatial resolution (to 15 m) of medium- resolution data. Moreover, almost all CLC changes that may lead to deforestation have been thoroughly investigated to look for forest management practices, in particular, sanitary clear-cuttings and planned logging that could lead to temporal deforestation.

2.3.2. Land Cover Changes Quantification

Land cover changes, defined as the transformation from one land cover class to another in two consecutive moments in time, were firstly categorised as land conversion or land modification following the definitions given by Lambin et al. [7,34]. Hence, land cover conversion means "the complete replacement of one cover type by another" and is measured by the shift from one CLC level 1 class to another (e.g., class 2—agricultural area to class 3—forests). These land cover changes result from such processes as urbanisation, deforestation (or afforestation), and agricultural expansion. On the other hand, land cover modification comprises all changes that "affect the character of the land cover without changing its overall classification" [7], e.g., extensification of agriculture, which is perceived as more beneficial to the environment.

This study aims to answer the following research questions:

1. What were the main land cover changes and land cover flows in the Polish Baltic coastal zone during the last three decades? Were they in line with the country average?
2. How did the changes in land cover affect the landscape in legally protected areas?
3. Is the spatial pattern of land cover changes clustered, dispersed, or random? Which regions are most affected by land cover changes?
4. How, positively or negatively, do changes in land cover affect the progress towards SDG?

The main trajectories of land cover changes were distinguished based on a slightly modified methodology introduced by Feranec et al. [35] that relies on an analysis of land cover transformation matrix during the analysed time period. Finally, the following land cover flows were identified to answer the above research questions:

LCF1—Urbanisation—refers to the conversion of agricultural land and forests into artificial surfaces (CLC classes 2.1, 2.2, 2.3, 3x, 4x to 1x).

LCF2—Expansion and intensification of agriculture—denotes conversion from natural and semi-natural areas (3.2, 3.3, 4.1, 4.2) into high-intensity agriculture (2.1; 2.2).

LCF3—Extensification of agriculture—states the modification from high to low-intensity forms of agriculture, namely from arable lands (2.1) and permanent crops (2.2) into pastures (2.3) or heterogeneous agricultural areas (2.4.).

LCF4—Afforestation—the re-creation of forest land, it generally comprises transformation from agriculture to forests.

LCF5—Deforestation—understood as the transition of forest land into non-forest land. Deforestation is predominantly associated with urbanisation (LCF1), i.e., transformation from forest to artificial surfaces, mainly 1.2, 1.3, and agricultural expansion, i.e., transformation of the forest into an agricultural area.

As the research focussed not only on land cover changes but also on their impact on achieving SDGs 2030, the following indices were calculated: land consumption rate (LConR), urban green space ratio (UGrR), forest cover ratio (FR), and the proportion of forest areas located within legally protected areas ($F_{LPA}P$). They were the basis for calculating SDGs' relevant indicators and contributed to monitoring progress goals 11 and 15. Primarily, these included target 11.3—towards the enhancement of inclusive and sustainable urbanisation, target 11.7—provide universal access to safe, inclusive, and accessible green and public spaces, as well as target 15.1—sustainable use of terrestrial ecosystems, in particular forests, and target 15.2—sustainable management of all types of forests, halt deforestation, restore degraded forests and substantially increase afforestation and reforestation globally.

Land consumption rate LConR, expressed in hectares per year, is defined as the ratio of the difference between the area occupied by artificial surfaces ($ArtS_{t+1}$) at the final year ($t + n$) and the initial year (t) to the number of years (n) between land cover inventories (Equation (1)):

$$LConR = \left[\left(\frac{AS_{t+n} - AS_t}{AS_t}\right)\frac{1}{n}\right]100\%. \quad (1)$$

The *LConR* relates directly to Goal 11 and allows us to calculate a measure of land use efficiency, which is defined as a ratio of land consumption rate to the population growth rate (11.3.1). The *LConR* index was computed for the Baltic Coastal zone in Poland, the Baltic coast regions (NUTS3), and protected areas.

The urban green space ratio in the year ($UGrR_t$) denotes the percentage share of an urban green area (CLC class 1.4) in total city acreage (Equation (2)):

$$UGrR_t = (UG_t/U_{area(t)}) \times 100\% \tag{2}$$

where UG_t—urban green area (i.e., CLC 1.4, and 3.x classes) at the year t; $U_{area(t)}$—urban area at the year t. With the CORINE Land Cover inventory, it allows comparing progress across cities towards the achievement of an optimal quantity of land allocated to public space (Goal 11, target 11.7 and indicator 11.7.1). The indicator was computed only for eight city counties, namely: Szczecin, Swinoujscie, Koszalin, Slupsk, Elblag, Gdansk, Gdynia, and Sopot, where the last three cities form the Trojmiejski (Tri-City) region.

The forest cover ratio (FR) is defined as the percentage of total forests and semi-natural areas (CLC level 2 cases 3.1 and 3.2) in the total land area at the time t. The measure predominantly applies to indicator 15.1.1., namely forest area ($F_{area(t)}$) as a proportion of total land area ($T_{area(t)}$), Equation (3):

$$FR = (F_{area(t)}/T_{area(t)}) \times 100\%. \tag{3}$$

The proportion of forest area located within legally protected areas ($F_{LPA}P_{(t)}$) at the analysed period t applies to 15.2.1 indicator—progress towards sustainable forest management (Equation (4)):

$$F_{LPA}P_{(t)} = (F_{area(t)}/LP_{area(t)}) \times 100\% \tag{4}$$

where $F_{area(t)}$—forest area at the time t; $LP_{area(t)}$—acreage of the legally protected area.

The spatial pattern of land cover changes was analysed by inferential statistics, namely Average Nearest Neighbour (ANN) and Ripley's K function, assuming that complete spatial randomness is a realisation of a Poisson point process (the null hypothesis). ANN measures the average distance from each point to its nearest point, which is compared to the distances between the nearest points and distances that would be expected based on chance. The average distance less than the average for a hypothetical random distribution indicates clustering, while greater distance indicates the dispersion of analysed features [36]. Contrary to ANN, the Ripley's K-function examines, instead of computing separating nearest neighbours, all inter-point distances. The function shows how to point pattern changes when the neighbourhood size increases or decreases. The variance stabilised Ripley K-function called the L function is generally used for data analysis (Equation (5)):

$$L(d) = \sqrt{\frac{A\sum_{i=1}^{N}\sum_{i=1,j\neq 1}^{N} k_{i,j}}{\pi N(N-1)}} \tag{5}$$

where d [illegible]

[illegible] of each central point of an individual land cover change patch. Then, the number of observed neighbouring centroids is compared to the number of central points that are expected based on an entirely spatially randomness. The advantage of Ripley's K function is the ability to analyse the pattern in the scale function. However, it should be remembered that due to edge effects, patterns are questionable at greater distances [37].

The density of land cover changes was portrayed by kernel density, choropleth, and graduated symbol maps. Kernel density calculates a magnitude-per-unit area from the centroid of land cover change polygons using a changing area as a weight factor to fit a smoothly shaped surface to each point [38].

Disparities of land cover flows were portrayed by Gini coefficient, due to Equation (6) [39]:

$$G(x) = \frac{\sum_{i=1}^{n}(2i - n - 1)x'_i}{\mu\, n^2} \quad (6)$$

of n ordered individuals with x'_i the size of individual i and $x'_2 < x'_3 < \cdots x'_n$, where: n—number of analysed regions, x'_i—the level of land cover flows in ascending order $< x'_1 < x'_2 < \cdots .. < x'_{n-1} < x'_n$, and μ—mean value.

Gini takes values from 0 to 1, with 0 showing perfect equality and 1 representing perfect inequality. The coefficient was used as an indicator of unequal distribution of chosen land cover flows in the analysed areas.

2.4. *Workflow*

The study was conducted in three consecutive phases: (1) preparation, (2) analysis, and (3) presentation of results. The preparatory phase comprised data acquisition, transformation to common Coordinate Reference System (CRS), and such geoprocessing operations as clip, intersect, and dissolve. It also established the level of details of land cover changes analysis, i.e., the entire Polish Baltic coastal zone, regional, and local for selected analysis. An important step in the preparatory phase was visual verification of CLC changes in all analysed time intervals.

The analytical phase relied on land cover net changes and land flows calculation over the four short periods (1990–2000, 2000–2006, 2006–2012, 2012–2018) and long-term changes over the past 30 years (1990–2018), as is presented in Figure 2. The results were related to the Polish Baltic coastal zone, regions, and legally protected areas. Based on the developed formulas (Equations (1)–(5)), the progress towards Sustainable Development Goals 2030 in the Polish Baltic coastal zone was assessed. Qualitative analysis showed the land cover change spatial pattern and density.

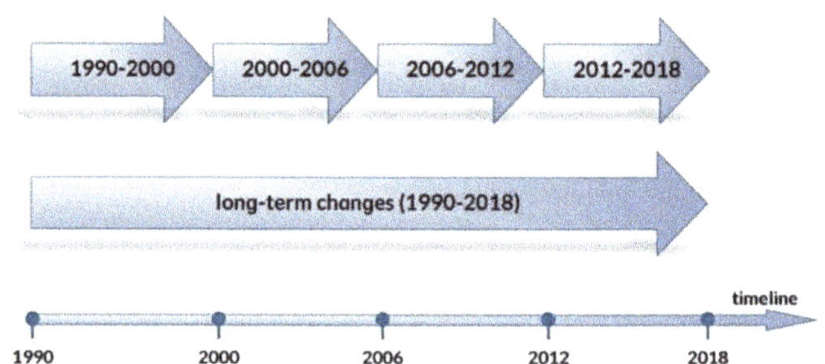

Figure 2. Short and long-term analysis of land cover changes; simplified diagram.

The final stage visualised the results using a set of thematic maps, mainly choropleth, charts, and tables as well as described and summarised the land cover changes across the Polish Baltic coastal zone.

3. Results

3.1. *CLC Change Data Verification*

CORINE Land Cover data were profoundly verified before making them publicly available. The verification that has been done both at the country level by national verification teams and at the European level by the European Environment Agency (EEA) technical team achieved at least 85% of thematic accuracy [24]. Moreover, the independent validation based on the LUCAS (The Land

Use/Cover Area frame Survey) Europe-wide data that had not been used as source data for CLC shows the total reliability of CLC2000 at level 87.0 ± 0.8% [26,40]. Moreover, as estimated by EEA [40], the highest class-level trustworthiness, above 95%, was obtained for water bodies and urban areas. Both arable land (2.1) and coniferous forest (3.1.1), the two largest CLC land cover classes, were assessed to have a high level of reliability, i.e., 90%, and 95%, respectively. However, validation was executed for CORINE land cover polygons, not the change polygons. So, there is no independent CLC change data assessment. Nevertheless, if the land cover classes were derived truthfully, we should not expect significant errors in the change database.

The verification of CLC change codes for the Baltic coastal zone confirms the high thematic accuracy of CORINE land cover change data for each analysed period, reaching a value of 95%. In particular, just a few errors were observed in land cover conversion, i.e., urbanisation, intensification of agriculture, afforestation, and deforestation. Predominantly, doubts in the attribution of the CLC change code were found where high-intensity agriculture (i.e., 2.1 and 2.2) was transformed into low-intensity forms, such as pastures (2.3) and mixed agricultural areas (2.4). They were detected on 11 polygons with a total area of 91 ha, which, given the total area of changes (1106.2 km^2), can be considered insignificant and not burdening the results of further analysis.

3.2. *Overview of Land Cover and Land Cover Net Changes*

Agricultural land accounts for the most significant share of the Polish Baltic coastal zone with 56.28%. It is followed by forests that occupy 35.39%. Artificial surfaces cover just 5.1%, while water bodies cover 2.59%, and wetlands cover merely 0.54% (see Table 5).

Table 5. The area in percent occupied according to the CORINE Land Cover (CLC) inventories.

CLC Code	CLC Nomenclature	1990	2000	2006	2012	2018	Net Changes in 1990–2018 [1]
1.1	Urban fabric	1.793	2.138	3.375	3.899	3.948	2.155
1.2	Industrial, commercial and transport units	0.466	0.470	0.515	0.614	0.656	0.190
1.3	Mine, dump and construction sites	0.057	0.071	0.132	0.167	0.243	0.186
1.4	Artificial, non-agricultural vegetated areas	0.231	0.245	0.297	0.344	0.346	0.114
2.1	Arable land	44.741	44.664	43.606	42.680	42.519	−2.222
2.2	Permanent crops	0.089	0.070	0.071	0.082	0.078	−0.010
2.3	Pastures	8.875	8.416	8.948	8.774	8.717	−0.157
2.4	Heterogeneous agricultural areas	8.105	7.494	5.369	4.999	4.975	−3.130
3.1	Forest	32.146	32.856	33.153	33.709	33.573	1.427
3.2	Scrub and/or herbaceous vegetation associations	0.288	0.377	1.333	1.506	1.714	1.426
3.3	Open spaces with little or no vegetation	0.136	0.114	0.106	0.098	0.098	−0.038
4.1	Inland wetlands	0.521	0.523	0.509	0.539	0.540	0.019
4.2	*Coastal wetlands* [2]	0.000	0.000	0.000	0.000	0.000	0.000
5.1	Continental waters	2.474	2.482	2.501	2.506	2.509	0.034
5.2	Marine waters	0.080	0.079	0.086	0.085	0.084	0.005
	Total	100	100	100	100	100	

[1] + increase; – decrease. 0—unchanged; [2] land cover class that does not occur in Poland.

The structure of agricultural land use on the Baltic coast exhibits considerable differences from region to region (Figure 3a), with an average share of 47.4%. The easternmost regions, Starogardzki and Elblaski, are mainly agricultural, with an agricultural land share of 67.3% and 62.3% respectively of which 55.1% and 51.0% [illegible] and Sopot), are occupied by agriculture. One of the characteristics of the Polish Baltic zone landscape is the high percentage of forests, scrubs, and herbaceous vegetation (Figure 3b). The average forest cover in the Baltic region amounts to 33.8% (11,852.75 km^2) and is higher than the national average equals of 29.5%. On average, there is 0.3 ha of forests per capita, which is 0.06 ha more than the national average, but still two times lower than the world average: 0.62 ha [41]. Forests occupy nearly half of the Slupsk region area, 48.1%, while in Koszalinski they occupy as much as 39.6%. A relatively small area, but still slightly more than 25%, of forested land exists in the Starogardzki region and Szczecin city, 26.2% and 25.1%, respectively.

Moreover, forests cover a relatively large part of the Trojmiejski region, 31.8%. This is about 2.5% more than the national average, which makes this agglomeration one of the greenest ones in Poland. The urbanisation level, i.e., the share of artificial surfaces, in the Polish Baltic regions is also significantly diversified, with the highest share of 44.0% in the Trojmiejski region and 36.4% in the city of Szczecin and the lowest one (Figure 3c) in the Slupski agro-forestry region. This diversity is also emphasised by the large difference between the mean and median values of urbanisation that amount to 13.4% and 4.6%, respectively.

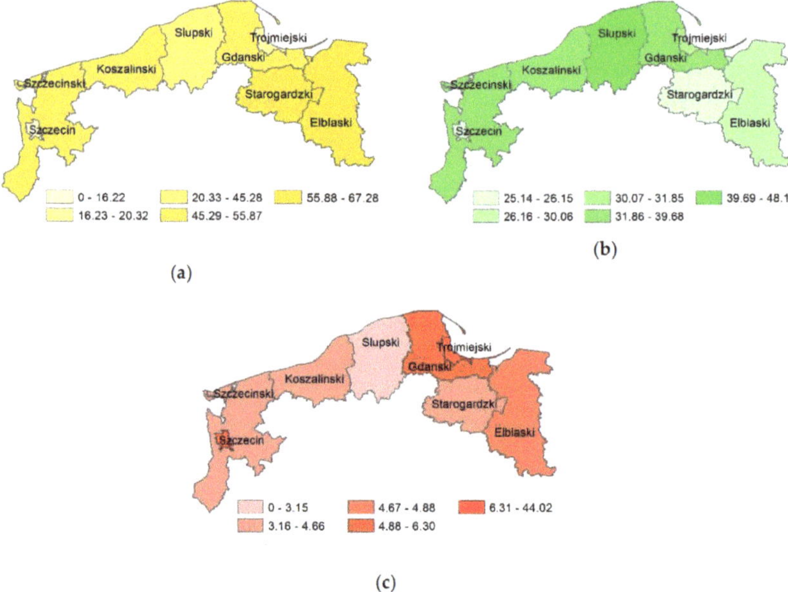

Figure 3. Polish coastal zone land cover structure in percent: (**a**) agricultural areas; (**b**) forest cover; (**c**) artificial surfaces.

The land cover changes in the Polish Baltic coastal zone comprise (1) gradual growth in urban areas, particularly transportation networks, commercial, industrial, and housing areas; (2) high (5.52%) loss of agricultural land, particularly arable land and pastures; and (3) a successive increase in forest cover associated with the agro-forestry programs implemented in Poland, and changes within the forest areas related to timber operations and forest renovation. A significant decrease in agricultural areas compensates for the overall increase in artificial surfaces and forests together with semi-natural lands (see Figure 4a). The growth of forests, shrubs, and semi-natural areas is slow and amounts to 0.1% per year, which after 28 years gives a 2.81% growth. The total increase in artificial surfaces was 2.65%, although the most significant growth occurred in 2000–2006 (1.4%), whereas in the remaining periods of CLC inventories, the gain in value remained significantly below 1% (Figure 4b). The percentages of various land cover categories in the increases of anthropogenic surfaces differ slightly with the highest proportion of land associated with roads and related infrastructure as well as construction sites (see Table 5). The loss of arable land follows a linear trend ($R^2 = 0.94$) with an annual decrease of 0.16%. The total loss of arable land in 1990–2018 equals 5.52%.

The net changes in the Baltic coastal regions also differ substantially (see Figure 4c). The Trojmiasto (Tri-city) agglomeration faced the highest loss of agriculture and at the same time the highest growth of urbanised lands. On the contrary, Szczecin, the second-largest city in the Polish Baltic region, has been afforested and developed at the expense of agricultural land. A relatively large number of agricultural areas have fallen in the Elblaski region. They have been afforested and taken over for transportation facilities and highways. The total land cover changes in the Polish Baltic coastal zone in 1990–2018 equal 3.22% (1078.14 km^2), whereas the annual land cover change takes the value of 0.12% (40.18 km^2). The amounts of total land cover changes, both in the acreage and number of patches, differ slightly between consecutive periods, being the highest in the last six-year period (Figure 4d).

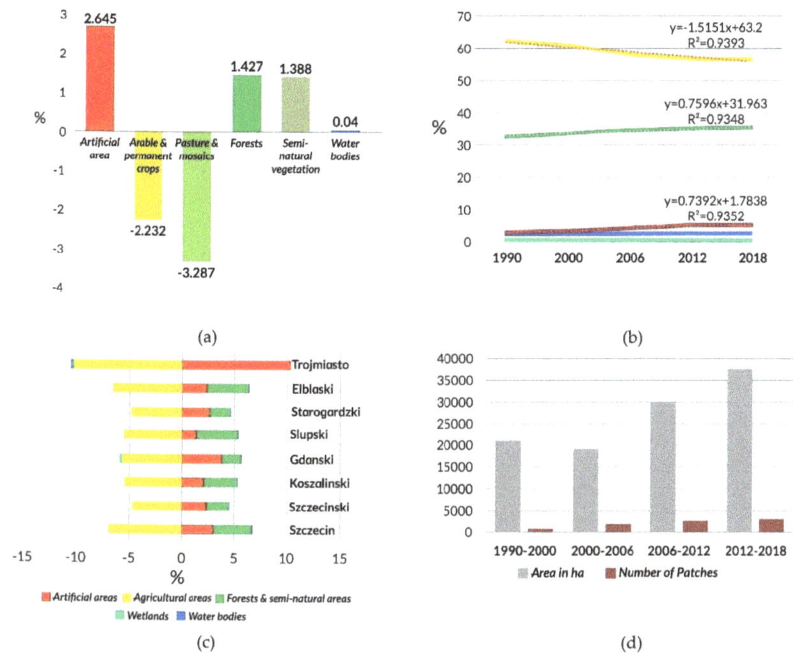

Figure 4. Land cover changes in the Polish Baltic coastal region: (**a**) net changes in percent 1990–2018; (**b**) percent of main CLC land cover types; (**c**) net changes in percent 1990–2018 by regions; (**d**) total area and number of patches.

The intensity of the long-term changes (from 1990 to 2018) varies considerably across the analysed area (Figure 5). Gdansk, the largest city on the Polish Baltic coast, has witnessed as much as 8.5% of land cover changes of the city's area. Moreover, the number of changes in the city amounted to 216, which results in the average area of a change polygon being 10.28 ha. Compared to 1990, land cover changed in 6.2% of the Gdanski county (the land county, neighbouring the city of Gdansk from the south), which makes it the second most affected county. Relatively small changes were recorded in typically small tourist cities such as Sopot (located near Gdansk, in the Trojmiejski region, in the eastern part of the Baltic zone) and Swinoujscie—placed on the west coast (0.3% and 0.0[illegible]

forests and scrub by 2.9%. In 1990, urban fabric areas covered 0.68% of protected areas, while in 2018, they covered as much as 1.98% with the highest increase observed in 2000–2006 (0.82%). The most considerable land cover changes affected agricultural land, as they covered a decline of 4.50% compared to 1990. The loss concerned all four agricultural classes of level 2 CLC with the highest decline of heterogeneous agriculture areas (2.38%) and arable land (1.79%). Pasture loss was 0.33% and permanent crops loss was merely 0.02%. Arable land was generally transformed into more extensive use, such as heterogeneous agriculture (CLC 2.4. class) or pasture (2.3 class), while 2.4 lands have very often been afforested. Forests in legally protected areas gradually increased their area by 1.53% compared to 1990, with the most significant changes taking place in 1990–2000 covering 1.21%, which was the result of the abandonment of agricultural land and reclamation of military areas that are assigned to the CLC class 3.2 (open spaces with little or no vegetation). However, a slight forest decrease (0.08%) was observed in 2018 compared to 2012. Most of these changes included forest clearing (class 3.2.).

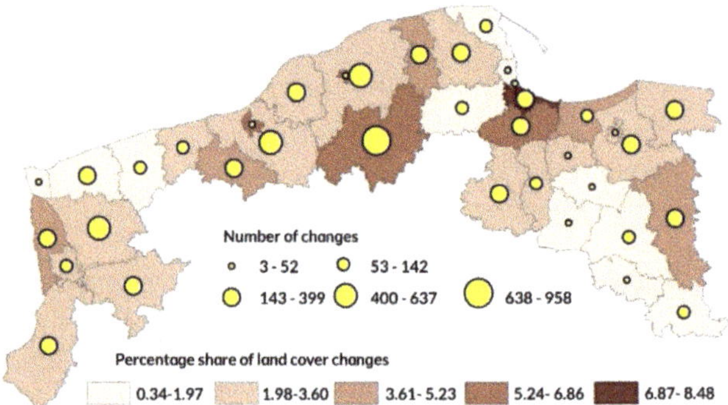

Figure 5. The percentage share of the land cover changes and the number of change polygons by counties (NUTS4 administration units in Poland).

3.3. *Spatial Pattern and Density of Land Cover Changes*

The spatial pattern of land cover changes over the Polish Baltic coastal zone in 1990–2018 was clustered, which is documented by the ANN statistics, particularly the NN ratio of 0.42 and the z-score of −95.18, indicating that there is a less than 1% likelihood that the clustered pattern is the result of random chance. Changes in land cover are also clustered in all analysed periods. The NN ratio did not differ much in subsequent analysed periods, assuming values less than 1, i.e., denoting a clustered pattern. In 1990–2000, the NN amounted to −0.33, in 2000–2006, it was −0.29, in 2006–2012, it took the value of 0.30, and in 2012–2018, this ratio equaled 0.39 (with p <0.001 for all statistics). Moreover, Ripley's K function graph (Figure 6) also shows a large concentration of land cover changes regardless of the scale, i.e., from the small, less than 20 km distances, to distances exceeding 100 km. Furthermore, the L (d) curve graph assumes a similar shape for changes in land cover in each of the analysed periods, indicating their clustered nature in close proximity, as well as throughout the entire study area.

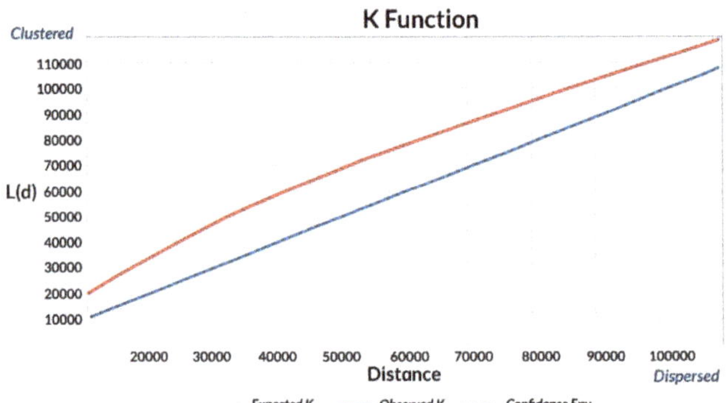

Figure 6. The spatial pattern of land cover changes in 1990–2018, in the relation of scale. Ripley's K function.

Based on the overall CLC changes data from 1990 to 2018, a kernel density analysis was carried out with a grid cell of 900 m and a radius of 9000 m. The results show that the density of land cover changes estimated by kernel density function is between 0.029 and 0.82 per square kilometres, and the average kernel density of the 7444 CLC changes varies from 0 to 0.67. The visualisation map (Figure 7) can demonstrate the characteristics of land cover changes spatial distribution, one of which is the CLC change concentration in the tri-city region, which is associated with a high intensification of artificial areas. The densification of land cover changes is also clearly visible in large forest complexes located in the regions Szczecinski (Goleniow Forest and Beech Forest), Koszalinski, and Slupski (Central Pomeranian Forests), and in the southern part of the Elblaski region (Elblag-Zulawy Forests).

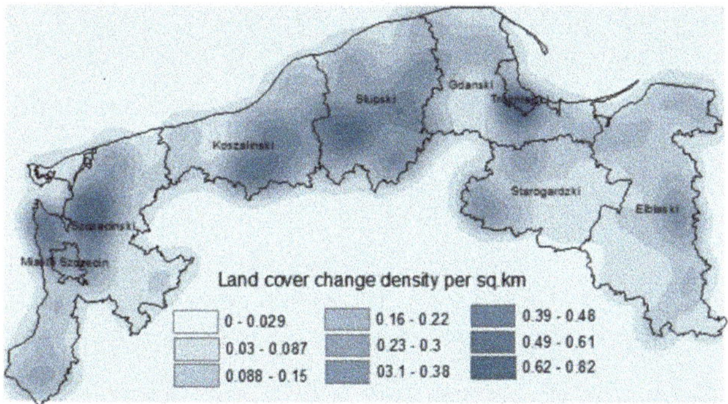

Figure 7. The density of land cover changes estimated by the kernel function.

3.4. *Land Cover Flows*

Urbanisation (LF1) is seen in Europe as one of the land cover flows that has the greatest environmental impact [35]. The rate of urban land cover types has been increasing continually, and the total value noticed for years 1990–2018 was 0.49 (Table 6). The marked growth of urban areas and the accompanying infrastructure occurred after 2006, which was caused by an increase in investments co-financed mainly from EU funds.

Table 6. Land cover flow in 1990-2018.

Land Cover Flow	1990–2000 %/km²	2000–2006 %/km²	2006–2012 %/km²	2012–2018 %/km²	1990–2018 %/km²
LCF1—Urbanization	0.059 1975.46	0.084 2812.52	0.164 5491.11	0.183 6127.27	0.49 16,406.36
LCF2—Intensification of agriculture	0	0	0.016 535.72	0.03 1004.47	0.046 1540.19
LCF3—Extensification of agriculture	0.091 3046.90	0.05 1674.12	0.009 301.34	0.007 234.38	0.157 5256.73
LCF4—Afforestation	0.208 6964.33	0.149 4988.87	0.301 10,078.19	0.23 7700.95	0.888 29,732.35
LCF5—Deforestation	0.035 1171.88	0.193 6462.10	0.346 11,584.90	0.599 20,055.94	1.173 39,274.82

and city beltways, was observed. When analysing the land cover flow for LCF1 depending on the area of the region, we can see that this is the equal distribution in terms of the size of the analysed area (Gini coefficient equal to 0.24772)—Figure 8a.

Urban sprawl also affects seaside resorts where new tourist buildings have been built, which is clearly visible on high-resolution images as shown in Dziwnow, which is a small city located in the west part of the Polish Baltic coastal zone. Particularly, the Dziwnow peninsula and the embankment of Lake Wrzosowisko were subjected to anthropological pressure after 2006 (Figure 9).

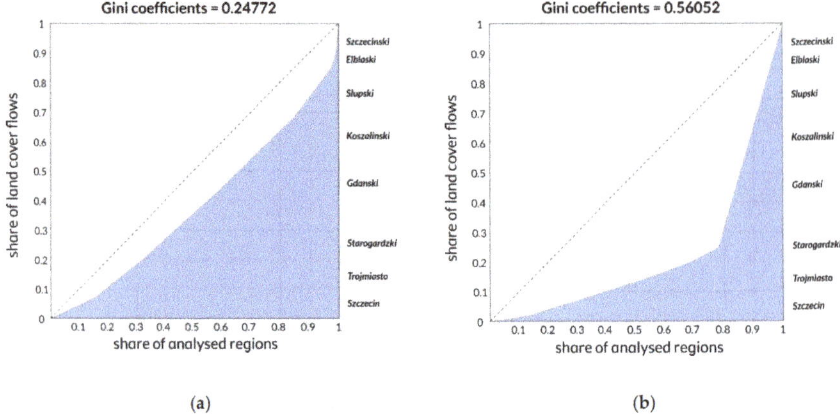

Figure 8. Gini coefficient for: (**a**) LCF1 and (**b**) LCF5.

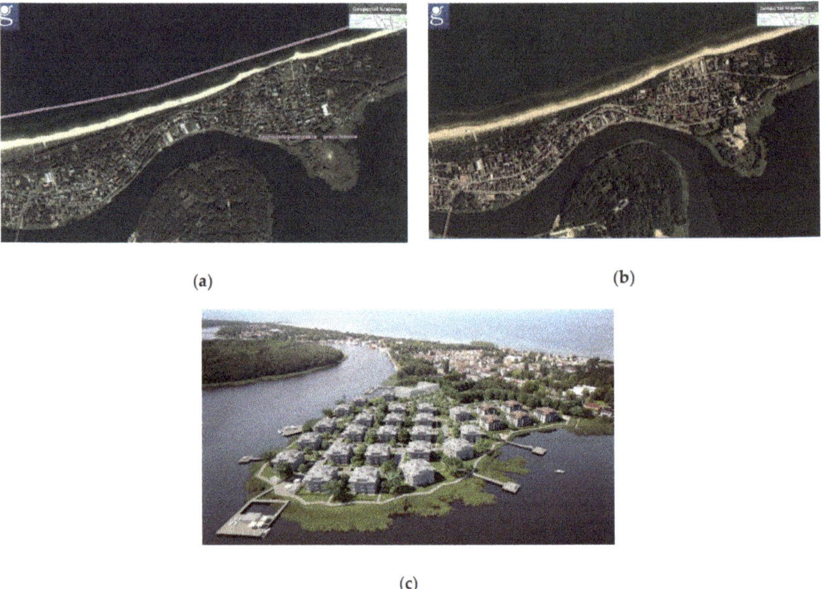

Figure 9. Dziwnow, the coastal resort in the Szczecinski region, 0.25 m aerial ortoimagery from geoportal.gov.pl: (**a**) 2005; (**b**) 2018; (**c**) planned, until 2020, development the Dziwnow peninsula.

Further changes concerned agricultural areas. In Europe, the fine-grained structure and associated biodiversity of traditional rural landscapes continue to be affected by land take, agricultural intensification, and farmland abandonment. However, for the analysed zone, a slight increase in the expansion and intensification of agriculture could be observed, especially for the Elblaski and Slupski regions in the last analysed period, i.e., 2006–2018 (Figure 10). For these areas, the process of revitalisation of rural areas began after 2010, which was supported by European Union programs.

Figure 10. Choropleth map of LCF2 and LCF4 in percent.

Along with the intensification of agriculture, changes in forest cover were noted, as the highest land cover flow in the Polish Baltic coastal zone (Table 6). The largest afforestation was noted in the Slupski, Elblaski, and Koszalinski regions, where the intensification of agriculture could be observed at the same time (Figures 10 and 11a).

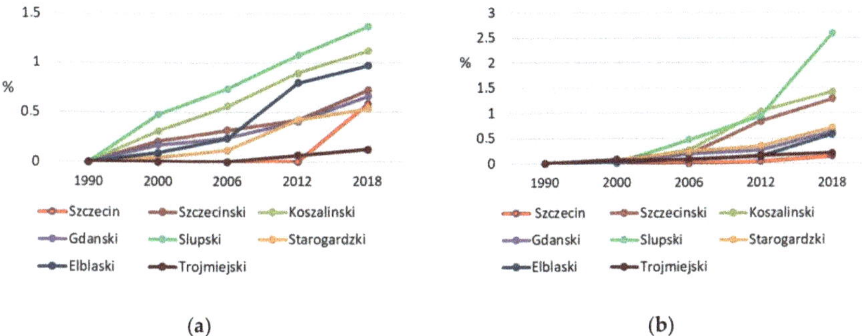

Figure 11. Land cover flows in 1990–2018 in percent: (**a**) LCF4, afforestation; (**b**) LF5, deforestation.

Based on the analysis of CLC data, the deforestation was also noted (LCF5). As based on the Gini coefficient (0.56052) as shown in Figure 8b, this land flow is not even in all areas studied. The deforestation has been most intensified in the last decade (Figure 11b). However, this not always does mean complete deforestation, e.g., for new transport infrastructure or housing. The increase in land cover flow consisting of the "disappearance" of forests is caused by the sustainable management of forests, i.e., the felling of diseased trees, in order to protect the woods against fire and to maintain the forest cover in good condition. Figure 12a–c present coniferous forest at various stages of growth, as captured by Maxar (DigitalGlobe) and Airbus satellites such as IKONOS, QuickBird, GeoEye,

11–12 August 2017 and registered in CORINE Land Cover in 2018. The largest (yellow) polygon cover an area of 3350 ha.

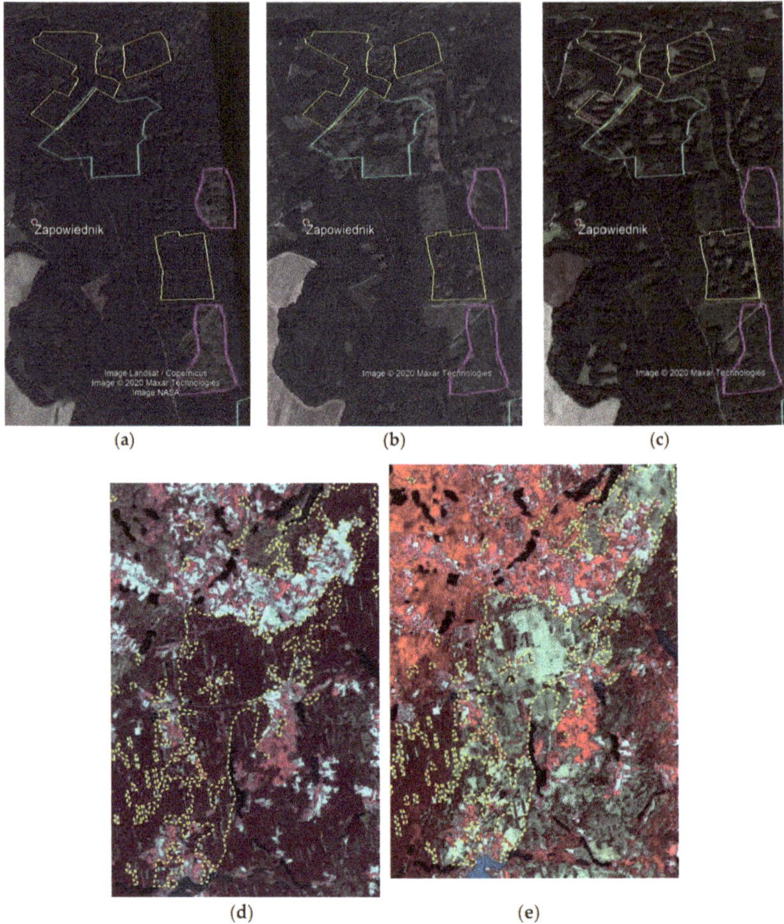

Figure 12. Coniferous forest in the Slupsk region: (**a**) 2006—without clear-cuttings; (**b**) 2012—green polygons indicate cutting conducted from 2006 to 2012; (**c**) 2018—new cutting marked in yellow; (**d**,**e**) Bytow and Lipusz districts of Bory Tucholskie Forest CIR compositions, before (**d**) and after hurricane (**e**) in 2017.

3.5. Progress towards SDGs 11 and 15

The long-term (28 years) land consumption rate (LconR) in the Baltic coastal zone amounts to 3.71%, with the total increase in artificial surfaces equal to 103.89%. The index took the highest value of 7.95% in 2000–2006, and the lowest, 0.56%, in 2012–2018. In the years 1990–2000, an annual rate of land consumption amounts to 1.48%. After the year 2006, the index decreased to 2.72%. The LconR indicator took the highest values for agriculture regions, namely Elblaski (6.6%), Gdanski (5.98%), and Starogardzki (5.71%), while the lowest one was noted in the Szczecinski region (0.04%) and the city of Szczecin (0.34%). In the legally protected lands, artificial surface area increased during 28 years (1990-2018) by 10,295.8 ha, i.e., by an average of 367.7 ha per year. In the year 1990, it amounted to 5691.42 ha, while in 2018, it was three times larger and took the value of 15,987.22 ha. The land consumption rate there amounts to 6.46% of the total HNV area. Nevertheless, LconR values in subsequent periods of CLC inventories differed significantly, assuming the lowest value of 0.28% in the latest period of 2012–2018, and the highest 15.12% in 2000–2006 (see Figure 13).

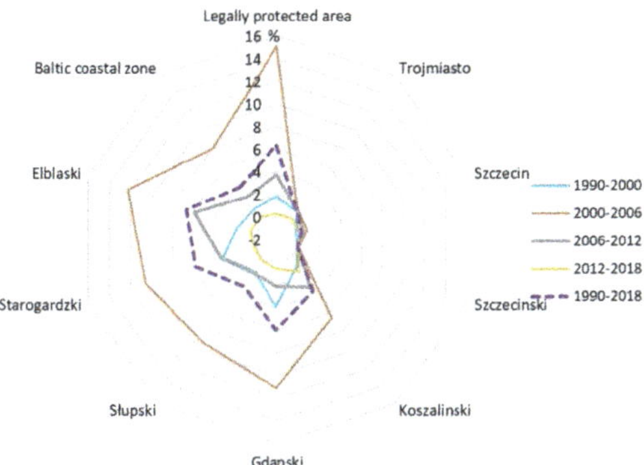

Figure 13. Long and short-term diversity of land consumption rate (LConR) across the Polish Baltic coastal zone and regions.

After 2006, the values of LconR in the Baltic coastal zone clearly decreased, reaching 3.75% in 2006–2012, and only 0.28% in 2012–2018. This should be considered as the positive effects of European and Polish legal regulations regarding land development and management towards sustainable goals. The large increase of the land consumption rate in 2000–2006 was mainly associated with the use of EU pre-accession funds for the development of new commercial centres, as well as the construction and reconstruction of road infrastructure, including the construction of animal culverts, marked pedestrian crossings, and bicycle paths.

The urban green ratio in the analysed cities varies significantly, having the lowest value in Swinoujscie (1.86%), the city located on the Odra delta, and the highest in Slupsk (11.31%), which is a small town located on the central Baltic coast. Slupsk and Sopot are the greenest cities in the Polish Baltic coastal zone, with the UGrR higher than 10% (Figure 14a). In contrast, urban green areas per capita take the highest value in Swinoujscie (80.0 m) and the lowest in Gdynia (16.4 m). In most cities, the size of urban green areas has increased since 1990 (Figure 14b). Elblag ranks first, increasing the urban green areas by 3.47%; in 1990, the urban greenery ratio was just 2.88%, ahead of only Swinoujscie and Koszalin. Then, it is followed by Gdansk, where urban greenery expands by 2.98%. Gdynia, Sopot, and Swinoujscie show relative stability of the urban green space ratio, with fluctuations below 0.2%. Only Szczecin reduced its urban greenery by 0.49%; a decrease was recorded in 2000, followed later by a very slow upsurge from 7.40% in 2000 to 7.75% in 2018.

Figure 14. The urban green ratio in the city counties: (**a**) UGrR in 2018; (**b**) UGrR changes in 1990–2018.

The forest cover ratio in the Baltic coastal zone has been steadily increasing, reaching 32.4% in 1990, while in 2018, it was 35.3%. The long-term (1990–2018) average annual increase in forest area was about 0.1%, i.e., 34 km^2. The index varies from 25.1% in Szczecin (city) to 48.1% in the Slupski region. In six regions, except for Slupsk and Starogardzki, its value exceeds 30%. In the period 1990–2018, the forest cover ratio (FR) increased slightly, from 3.56% in Szczecin to 1.84% in the Gdansk region, while only a minimal decline of 0.14% was observed in the Trojmiejski region.

The proportion of forest areas located within legally protected areas, or the $F_{LPA}P$ index, remained at 48.76% in 2018, gradually, increasing its value compared to 1990 (see Figure 15). The highest (88.8%) and most stable in the analysed period, the value of the $F_{LPA}P$ indicator, is observed in the Trojmiejski region, where the tri-city landscape park and Oliwa forests are located. The least, albeit 25% of forests in legally protected areas, are observed in the Szczecin region.

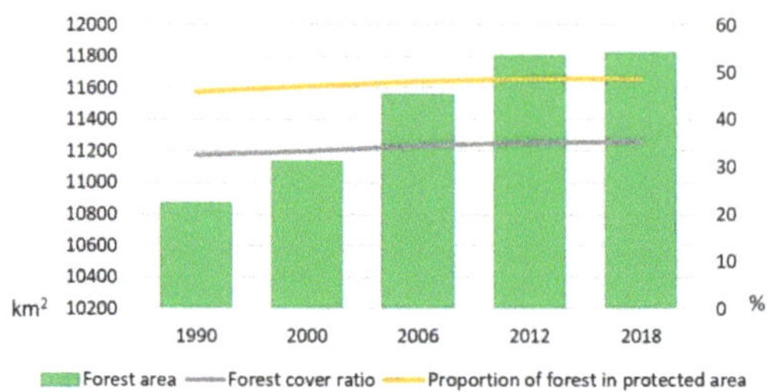

Figure 15. Forest area in km^2 and forest cover in protected areas.

4. Discussion

Changes in land cover because of civilisation processes are inevitable and include mainly urban sprawl, transport networks densification, the intensification of agriculture, and changes related to forest management. They are dynamic, widespread, and accelerating, and they are mainly driven by natural phenomena and anthropogenic activities [42,43]. By and large, land cover changes are increasingly recognised as an essential driver of environmental change on both spatial and temporal scales [34]. Current land cover changes are triggered by economic development, investment, agricultural policy, and environmental protection policy. An important factor influencing the changes is the rise of consumption resulting from the growth in the wealth of society [44]. In Poland, as in other post-Communist countries, the land cover has changed since 1989, after the collapse of the Soviet Union and the sociopolitical transformation of Central and Eastern European countries [17,18,45].

The CLC database was developed using satellite remote sensing data, i.e., Landsat, Indian Remote Sensing (IRS) Satellite, and Sentinel, which are not VHR data. However, not all land cover changes can be captured using medium-resolution images that have been used to interpret changes within the EU CORINE Land Cover project. What is more, the spatial resolution of CLC change data, i.e., 5 ha as a minimum mapping unit with a minimum width of 100 m, allows the storage of fragmented landscapes in the form of mixed classes (e.g., 2.4 or 3.2), which ultimately leads to generalisation. Therefore, CORINE land cover inventories are dedicated to analysis on a regional or national scale, starting from 1: 75,000 [46].

CLC data show that total land cover changes increased from the 1990–2000 period to the 2012–2018 period. The main land cover trends (in the 1990–2012 period) and their environmental impacts according to European land accounting [47] are as follows:

- urban and infrastructure expansion resulting in a loss of productive soils and landscape fragmentation;
- continuous decrease of agricultural areas and, as a consequence, farmland abandonment and biodiversity loss;
- intensification of forest land use, leading to a declining quality of forest ecosystems.

These trends are observed in Europe, Poland [46], and the Polish Baltic coastal zone. Land cover changes in Poland amount to 0.93%, 0.53% and 0.99% respectively in 1990–2000, 2000–2006, and 2006–2012 [43,44], which gives a 0.10% (3348.2 ha) annual land cover change of the total area in the years 1990–2006 and 0.16% for the period 2006–2012. The Polish coastal zone faced slightly slower changes respectively: 0.63%, 0.59%, 0.90% in the subsequent changes, and 1.13% in the last period of 2012–2018.

Urban expansion is a constant trend across all regions in the Baltic zone, reaching the highest value in the Trojmiejski region (4.7%) and the lowest in Slupski (0.2). This process has been more intensive since 2000, which is related to Poland's pre-accession declaration to the EU and financing structural investments. Although changes in land cover between 1990 and 2018 in the Baltic zone showed that agricultural areas changed most often, changes in agriculture structure, denoted as LF3, covered just 5357.2 ha, which is 0.16% of the Baltic coastal zone area. Conversion from natural and semi-natural areas to high-intensity agriculture, i.e., the expansion and intensification (LF2) of agriculture, was visible only in the period 2006–2012, but its size amounts to 0.05% (1674.1 ha) in 1990–2018. Land cover changes related to forest management remain the largest in terms of the total turnover of 2.06% of the total analysed area, including afforestation by 0.89% and deforestation by 1.17%. The results indicate that this process was more intense in 2012–2018, which was mainly due to the hurricane on the night of 11–12 August 2017. This hurricane destroyed about 26,000 ha of forests [21], of which approximately 2000 ha are in the area under analysis. The increase in the forest cover of the country is a visible result of the implementation of the National Program for Extending Forest Cover (KPZL) adopted in 1995. Its intention was to increase Poland's forest cover to 30% in 2020 and 33% in 2050.

The changes over land use in the Polish Baltic zone in the last three decades only slightly follow the tendency in the neighbouring Baltic countries, such as Germany, Lithuania, and Estonia, where they were much more intensive [10,18]. The results and tendencies obtained in the Polish Baltic zone are much more similar to other coastal countries such as Malta, Cyprus, Bulgaria [46].

Changes in land cover triggered by urbanisation, landscape fragmentation, and intensification of agriculture have been recognised and discussed by many scientists [48–50] as those that deeply affect biodiversity and human life, especially in protected areas. Legally protected areas located in the Polish coastal zone were also affected by land cover changes, although the total size of changed areas during 1990–2018 was mainly associated with activities aimed at the protection of nature in these areas in compliance with European and national provisions.

Trends and the rate of land cover changes force local and regional spatial planning authorities to identify harmful land cover flows and develop a policy that prevents their further growth. According to the National Strategy of Regional Development [51] and EU's territorial cohesion policy for the Baltic Sea region [19], the spatiotemporal dynamics of land cover changes should be monitored on a regular basis, at local and regional levels, since only continuous land use monitoring can ensure appropriate, sustainable territorial management, and development. In Poland, according to Geodetic and Cartographic law, the duty to monitor land cover/land use changes belongs to the responsibilities of the Voivodeship self-government authorities [52]. However, as noted by Noszczyk [53], monitoring activities involve general observations of the current land use structure in comparison with the previous

for sustainable development potential and counteracting marginalisation. The most important ones include low transport accessibility and a low availability of goods and services, shaping development capabilities for inhabitants [51]. The analysis shows that industrial, commercial, and transport areas increased slightly in the Baltic area, reaching an average value of 0.7% in 2018 and a growth of 0.02% compared to 2012, i.e., two years after the National Regional Development Strategies had entered into force. The highest rises were observed in the Trojmiejski region (0.25%) and the city of Szczecin (0.11%), while the lowest ones (0.01%) were observed in the Koszalinski and Slupski regions, i.e., those with the worst road and rail access to the center of the voivodeship [54]. However, it is worth noting that large forest cover (35.4%) and a significant share of legally protected areas (20.7%) prevent the free development of commercial areas and road networks, which may contribute to reducing the development of tourism, and thus reduce the income of the population and public administration. Therefore, pursuant to the National Strategy of Regional Development [51], local and regional policies should complement the environmental policy for the comprehensive protection and conservation of nature in the region, as well as improving the use of endogenous regions' potential in order to increase economic growth. In such a context, the monitoring of regional development plays a considerable role, in particular, monitoring land cover changes and flows. It is assumed that the EU's agricultural and environmental policy, together with national environmental protection, is an important element shaping land cover structure in the Polish Baltic coastal zone. Of utmost importance is the Act on the Protection of Agricultural and Forest Land [55], which requires the consent of the Minister of Agriculture and Rural Development for agricultural and forest land consumption. However, this national legislation does not apply to the transformation of agricultural land into non-agricultural land within the city limits.

The responsible and sustainable management of coastal zones requires access to up-to-date, reliable spatial data stored in a spatial information system. Evidence-based remotely sensed land cover data at a fine scale is of utmost importance for public administration as an essential tool for managing and monitoring these areas. The CORINE land cover datasets of an agreed and harmonised nomenclature have been available to all of Europe since 1990 for subsequent six-year periods, and as such, they constitute an invaluable source of spatiotemporal data for monitoring the state and changes of landscape, particularly land accounts according to EU and national regulations [25,56]. Moreover, recently, the EEA has established a project aimed at governing coastal zone areas, in particular to monitor landscape dynamics in coastal zones in a spatially explicit manner. The first release thematic hotspot products will consist of land cover/land use datasets for 2012 and 2018 as well as a change layer between 2012 and 2018. The products are going to provide land cover status and changes every six years. The first delivery is expected by the end of 2020 [57]. As [47] noted, regular inventories of land cover changes are vital to assess the reasons and consequences of natural and artificial processes, identify trends, maintain ecological balance, and take these factors into account in decision-making processes.

Researchers' interest towards attaining the 17 SDGs, agreed in September 2015, and set up in the 2023 Agenda for Sustainable Development is still a hot topic in many scientific fields. The indicators for SDGs achievements are grouped into three tiers, according to methodological clarity and data accessibility. Indicator 11.3.1 has been classified as tier II, although, as noted by Nicolau et al. [58] it is not conceptually clear, as the agreed and precise definition of urban land is still missing. The elaborated land consumption rate (LconR) is based on a remotely sensed urban area and its relative growth, and as such, it may be considered as an input component to 11.3.1 indicator calculation. Indicators 15.1.1 and 15.2.1 belong to tier I (conceptually clear, methodology and standards are available, and data are regularly produced by most of the countries). However, Liu et al. [59] found that the 15.2.1 progress towards sustainable forest management required further deconstruction and analysis of its definition. Moreover, they introduced five sub-indicators, among them a sub-indicator "proportion of forest area located within the legally established area", which was used in the presented study.

5. Conclusions

The investigation of land cover and its changes over time and space play an essential role in understanding socioeconomic and physical phenomena at any level. The analysis of coastal areas is particularly important because of their economic and social significance for all countries. The Polish Baltic coastal area is dominated by agro-forestry landscape, with a small share (5.19%) of urbanised surfaces. Due to the nature of the land cover and the strategic geolocation, analysing the land cover changes in this area is very important for environmental, economic, and political purposes.

Total land cover changes in the Baltic coastal zone reveal a valuable trend in national territorial protection policy. The analysis showed a dispersed spatial pattern of land cover changes in the Baltic coastal regions in each of the analysed periods. That indicates that land cover changes have affected every Baltic region. However, the scale of changes varies depending on the region, being the highest in the Trojmiejski (tri-city) and the lowest in Szczecinski. The structure of land cover changes in the Polish coastal zone of the Baltic Sea is intricate and exhibits marked differences between regions. Urbanisation mostly affected the Trojmiejski urban region and small, mainly touristic cities located along the coastal zone. The intensification and extensification of agriculture (LF2 and LF3) dominate in the regions of Elblaski, Slupski, and Koszlinski, where after the year 1990, as a result of the abandonment of cooperative farms, the structure and type of agricultural production changed. These regions also saw the marked changes in forests—on the one hand, clear-cutting, harvesting of trees, and forest damage due to natural hazards (i.e., hurricanes), and on the other hand, afforestation in accordance with national and regional environmental and agricultural policies.

Two distinct periods were observed in land cover changes and flow. The first, after the fall of socialism and before Poland's accession to the EU (1990–2006), was characterised by slower changes, consisting mainly in the abandonment of agricultural land and afforestation. Secondly, in the years 2006–2018, there was a visible intensification of urbanisation due to the availability of EU structural and cohesion funds. The expansion of urban and related infrastructure was accompanied by deforestation.

Our findings show that agricultural land loss, at an average at 6612.8 ha per year, was the dominant land cover net change in the years 1990–2018. It was followed by a steady growth of the forest and semi-natural area, by an average of 3366.2 ha annually. Afforestation is the result of planting new trees as well as a natural expansion of forests on abandoned agricultural land (e.g., CLC change from 2.4 to 3.1). The enlargement of the artificial surface was the third observed trend of land cover changes, with a total size of 88,560.0 ha (2.65%) and annual growth of 3162.9 ha.

The analysis of land cover changes and calculated on its basis indices monitoring the achievement of SDGs by 2030 indicate an increase in land consumption rate and steady increase in the forest cover ratio, and in most of the cities in the analysed area, a growth of urban green ratio.

As presented in this article, land cover changes can be easily observed with the CLC database. Information on the land cover was derived from computed-aid visual interpretation based on optical images (Landsat 4/5, 7, and 8, SPOT 4/5, IRS-P6, RapidEye and Sentinel 2) in the visible and near-infrared spectrum, which allows the creation of false and true-colour compositions, and reached the 95% of CLC thematic reliability.

Temporal trajectories of apparent land cover changes based on the multi-decadal CLC inventories provide valuable information for regional and national authorities that force them to make transparent (based on data) decisions. From the administrative, research, and technical points of view, the potential of CORINE land cover data is unquestionable, since it consistently stores easily accessible spatiotemporal data. This fact is reflected in the INSPIRE data specification for land cover [60], where the CLC is shown as one of the sets where technical and semantic standardisation has been achieved, constituting one of the most harmonised European datasets. Therefore, it can be stated that CORINE may be a key element of future studies covering land cover changes related to the goals of sustainable development across the world.

Funding: This research was funded by Military University of Technology in Warsaw, Faculty of Civil Engineering and Geodesy, Institute of Geospatial Engineering and Geodesy.

Acknowledgments: Corine Land Cover data were obtained from the Chief Inspectorate of Environmental Protection CLC web page http://clc.gios.gov.pl/ on 11 December 2019. Boundaries of NUT3 were gained from the Eurostat GISCO Database web page https://ec.europa.eu/eurostat/web/gisco/geodata/reference-data/administrative-units-statistical-units/nuts. Data on the administrative division was gained from the Head Office of Geodesy and Cartography.

Conflicts of Interest: The authors declare no conflict of interest.

Appendix A

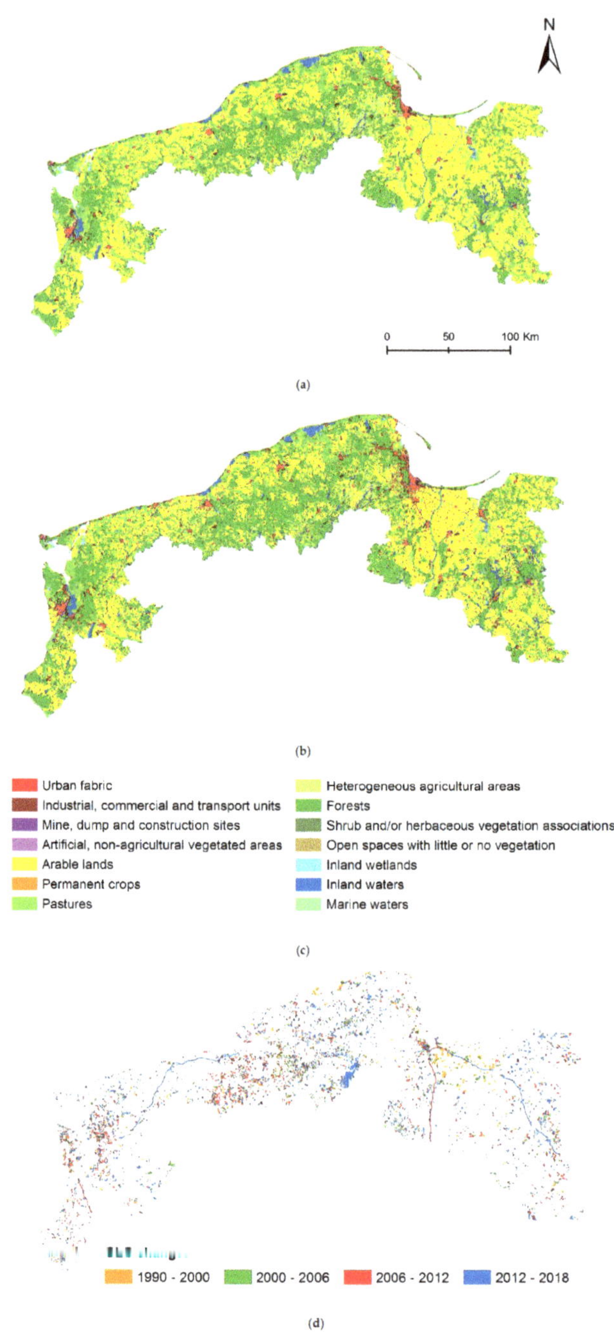

Figure A1. CORINE Land Cover types in Polish Baltic coastal area in (**a**) 1990; (**b**) 2018; (**c**) CLC classes—level 2; (**d**) CLC changes in 1990–2018.

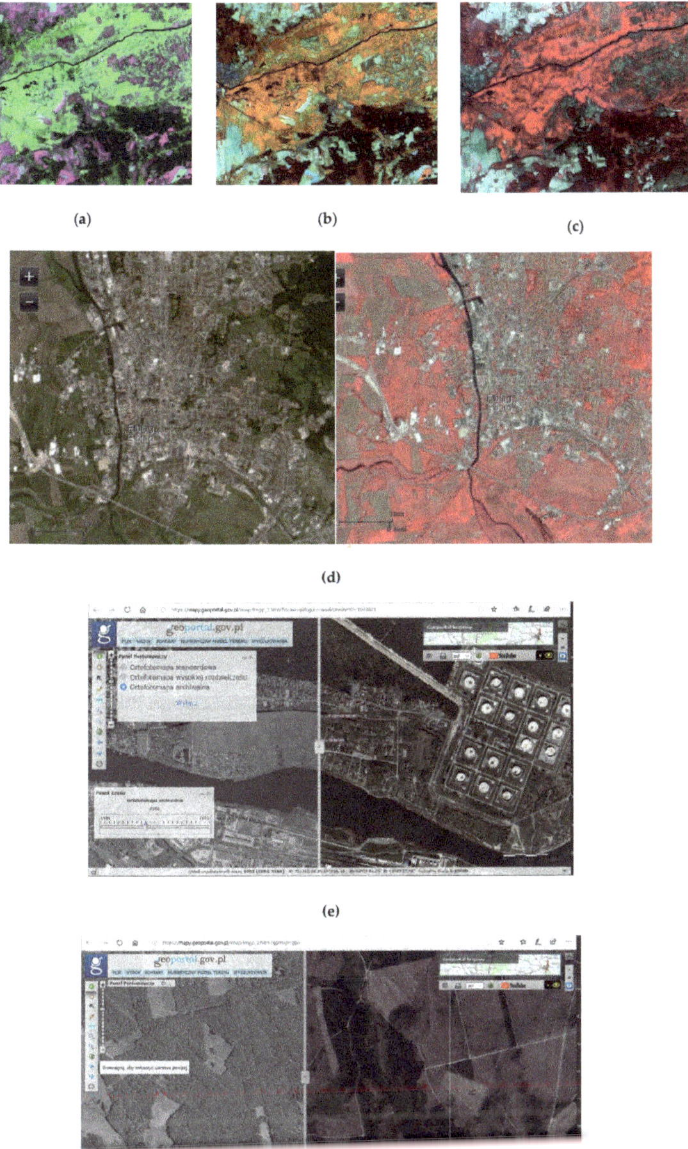

Figure A2. Satellite images and aerial ortorectified data supporting CLC change data checking (**a**) Landsat 5 false colour composition; (**b**) Landsat 7 false colour composition; (**c**) pan-sharpened Landsat 8 false colour composition; (**d**) High-resolution mosaic of Sentinel-2 (10 m) in true (bands 4,3,2) and false colour (bands 8,4,3) available by via Copernicus services (Elblag); (**e**) Geoportal services of archive and new aerial ortoimagery (Gdansk); (**f**) Geoportal services—forest damage in the Slupski region after the hurricane on 11–12 August 2017.

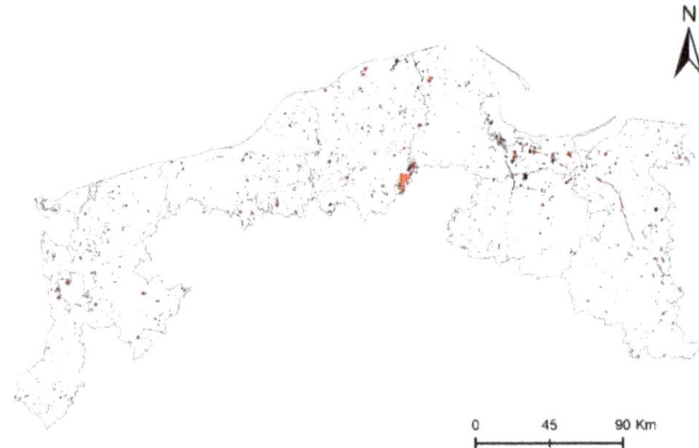

Figure A3. Red marked CLC changed polygons used for data verification.

References

1. Strandberg, G.; Kjellström, E. Climate Impacts from Afforestation and Deforestation in Europe. *Earth Interact.* **2019**, *23*, 1–27. [CrossRef]
2. Sagan, C.; Toon, O.B.; Pollack, J.B. Anthropogenic Albedo Changes and the Earth's Climate. *Science* **1979**, *206*, 1363–1368. [CrossRef]
3. Chun, J.; Kim, C.-K.; Kang, W.; Park, H.; Kim, G.; Lee, W.-K. Sustainable Management of Carbon Sequestration Service in Areas with High Development Pressure: Considering Land Use Changes and Carbon Costs. *Sustainability* **2019**, *11*, 5116. [CrossRef]
4. Vitousek, P.M.; Mooney, H.A.; Lubchenco, J.; Melillo, J.M. Human Domination of Earth's Ecosystems. *Science* **1997**, *277*, 494–499. [CrossRef]
5. Sala, O.E.; Chapin, F.S.; Armesto, J.J.; Berlow, E.; Bloomfield, J.; Dirzo, R.; Huber-Sanwald, E.; Huenneke, L.F.; Jackson, R.B.; Kinzig, A.; et al. Global Biodiversity Scenarios for the Year 2100. *Science* **2000**, *287*, 1770–1774. [CrossRef] [PubMed]
6. Trimble, S.W.; Crosson, P.U.S. Soil Erosion Rates-Myth and Reality. *Science* **2000**, *289*, 248–250. [CrossRef]
7. Lambin, E.F.; Geist, H.J.; Lepers, E. Dynamics of Land-Use and Land-Cover Change in Tropical Regions. *Annu. Rev. Environ. Resour.* **2003**, *28*, 205–241. [CrossRef]
8. Dorocki, S.; Raźniak, P.; Winiarczyk-Raźniak, A. Changes in the command and control potential of European cities in 2006–2016. *Geogr. Pol.* **2019**, *92*, 275–288. [CrossRef]
9. Esposito, M.A.; Perdigao, V.; Christensen, S.; Di Cura, A. *The LaCoast Atlas: Land Cover Changes in European Coastal Zones*; EU—European Commission Publication Office, Joint Research Centre: Milan, Italy, 2000.
10. Antso, K.; Palginõmm, V.; Szava-Kovats, R.; Kont, A. Dynamics of Coastal Land Use over the Last Century in Estonia. *J. Coast. Res.* **2011**, *SI 64*, 1769–1773.
11. Neumann, B.; Ott, K.; Kenchington, R. Strong sustainability in coastal areas: A conceptual interpretation of SDG 14. *Sustain. Sci.* **2017**, *12*, 1019–1035. [CrossRef]
12. Merkens, J.-L.; Reimann, L.; Hinkel, J.; Vafeidis, A.T. Gridded population projections for the coastal zone under the Shared Socioeconomic Pathways. *Glob. Planet. Chang.* **2016**, *145*, 57–66. [CrossRef]
13. Merkens, J.-L.; Vafeidis, A. Using Information on Settlement Patterns to Improve the Spatial Distribution of Population in Coastal Impact Assessments. *Sustainability* **2018**, *10*, 3170. [CrossRef]
14. Neumann, B.; Vafeidis, A.T.; Zimmermann, J.; Nicholls, R.J. Correction: Future Coastal Population Growth and Exposure to Sea-Level Rise and Coastal Flooding—A Global Assessment. *PLoS ONE* **2015**, *10*, e0131375. [CrossRef]
15. Eurostat. Eurostat regional yearbook 2011, chapter 13 Coastal Region. In *Eurostat Regional Yearbook 2011*; Publications Office of the European Union: Luxembourg, 2011; pp. 170–183.

16. Institute of Meteorology and Water Management. Assessment of Actual and Future Climate Changes on Polish Coastal Zone and Its Ecosystem. Available online: https://nfosigw.gov.pl/download/gfx/nfosigw/pl/nfoekspertyzy/858/210/1/2014-424.pdf (accessed on 19 December 2019).
17. Fortuniak, K. Stochastyczne i deterministyczne aspekty zmienności wybranych elementów klimatu Polski. *Acta Univ. Lodz. Folia Geogr. Phys.* **2000**, *4*, 1–139.
18. Neumann, T.; Schernewski, G.; Hofstede, J. *Global Change and Baltic Coastal Zones*; Coastal Research Library; Springer: Dordrecht, The Netherlands, 2011; Volume 1. [CrossRef]
19. Zaucha, J. *Territorial Cohesion—Baltic Sea Region Examples*; Baltic 21 Series; s.Pro Sustainable Projects: Berlin, Germany, 2011; Volume 2.
20. Eurostat. *Methodological Manual on Territorial Typologies—2018 Edition*; Publications Office of the European Union: Luxembourg, 2019. [CrossRef]
21. GUS. Statystyka Regionalna—Regional Statistic. Available online: http://stat.gov.pl/statystyka-regionalna/jednostki-terytorialne/klasyfikacja-nuts/ (accessed on 13 November 2019).
22. Tylkowski, J. The tendencies of bioclimatic conditions changes and dynamics occurrence of thermally stimulus weather events in the Polish Baltic coastal zone. *J. Educ. Health Sport* **2017**, *7*, 467–480. [CrossRef]
23. Pruszak, Z.; Zawadzka, E. Vulnerability of Poland's Coast to Sea-Level Rise. *Coast. Eng. J.* **2005**, *47*, 131–155. [CrossRef]
24. Bossard, M.; Feranec, J.; Otahel, J. *Technical Report No 40/2000 CORINE Land Cover Technical Guide—Addendum 2000*; EEA: Copenhagen, Denmark, 2000.
25. Bielecka, E.; Jenerowicz, A. Intellectual Structure of CORINE Land Cover Research Applications in Web of Science: A Europe-Wide Review. *Remote Sens.* **2019**, *11*, 2017. [CrossRef]
26. GIOŚ. CORINE Land Cover-CLC. Available online: http://clc.gios.gov.pl/ (accessed on 30 November 2019).
27. Główny Urząd Geodezji i Kartografii. Geoportal Infrastruktury Informacji Przestrzennej. Available online: Geoportal.gov.pl (accessed on 10 February 2020).
28. Copernicus Proramme. Copernicus Land Monitoring Service. Available online: https://land.copernicus.eu/imagery-in-situ (accessed on 10 February 2020).
29. European Space Agency. Copernicus Open Access Hub. Available online: https://scihub.copernicus.eu/ (accessed on 10 February 2020).
30. U.S. Geological Survey. Landsat Data Access. Available online: https://www.usgs.gov/land-resources/nli/landsat/landsat-data-access (accessed on 20 April 2020).
31. European Commission Eurostat. NUTS. Available online: https://ec.europa.eu/eurostat/web/gisco/geodata/reference-data/administrative-units-statistical-units/nuts (accessed on 10 December 2019).
32. European Parliament. *Regulation (EC) No 1059/2003 of the European Parliament and of the Council of 26 May 2003 on the Establishment of a Common Classification of Territorial Units for Statistics (NUTS)*; OJ L 154; The European Parliament: Brussels, Belgium, 2003; pp. 1–41.
33. Sun, W.; Chen, B.; Messinger, D.W. Nearest-neighbor diffusion-based pan-sharpening algorithm for spectral images. *Opt. Eng.* **2014**, *53*, 013107. [CrossRef]
34. Lambin, E.F.; Rounsevell, M.D.A.; Geist, H.J. Are agriculturala land-use model able to predict changes in land-use intensity. *Agric. Ecosyst. Environ.* **2000**, *82*, 321–331. [CrossRef]
35. Feranec, J.; Jaffrain, G.; Soukup, T.; Hazeu, G. Determining changes and flows in European landscapes 1990–2000 using CORINE land cover data. *Appl. Geogr.* **2010**, *30*, 19–35. [CrossRef]
36. Ripley, B.D. Th

38. Silverman, B.W. *Density Estimation for Statistics and Data Analysis*; Chapman and Hall/CRC: New York, NY, USA, 1986.
39. Gini, C. On the Measure of Concentration with Special Reference to Income and Statistics. *Color. Coll. Publ. Gen. Ser.* **1936**, *208*, 73–79.
40. European Environment Agency. *The Thematic Accuracy of Corine Land Cover 2000: Assessment Using LUCAS (Land Use/Cover Area Frame Statistical Survey)*; EEA Technical Report No 7/2006; EEA Publication Office: Copenhagen, Denmark, 2006; p. 85. Available online: https://www.eea.europa.eu/publications/technical_report_2006_7 (accessed on 5 May 2020).

41. GUS. *Forestry*; Statistical Publishing Establishment: Warsaw, Poland, 2017. Available online: www.stat.gov.pl (accessed on 12 December 2019).
42. Agarwal, C.; Green, G.M.; Grove, J.M.; Evans, T.P.; Schweik, C.M. *A Review and Assessment of Land-Use Change Models: Dynamics of Space, Time, and Human Choice*; General Techical Report NE-297; U.S. Northeastern Research Station, Forest Service, Department of Agriculture: Newtown Square, PA, USA, 2002; p. 61.
43. Guerra, C.A.; Rosa, I.M.D.; Pereira, H.M. Change versus stability: Are protected areas particularly pressured by global land cover change? *Landsc. Ecol.* **2019**, *34*, 2779–2790. [CrossRef]
44. Matyka, M. Analiza regionalnego zróżnicowania zmian w użytkowaniu gruntów w Polsce. *Pol. J. Agron.* **2012**, *10*, 16–20.
45. Woch, M.; Konarski, M. *Ewolucja Prawa Miejscowego Jednostek Samorządu Terytorialnego w Polsce*; Wydawnictwo Marek Woch: Warszawa, Poland, 2014.
46. Bielecka, E.; Ciołkosz, A. Land Cover Structure in Poland and its changes in the last decade of 20th Century. *Ann. Geomat.* **2004**, *2*, 81–93.
47. EEA. *Landscapes in Transition An Account of 25 Years of Land Cover Change in Europe*; EEA Report No 10/2017; Publications Office of the European Union: Luxembourg, 2017. Available online: https://www.eea.europa.eu/publications/landscapes-in-transition (accessed on 28 November 2019).
48. Hosciło, A.; Tomaszewska, M. CORINE Land Cover 2012—4th CLC inventory completed in Poland. *Geoinf. Issues* **2014**, *6*, 49–58. [CrossRef]
49. Walz, U. Monitoring of landscape change and functions in Saxony (Eastern Germany)—Methods and indicators. *Ecol. Indic.* **2008**, *8*, 807–817. [CrossRef]
50. Łowicki, D.; Walz, U. Gradient of Land Cover and Ecosystem Service Supply Capacities—A Comparison of Suburban and Rural Fringes of Towns Dresden (Germany) and Poznan (Poland). *Procedia Earth Planet. Sci.* **2015**, *15*, 495–501. [CrossRef]
51. Ministry of Regional Development. *National Strategy of Regional Development 2010–2020. Regions, Cities, Rural Areas*; Ministry of Regional Development: Warsaw, Poland, 2010.
52. Journal of Laws. *Geodetic and Cartographic Law*; Dz.U. 2016, poz. 1629; Law Office of the Polish Parliament: Warsaw, Poland, 2016.
53. Noszczyk, T. Land Use Change Monitoring as a Task of Local Government Administration in Poland. *J. Ecol. Eng.* **2018**, *19*, 170–176. [CrossRef]
54. Bański, J.; Degórski, M.; Komornicki, T.; Śleszyński, P. The delimitation of areas of strategic intervention in Poland: A methodological trial and its results. *Morav. Geogr. Rep.* **2018**, *26*, 84–94. [CrossRef]
55. Journal of Laws. *Act on the Protection of Agricultural and Forest Land*; Dz.U. 1995 nr 16 poz. 78; Law Office of the Polish Parliament: Warsaw, Poland, 2017.
56. Gibas, P.; Majorek, A. Analysis of Land-Use Change between 2012–2018 in Europe in Terms of Sustainable Development. *Land* **2020**, *9*, 46. [CrossRef]
57. European Environment Agency. Thematic Hotspot Monitoring in Coastal Zones. New Product Update 2019–20. August 2019. Available online: https://land.copernicus.eu/local/coastal-zones (accessed on 17 February 2020).
58. Nicolau, R.; David, J.; Caetano, M.; Pereira, J. Ratio of Land Consumption Rate to Population Growth Rate—Analysis of Different Formulations Applied to Mainland Portugal. *ISPRS Int. J. GeoInf.* **2018**, *8*, 10. [CrossRef]
59. Liu, S.; Bai, J.; Chen, J. Measuring SDG 15 at the County Scale: Localisation and Practice of SDGs Indicators Based on Geospatial Information. *ISPRS Int. J. GeoInf.* **2019**, *8*, 515. [CrossRef]
60. INSPIRE Thematic Working Group Land Cover. D2.8.II.2 INSPIRE Data Specification on Land Cover—Technical Guidelines. 2013. Available online: https://inspire.ec.europa.eu/id/document/tg/lc (accessed on 14 March 2020).

MDPI

Article

Determining Long-Term Land Cover Dynamics in the South Baltic Coastal Zone from Historical Aerial Photographs

Andrzej Giza [1,*], Paweł Terefenko [1], Tomasz Komorowski [2] and Paweł Czapliński [3]

1 Institute of Marine and Environmental Sciences, University of Szczecin, 70-383 Szczecin, Poland; pawel.terefenko@usz.edu.pl
2 Institute of Management, University of Szczecin, 71-004 Szczecin, Poland; tomasz.komorowski@usz.edu.pl
3 Institute of Spatial Management and Socio-Economic Geography, University of Szczecin, 71-101 Szczecin, Poland; pawel.czaplinski@usz.edu.pl
* Correspondence: andrzej.giza@usz.edu.pl; Tel.: +48-91-444-24-32

Abstract: Coastal regions are dynamic environments that have been the main settlement destinations for human society development for centuries. Development by humans and environmental changes have resulted in intensive land cover transformation. However, detailed spatiotemporal analyses of such changes in the Polish Baltic coastal zone have not been given sufficient attention. The aim of the presented work is to fill this gap and, moreover, present a method for assessing indicators of changes in a coastal dune environment that could be an alternative for widely used morphological line indicators. To fulfill the main aim, spatial and temporal variations in the dune areas of the Pomeranian Bay coast (South Baltic Sea) were quantified using remote sensing data from the years 1938–2017, supervised classification, and a geographic information system post-classification change detection technique. Finally, a novel quantitative approach for coastal areas containing both sea and land surface sections was developed. The analysis revealed that for accumulative areas, a decrease in the land area occupied by water was typical, along with an increase in the surface area not covered by vegetation and a growth in the surface area occupied by vegetation. Furthermore, stabilized shores were subject to significant changes in tree cover area mainly at the expense of grass-covered terrains and simultaneous slight changes in the surface area occupied by water and the areas free of vegetation. The statistical analysis revealed six groups of characteristic shore evolutionary trends, of which three exhibited an erosive nature of changes. The methodology developed herein helps discover new possibilities for defining coastal zone dynamics and can be used as an alternative solution to methods only resorting to cross sections and line indicators. These results constitute an important step toward developing a predictive model of coastal land cover changes.

Keywords: land cover; dune coast; air photograph; South Baltic Sea

Citation: Giza, A.; Terefenko, P.; Komorowski, T.; Czapliński, P. Determining Long-Term Land Cover Dynamics in the South Baltic Coastal Zone from Historical Aerial Photographs. *Remote Sens.* **2021**, *13*, 1068. https://doi.org/10.3390/rs13061068

Academic Editor: Feng Ling

Received: 10 February 2021
Accepted: 10 March 2021
Published: 11 March 2021

1. Introduction

Coastal zones are highly [illegible]

[illegible] mid-twentieth century, increased cyclonal activity has been observed in winter periods in the North Atlantic due to global climate change [2,3]. The number of extreme storm surges in the Baltic Sea is increasing steadily [4,5], which along with milder winters and limited ice cover, further exacerbates coastal erosion [6]. The observed rates of change in coastlines, particularly in case of erosion, have been a significant concern for communities inhabiting littoral zones. Appropriate management closely linked to a balanced development of littoral areas is becoming a major challenge for the research community owing to its role in the knowledge-based shaping of environmental policies [7]. The study of long-term changes in land cover is important, since it supplements the knowledge of coastal zone dynamics by analyzing vegetation cover, which is a crucial element stabilizing shore ecosystems. Changes occurring in the coastal zone landscape and its ecosystems are critical in terms of spatial planning and determining natural environment conditions.

In recent years, remote sensing (RS) and Geographic Information Systems (GIS) have gained in popularity, in assessing the spatial and temporal dynamics of coastal zones, as powerful and cost-effective tools [8–10]. RS methods have been particularly useful in analyzing land cover changes. Satellite images constitute a basic source of data, providing information on environmental conditions over a far greater area than that acquired by aerial imaging. In turn, aerial photographs are a valuable source of information regarding environmental conditions at a given moment with a high degree of accuracy. RS methods and spatial information systems are valuable when documenting and measuring changes to the terrain. Adopting a dynamic perspective toward the landscape promotes mutual human and nature interactions [11]. Furthermore, remote sensing tools are successfully applied in studies for detection changes in shallow nearshore areas [12–14]. The use of archival photos taken using monochromatic technology offers many possibilities, as confirmed by numerous studies [15–21]. This technology requires from the photograph interpreter significant experience and knowledge of the physical and geographical conditions of the analyzed area. Aerial photographs are also employed for monitoring changes in seashore dynamics; however, such studies are conducted by analyzing the location of the dune or cliff base line in individual years. Currently, precise measurements are commonly made using satellite navigation, taking advantage of real time kinematic GPS to generate transverse profiles of the coastal zones. For larger areas, light detection and ranging (known as LiDAR) technology has been successfully employed [22–25].

The South Baltic land cover changes and their dynamics have long been a matter of widespread research interest. The first detailed study of the South Baltic coast concerned the Island of Wolin's flora [26]. The paper listed species and their locations, as well as plant communities of many different species, both common and rare. Short and Hesp [27] created a classification of dune shores based on the percentage of vegetation cover, in which the morphological development was determined by the vegetation affecting eolian processes and form creation [28]. Coast development on the tideless Baltic Sea shores, and the possibilities for forecasting in the area, were also explored by Musielak and Furmańczyk [29]. Furthermore, Dudzińska-Nowak analyzed dune/cliff base changes between 1938–1996 within the Rewal Commune [30], while the studies conducted by Łabuz regarding the impact of anthropogenic activity on dune vegetation cover on the Island of Wolin between Międzyzdroje and Świnoujście confirmed the natural trends of vegetation cover development in that area [31,32]. Thus far, most studies have been time- or site-specific. Spatiotemporal changes in land cover have not been sufficiently addressed. The most extensive study to date has been presented by Bielecka et al. [33]. Her paper provides the first comprehensive quantitative and qualitative analysis of land cover changes and their characteristics in the Polish Baltic coastal zone over the last three decades. It does not, however, offer local-scale information, since the analysis' scale is adjusted to the resolution of the European CORINE Land Cover database.

Therefore, to fill this gap the present study offers analyses of the coastal dynamics on a local scale, focusing on long-term land cover changes. A new reproducible quantitative approach using basic field analysis to define the evolutionary trends of individual sections of a dune coast is proposed. For this purpose, a spatial database of archival aerial photos was developed and subsequently used to analyze changes in land cover. Finally, models of seashore evolutionary trends were created based on the results of the land cover condition analysis. These methodologies constitute an important step toward developing a predictive model of coastal land cover changes. Furthermore, the implementation of this new approach will enable researchers to test scientific hypotheses stating that the analysis of land cover changes in enumeration units could be an alternative for widely used morphological line indicators in a coastal dune environment.

2. Materials

2.1. Study Site

The Baltic Sea is a non-tidal, semi-enclosed, and shallow body of brackish water with a coastline divided between 14 countries. Poland has a 500-kilometer-long diversified coast [34]. In several locations along the coast there are sections of cliffs that account for 18% of the entire coastline [5]. Apart from small alluvial sections (less than 3% of total), the remainder is a spit- and barrier-type coast, with dunes ranging in height from less than 2 to 49 m.

Generally, those dunes are subject to erosion along the whole Polish coast, and thus usually long sections of the dune coast have to be reinforced with some technical protective infrastructure. Several locations that are exceptions to this, such as the Spit of the Świna River or Świnoujście region (both in the western part of the Polish coast), experience shore accumulation. These unique sites were chosen as the subject-matter of our land cover change analysis. The research area encompasses the western part of the Polish coast with several regions of dune coast located within the administrative limits of the West Pomeranian Voivodeship. It comprises a stretch from the city of Świnoujście to Mrzeżyno sea resort (Figure 1). To ensure accuracy and verifiability, and to allow for a comparison of our results with those of previous studies, we narrowed the scope of this study to the dune-covered coast. Implementing such limitation rules resulted in several exclusion areas of varying sizes.

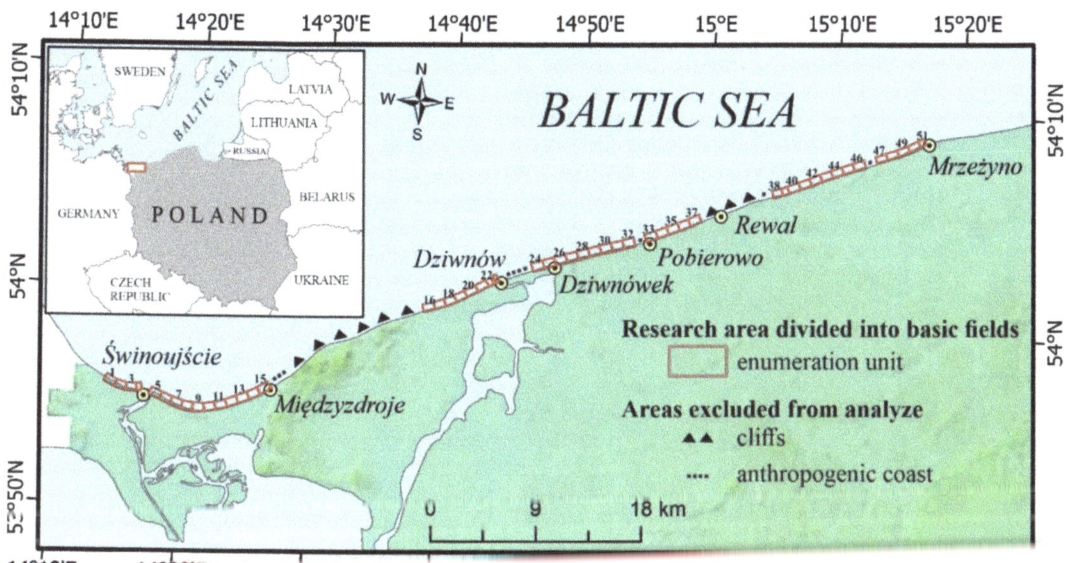

The first excluded fragments represent coastal cliffs. The decision to drop these parts of the coast was due to the substantial differences in height occurring between the cliff edge and base. Accounting for these differences would require the application of different analytic techniques to determine land cover changes.

Furthermore, considering that the analyzed terrain comprised areas where natural processes of shore and land cover evolution occurred, the analysis also excluded all coastal segments comprising river estuaries (the mouths of the Świna and the Dziwna rivers) with their breakwaters, as well as areas modified by intensive construction. Moreover, the exclusion group comprised areas with strong shore fortifications, heavy-duty concrete bands, shore reinforcements with artificial dunes and plantings, and places featuring anthropogenic changes located in the direct vicinity of the dune. Finally, human-modified spaces such as military grounds were also excluded from the analysis. As substantial changes in the area concerned were caused by military activity, they were masked in the 1973 aerial photograph that was available to us, thus preventing comparisons with other photos that we had.

The climatological conditions characterizing the investigated area are controlled by atmospheric pressure patterns and highly influenced by the Atlantic Ocean. The wave energy in the investigated area is medium, with an average significant wave height calculated for the 1958–2001 period ranging from 0.37 m in summer to 1.04 m during winter months. Offshore waves are mainly driven by westerly and southwesterly winds [35].

In recent decades, the highest absolute amplitude of sea level changes in the study area was recorded in 1984 and reached a value of 2.79 m, while the most extreme storm surge (a combination of the sea level and the waves) occurred in November 1995, with a water level of +1.61 m above mean. However, extreme value analyses showed that a 100-year storm surge in the western part of the Polish coast could reach +1.71 m above mean, and a 500-year event would exceed 2 m [5].

The coastal zone itself appears not to be crucial to the national economy, as the maritime sectors provide employment to less than 1% of the total working population. Additionally, the region is rather scarcely populated. The population of the municipalities enjoying direct access to the sea accounts for less than 3% of Poland's entire population. The average population density in the municipalities in the investigated areas is a maximum of 95 persons per km^2, far below the national average of 123 [33]. However, the coast is by far the most popular vacation destination in Poland. It is estimated that the coastal zone attracts approx. 35% of all tourist traffic in Poland during July-August. This corresponds to an average of 71 tourists per 100 residents during the summer [5]. Furthermore, the Baltic Sea coastal zone boasts more than a 20% share in the country's high nature value areas, which attract tourism development and improve the residents' quality of life [33].

2.2. Data

In this study, five series of RS images from 1938, 1951, 1973, 1996, and 2017 were used. The 1938 photomap is in the worst condition due to its age. Although it contains a map graticule along with location names and object descriptions, those elements had a limited impact on the image interpretation outcome. Archival panchromatic aerial photographs from 1951 and 1973 also exhibit signs of damage, such as scratches and discoloration, which directly affects image clarity, although the photographs from 1951 offer satisfactory brightness and clarity, while those from 1973, despite their smaller scale, provide a better contrast. The photographs taken in 1996 are in the best technical condition, as they bear no signs of use and are the most suitable for interpretation. Finally, digital color orthophotomaps from 2017 were obtained from the Head Office of Geodesy and Cartography.

All the 117 aerial photographs analyzed had similar scales and pixel sizes, constituting a valuable comparative material for the detailed analysis of land cover changes. The numbers of images per series, as well as their scales, representing the spatial range of the image, their scan resolutions, and the resulting spatial resolutions (pixel sizes) of the digital versions are presented in Table 1.

Table 1. Archival aerial photograph details.

Year	Count	Scan Resolution	Scale	Spatial Resolution
2017	42	Digital version	1:5 000	0.25 m
1996	26	1200 dpi	1:26 000	0.5 m
1973	18	1200 dpi	1:28 000	0.75 m
1951	20	1200 dpi	1:22 000	0.15 m
1938	11	600 dpi	1:25 000	1 m

3. Methodology

The methodology comprised three basic steps: image processing, land cover detection, and statistical analysis of the changes. The methodology workflow is presented in Figure 2.

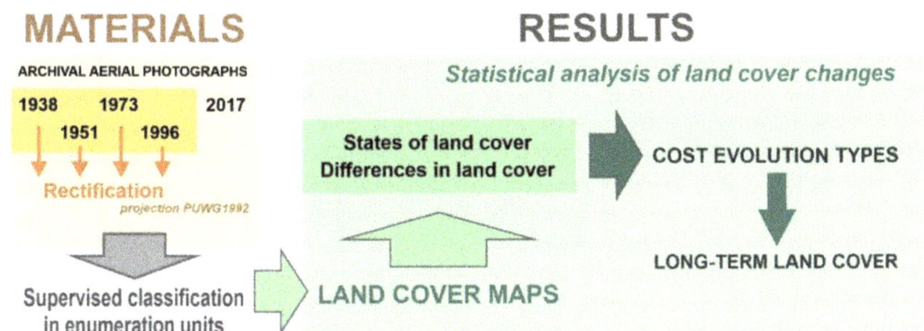

Figure 2. Methodology workflow.

3.1. Image Processing

A photomap from 1938 and archival aerial photographs from 1951 and 1973 were subjected to geometric correction (rectification) into a system of coordinates of the orthophotomap compiled with the use of color photographs taken within the scope of the PHARE program in 1996, which organized all the data in a uniform Państwowy Układ Współrzędnych Geodezyjnych 1992—Polish National Coordinate System 1992 (PUWG 1992) layout. The process involved combining each image with a system of coordinates using n-th degree polynomials [36,37]. Rectification was performed using first degree polynomial transformation in Erdas Imagine 8.5. The purpose of the method is to ensure that an unprocessed image can provide a suitable representation.

The parameters defining a mathematical equation between the previous and the new image reference frame are defined using the ground control points (GCPs), which are points on an output image that correlate to points on the input image to facilitate transformation. The minimum number of ground control points required for the first-degree polynomial is three [38]. However, during rectification, many GCPs were marked and distributed evenly over the image surface, including the marginal areas. Overall, 143 points were placed over the entire surface. The process was performed for each photograph series, beginning with the most recent photos, i.e., those taken in 1973, followed by those dating from 1951 and 1938.

Image rectification is crucial due to the impact of the cartometricity of the processed image on the subsequent analyses. As a result of rectification, the accepted root-mean-square error (RMSE) value in the case of the available material ranged between 2 and 5 m, while the greatest errors were observed for the photomap of 1938. This was so because the image provided for our analysis was processed into a digital form with the use of a flat

3.2. Land Cover Detection

For a detailed analysis of land cover changes, an approx. 50-km-long area of dune sections of the coastline was designated as shown in five photograph series spanning the following periods: 1938–1951, 1951–1973, 1973–1996, and 2017–1996. ArcGIS Pro 2.6 software was used, and the entire area of interest was divided into basic fields (Figures 1 and 3). Each such subarea represented a 1-km-long and 300-m-wide coastal fragment. On several occasions, basic field lengths were slightly shortened to border areas excluded from the analysis. In total, 51 basic fields were extracted and analyzed (Figure 3).

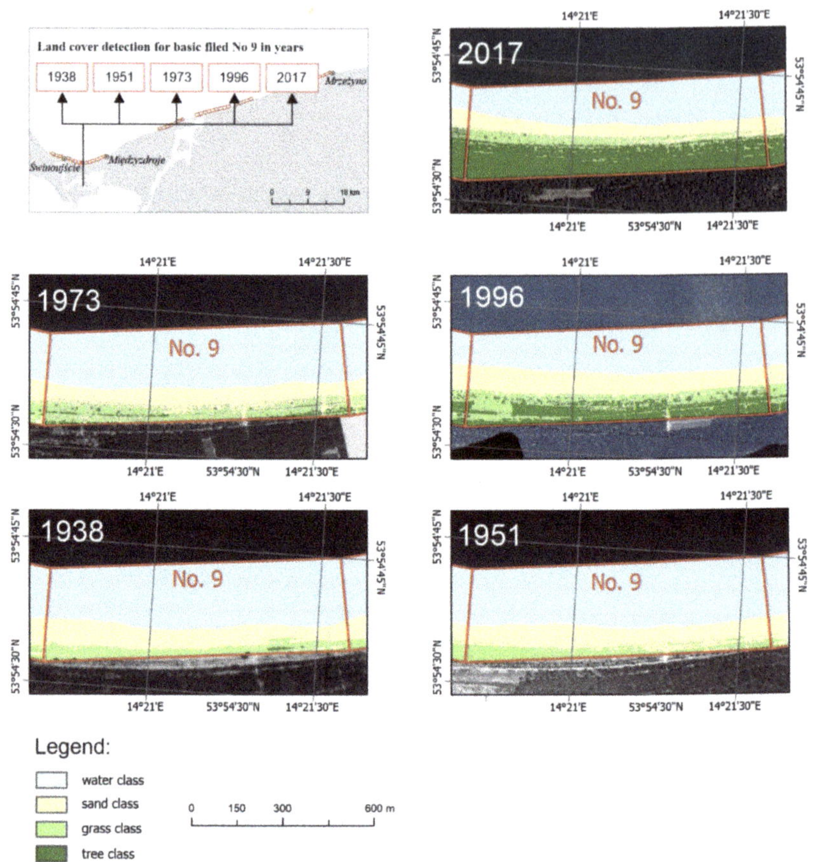

Figure 3. An example of land cover detection for basic field no. 9.

Aerial photograph interpretation was based on Stone [39], who discussed the images of individual plant species captured in photos. How vegetation cover is represented in aerial photographs depends on a multitude of factors, which hinders interpretation and limits the possibility of using standard keys. Vegetation cover constitutes an environmental component that is typically clearly visible, even in panchromatic aerial photographs. However, it is not always distinctly interpretable or identifiable [40].

Within the scope of the supervised classification of the archival aerial photograph series, various states of land cover divided into classes were obtained separately for each analyzed area. Firstly, basic fields were divided into two smaller fragments, isolating an area comprising the sea and the beach, as well as another area comprising the beach, dune vegetation, trees, and shrubs. This particular division stemmed from the fact that in 8-bit photographs, the pixel phototones for water-covered and vegetation-covered areas are remarkably similar. Therefore, to ensure the correct interpretation and classification of archival photograph details, the separate class of "water" had to be singled out.

Once the class of "water" was distinguished, the remaining objects were classified in terms of recognizable properties. Direct object properties were considered, such as image shape and size, phototone, color, texture, and structure. Indirect characteristics were also important, as they indicated the presence of objects or their properties, such as own shadow, shadow cast, topographic location, and pixel values related to other landscape elements, which were verified on the basis of photograph image structure and texture [41].

Identification features may also be described using tools provided by image processing systems. In a digital image, texture is understood as the superficial pixel distribution of various degrees of brightness values, accounting for the regularities and mutual relationships occurring in that distribution [42]. In object recognition, the influence of an interpreter's subjective assessment may be significantly decreased. A digital image record allows for the precise definition of the nature of tone changeability and for the capture of subtle differences in the reflection of the individual ranges of electromagnetic radiation [41].

A supervised classification was done to locate specific areas within the image that show homogenous categories of the land cover types. Object base classification was used with a manually created training sample for each image over identifiable cover types. The process involved using the Training Samples Manager tools in the ArcGIS PRO software in order to set the neighborhood mode, the geographic constrains of the pixels, and the spectral Euclidean distance. Multiple areas were drawn and grouped for a single category. Those areas represented the best match of pixels. The process was repeated for every known category on the image. The supervised classification used the maximum likelihood parametric classification rule.

When interpreting vegetation cover, we were able to differentiate between two basic groups: tree vegetation (with shrubs) and grasses. Non-forest formations can be discerned through changes in texture, structure, tone, and color to determine the different species; the density of vegetation cover itself is essential as well. Grass vegetation is typically smooth and nearly uniform in color, while coastal grasses are lighter than other vegetation. Eventually, the following land cover classes were specified:

- tree class
- grass class
- sand class
- water class

These classes have inclusive names to account for a wide range of species. The term "grass" indicates low-growing vegetation up to a height of approximately 0.5 m, comprising various plant communities. The term "tree" indicates vegetation growing higher than 0.5 m and contains shrubs and young trees. Both these classes were clearly distinguishable in our aerial photographs.

This stage was also important for collecting the testing sites of the study area. These are very important to determine the accuracy assessment for each classification algorithm and to check the validation of classifications. Usually, whenever possible, the testing sites should be represented as ground control points collected from the field of the study area. Because the presented work is based on the analysis of historical data, such control points (one for each of 4 classes) has been collected using office work from each of the 117 analyzed photos.

The final step of the image classification was the accuracy assessment stage. This process was performed as an estimation with the aid of a remotely sensed dataset for classification conditions, and it is useful for the evaluation of the classification approach. [illegible]

classification was 89.5%. The values were different for each series of photographs, reaching 82.43% for 1938, 90.25% for 1951, 92.22% for 1996, and 93.12% for 2007, respectively.

3.3. Statistical Analysis

Having identified and measured the area covered by the individual classes, the results were recalculated from real values expressed in square meters to relative values expressed as a percentage share of each class. These shares allowed for a detailed analysis of land cover changes without comparing the sizes of basic field surfaces in relation to their actual surfaces. Consequently, a change inferred from the share size of a given class could be analyzed both for an individual basic field and the entire analyzed section of the coast extending from Świnoujście to Mrzeżyno.

All statistical calculations and analyses were realized within Statistica 12.5 software. Firstly, in order examine changes over time, these land cover classes were subjected to statistical analysis. First, several standard descriptive statistics, histograms, and standard box plots were used to indicate variability in the land cover classes. The box plot, also called box-whiskers plot, was used in order to present the variability of the land cover classes. Values larger than the sum of the third quartile and three quarter deviations (Q_3+3 (Q_3- Q_1)/2), as well as those less than the difference of the first quartile and three quarter deviations (Q_1+3 (Q_3- Q_1)/2), are considered to be atypical, i.e., outliers [44].

Further on, the share values of the individual land cover classes were subjected to cluster analysis, which is used to arrange objects in groups according to the degree of relation [44]. This method is most helpful in the exploratory phase of research, where no hypothesis has yet been formulated. Therefore, cluster analysis does not constitute a statistical test but helps group objects according to their similarities. In the present research, it was used to cluster two different types of data. First of all, it was employed to distinguish groups of basic fields with similar shares of land cover classes. Second, it was used to examine and create a clustering process of similar patterns in land cover changes between the time periods studied.

To this effect, the Euclidean distance—the geometric distance in a multi-dimensional space—was used to determine the distance between the new clusters. The Ward's method that was used also assumes that every object constitutes a separate cluster. At every subsequent stage, groups are created by combining objects and clusters formed at earlier stages. Consequently, the objects were combined into ever-growing clusters until a sufficient level of grouping was achieved. The method employs variance analysis to estimate the distance and aims to minimize the sum of the deviation squares inside clusters. At every stage, of every possible pair of clusters that can be combined, the cluster that eventually yields the minimum diversification is selected [45]. The measure of such diversification in relation to average values is the error sum of squares, which is defined by the following formula:

$$ESS = \sum_{i=1}^{k} (x_i - x)^2 \quad (1)$$

where:

x—the base value of the variable constituting the segmentation criterion,

x_i—value of the variable constituting the segmentation criterion for the i-th object,

k—number of objects in a cluster.

Linear regression, being the simplest regression variant, was employed to determine the development trends of individual classes of land cover over time. It assumes that there exists a linear relationship between an explained variable and an explaining variable. In linear regression, it is presumed that an increase of one variable is accompanied by the growth or decrease of the other variable. A regression function takes on the form of a linear function:

$$y = bx + a \quad (2)$$

where:

a and b are regression estimators.

Linear regression analysis is used to determine such coefficients with which the model could best predict the value of a dependent variable. Such approach ensures that the estimation error is as small as possible. It also enables us to set a development trend of land cover class changes which could be used for defining model data in further analyses [44].

4. Results

4.1. Descriptive Analysis of Land Cover

The percentage shares of the individual classes are presented in the form of a stacked column chart (Figure 4). Characteristic arrangements were observed for areas 1 to 15, which comprised the section stretching from Poland's western border through the mouth of the Świna River to Międzyzdroje. This section exhibited an accumulative character with a fragment of a stabilized shore lying to the west of Międzyzdroje (the areas 13 to 15). In the aerial photographs from 1938 and 1951, areas 1 to 12 revealed a wide beach along with grass vegetation covering the territory; there were no visible trees or shrubs there, though. Trees and shrubs in those areas appeared as late as 1973, and their surface area expanded greatly in 1996. This territory clearly implied accumulative processes, occurring particularly in areas 5, 6, and 7, lying to the east of the breakwater at the mouth of the Świna River. In area 5, water occupied 66.2% in 1938 and only 36.3% in 1996, whereas in area 6 water area shrank from 53.9% in 1938 to 28.3% in 1996 (Table S1).

The areas between 11 and 15 constituted a section of a stabilized shore. The accumulation found in that stretch of the coast was slight, but a natural succession of vegetation cover entering new areas was observed. In areas 13 and 14, trees and shrubs were recorded in 1938 and 1973. The shore was stabilized, and the water surface area was virtually unchanged, accounting for approximately 60%. The proportions of the remaining land cover classes kept changing. An expansion of tree and grass cover at the expense of surface clear of vegetation could be observed. Areas 16 to 37 were clearly more stable in terms of land cover change dynamics (Tables S2 and S3). Between areas 16 and 23 (Tables S1 and S2), lying to the west of the mouth of the River Dziwna, the water surface area exhibited minimal to no expansion, while changes in the land cover structure were noted. Such changes were particularly visible between the periods 1938–1951 and 1951–2017. During the former, the surface clear of vegetation expanded along with an increase in the beach width. During the latter, a rise in grass vegetation became evident, followed by the appearance of trees and shrubs, which were visible in those areas in 1996.

The portion of the coast located between Dziwnów and Trzęsacz, comprising the areas from 24 to 37, was a stabilized shore fragment. Water and beach surface areas remained the same, and the tree area increase became noticeable with a simultaneously shrinking surface occupied by grass.

In areas 38 to 51 (Pogorzelica-Mrzeżyno), the water surface area advanced, which implicated that erosion processes were underway (Table S3). Significant areas clear of vegetation were observed, specifically in 1938 and 1951 (Figure 4). In subsequent years, a rise in the surface occupied by trees, shrubs, and grasses was observed.

The investigated status of land cover, with account taken of particular classes within the analyzed coastal sections, revealed that the terrains extending in the direct vicinity shared similar proportions of land cover structure. The dynamics of the changes occurring in those terrains were also similar. Additionally, it was noted that the territories excluded from the research, such as river estuaries or cliff-coast sections, were naturally distinct areas with different land [illegible]

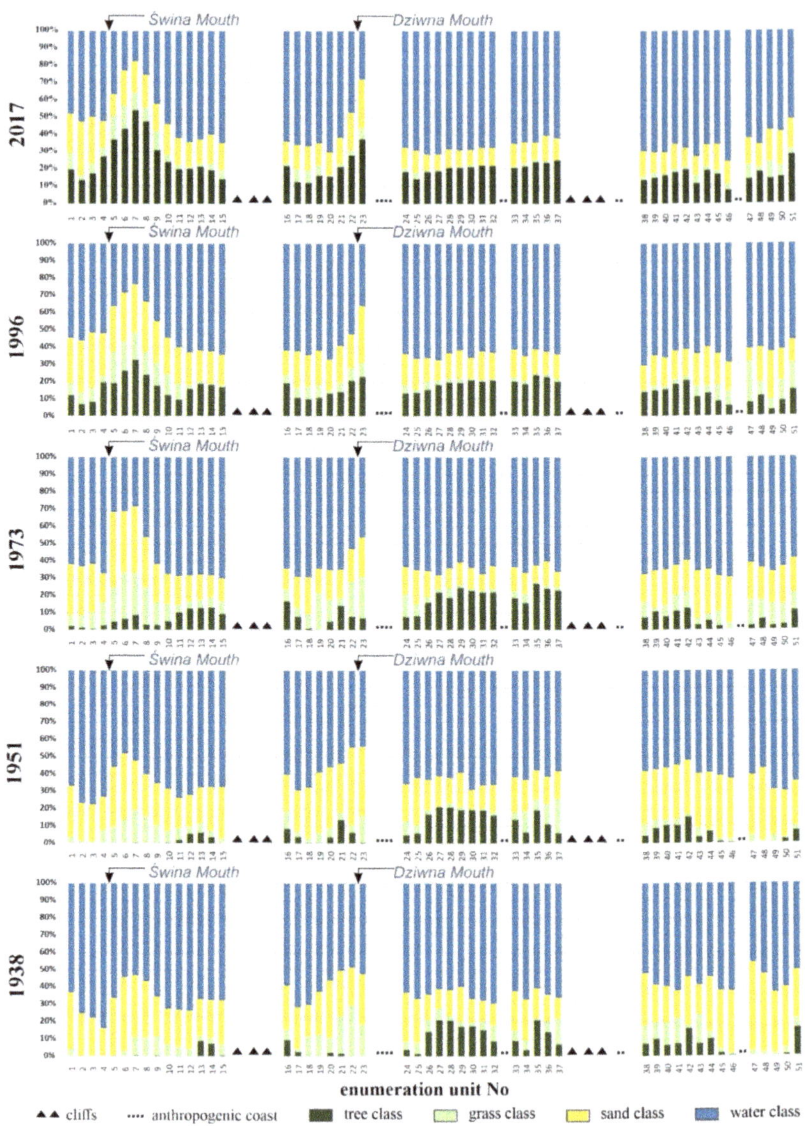

Figure 4. Percentage share of each class.

4.2. Differences in Land Cover

To achieve a detailed qualitative change analysis, calculations were performed to identify differences in the surface areas according to land cover class for each area between the analyzed years.

Between 1938 and 1951, a drop in the area of land clear of vegetation was observed along the entire analyzed coast fragment with a simultaneous rise in grass-covered and tree-covered terrains. A growth in the water-covered area became noticeable as well. The section of the coast encompassing areas 1–9 constituted an exception to the aforementioned trend. In that terrain, a significant loss of water area of more than 20% was observed, with a simultaneous expansion of the area occupied by other land cover classes (Figure 5).

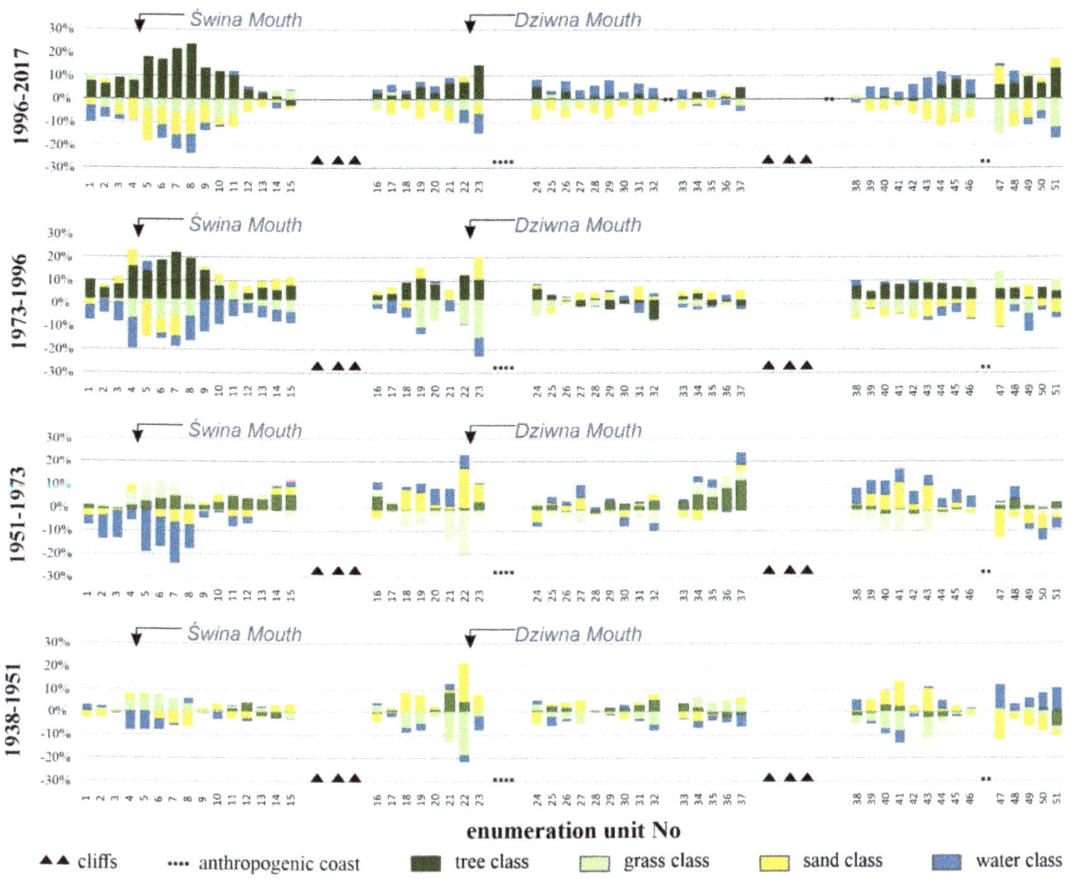

Figure 5. Percentage changes in land cover.

The years 1951–1938 comprised a period during which the tree class cover grew, amounting to 24% in relation to 1938 for area 7 (Figure 5). A simultaneous drop in the area of land covered by grass was recorded. The water-covered area in the individual areas of the entire analyzed coast section fell as well.

The accumulation process in that section was less intensive owing to the currents running alongside the shore and transporting sandy material to the terrains lying further to the west. The lack of beach expansion resulted in limited opportunities [illegible]

As a result of our analyses of the land cover changes occurring between 1938 and 1996, vegetation succession was confirmed for the investigated fragment of the coast, mainly represented by an increasing share of the tree and shrub land cover classes. The first section of the investigated length of coast extended between Świnoujście and Międzyzdroje (areas 1–15), which had an accumulative character suitable for vegetation succession, and where the most substantial vegetation spread was observed.

The second fragment comprised areas 16 to 23, where slight changes were recorded regarding the surface area occupied by water and an increasing area covered by vegetation.

The largest change was observed for the tree class, with an increase of 20% from 1938 to 1996 in area 22.

The third section included areas 24 to 37, where the water surface area had been dominant since 1938, and accounted for approximately 60% of the total land cover. The changes occurring along this fragment of coast involved the expansion of the surface overgrown by trees and shrubs and a decrease in the share of grasses, owing to the lack of opportunities for them to inhabit new grounds. No beach area increments were noted here.

The last fragment of coast comprised areas 38 to 51, where a consistent tendency for the water surface area to expand and a persistent substantial share of the area free of vegetation were observed. A rising share of grass and trees was recorded. This analysis demonstrated that there was a significant correlation between the water surface area and the area free of vegetation, on the one hand, and the changes in the area occupied by vegetation, on the other, which suggested that the structure of the land covered by vegetation was influenced by coastal dynamics.

The relationship between the area covered by water and the area free of vegetation was evident in our subsequent analyses. Where the water surface exhibited rapid changes (10% or more), an inversely proportional change in the area free of vegetation was noticeable.

It is characteristic of accumulative areas that decreases in water-covered terrain and expansion of terrain free of vegetation are accompanied by a growth in the area carrying vegetation. For stabilized shores, tree growth occurs evidently at the expense of grassy areas, with simultaneous small variations in the areas occupied by water and free of vegetation.

4.3. Statistical Analysis of Land Cover Changes

Finally, the basic fields investigated were analyzed in terms of similarities both in the percentage of the surface occupied by the individual land cover classes and in differences of land cover between the analyzed years. Cluster analysis using Ward's method [46] was implemented, and the results are presented in the tree diagram below (Figure 6).

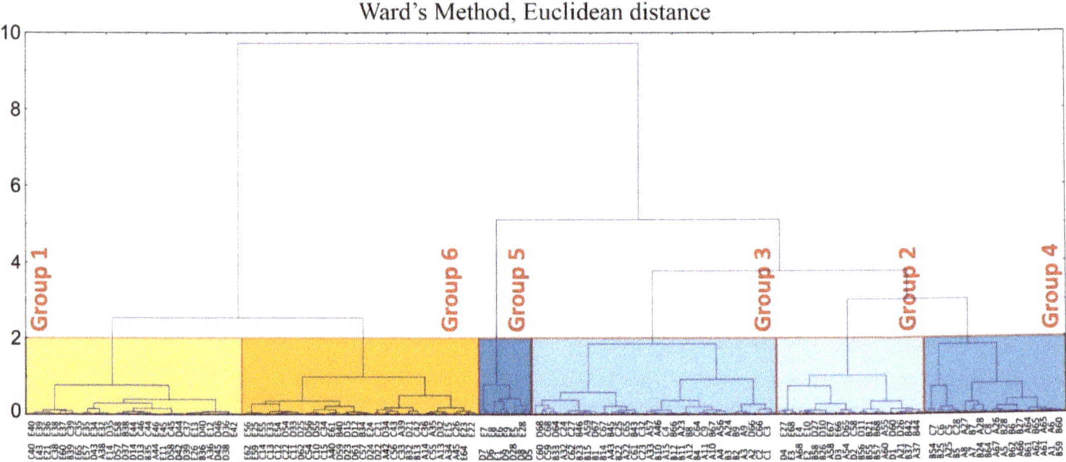

Figure 6. Land cover grouping based on the percentage shares of the individual classes via cluster analysis using Ward's method for all the years in question. Identifier format: YB, where Y—year (A—1938, B—1951, C—1973, D—1996, E—2017), B—basic field number (from 1 to 51).

Firstly, similarities between the percentage shares of the individual classes were investigated for each year under analysis. According to the measure of the Euclidean distance, areas closest to one another exhibit the smallest binding distance, which becomes increasingly small as differences between individual areas grow. The above diagram (Figure 6) shows that the areas fell within six characteristic groups. The assignment of the individual areas to the given group in each year was analyzed, and the shares of the particular land cover classes were examined. Furthermore, the average value of the given class in each group was determined. To facilitate the understanding of the spatial distribution of the identified clusters, a diagram was developed that presents the assignment of each single basic field to the specific identified cluster for all the years in question (Figure 7).

Figure 7. Diagram showing the assignment of all the areas to groups based on Ward's method.

One characteristic of Group 1 was the approximate 4% and 15% shares of grass and beach surfaces, respectively. Group 2 comprised areas that featured a wide beach, occupying a surface of more than 20%. In Group 3, the approximately 9% share of grass vegetation was a typical occurrence, where it may have been proof of the evolution of a dune that was invaded by grass. Group 4 typically comprised accumulative areas with a slight share of water, a very wide beach, a large proportion of grass in the land cover, and a substantial area taken over by trees. Group 5 was characterized by the approximate shares of 12% and 15% of grass and beaches, respectively, and small areas covered by water. Group 6 was composed of areas where erosion prevailed, grass was virtually non-existent or had a scant share, and the share of trees and shrubs reached approximately 21%.

Based on the spatial distribution of the similarity groups, we were able to identify the variability of the different cluster changes over time. The visible transitions of one type into another in different years constituted a layout that corresponded to the natural coastal structure. In turn, during the analysis of the coast types over time, it was observed that a portion of the areas remained in the same group; for example, area 26 had shown the same coast type since 1938 that was assigned to Group 1. Transitions between the groups could be observed as well. Namely, area 21 once belonged to Group 4 (in 1938), Group 2 (in 1951), Group 1 (in 1973), Group 2 (in 1996), and Group 6 (in 2017), indicating that it experienced significant land cover changes. Overall, 22 areas exhibited the most frequent transformations in structure over time, transitioning between three groups. For 18 areas, a transition between two groups was recorded, with four areas moving between 12 groups.

Furthermore, to better visualize the similarity groups distinguished using Ward's analysis, an average, a minimum, and a maximum of areas were calculated based on all the basic field falling within each cluster. The final averaged land cover characteristic of the clusters is shown in Figure 8, and the extreme values (min. and max.) are shown in Table 2.

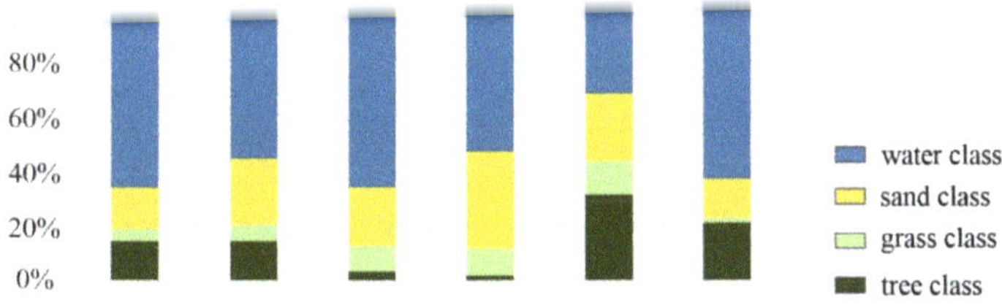

Figure 8. Average land cover areas in Groups 1–6 classified according to Ward's method.

Table 2. Land cover characteristic in Groups 1–6 classified according to Ward's method.

	Group 1			Group 2			Group 3			Group 4			Group 5			Group 6		
	Mean	*Max*	*Min*	*Mean*	*Max*	*Min*	*Mean*	*Max*	*Min*	*Mean*	*Max*	*Min*	*Mean*	*Max*	*Min*	*Mean*	*Max*	*Min*
Water class	66%	75%	36%	55%	62%	47%	66%	84%	53%	53%	66%	28%	32%	48%	17%	63%	67%	60%
Sand class	15%	23%	8%	24%	32%	15%	21%	30%	8%	35%	51%	22%	24%	33%	16%	14%	18%	9%
Grass class	4%	14%	0%	6%	12%	2%	9%	24%	0%	10%	29%	1%	12%	18%	7%	2%	4%	0%
Tree class	15%	37%	5%	15%	28%	7%	4%	15%	0%	2%	9%	0%	31%	54%	18%	21%	27%	17%

The second part of our cluster analysis, again conducted using Ward's method, focused on similarities in differences of land cover values identified between the individual periods. In the subsequent stage of the study, all the differences observed for the individual periods were subjected to a cluster analysis, and the degree of the relationships between them was examined. The relationships were determined on the basis of the percentage shares of the given classes of land cover in a basic field, much as they were in the case of the classification trees.

Six groups were distinguished based on these results, as shown in Figure 9. The spatial distribution of the land cover share differences between the similarity groups along the coast is shown in Figure 10.

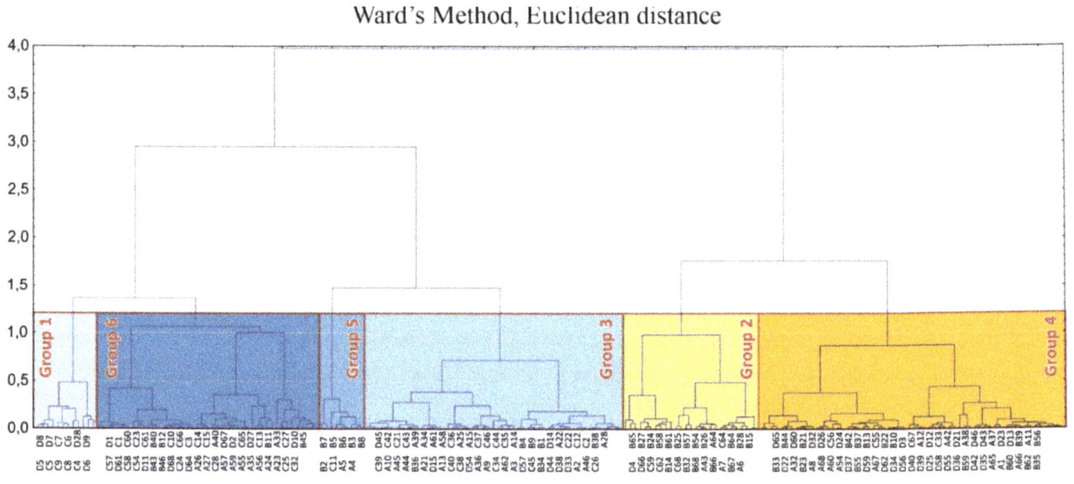

Ward's Method, Euclidean distance

Figure 9. Grouping of differences in land cover in the areas based on a cluster analysis using Ward's method for all the areas and all the years in question. Identifier format: YB, where Y—year (A—1973–1996, B—1951–1973, C—1938–1951, D—1996–2017), B—basic field number (from 1 to 51).

Figure 10. Areas in Groups 1–6 classified according to Ward's method.

As was the case of the land cover analysis, the similarity classes of land cover changes were described using an average, minimum and maximum of area calculations. An averaged cluster land cover characteristic is shown in Figure 11, and the extreme values (min. and max.) are shown in Table 3.

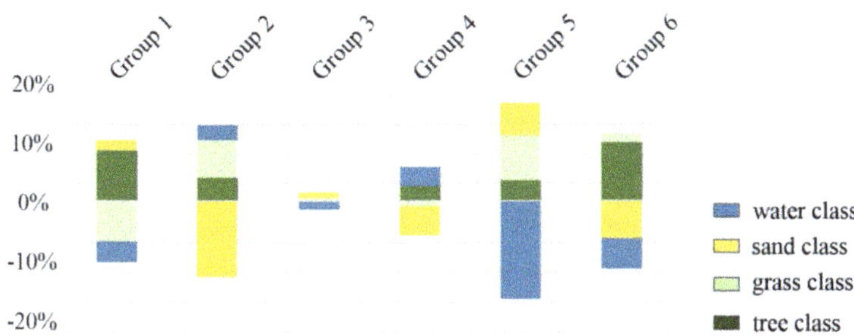

Figure 11. Average differences in land cover areas across Groups 1–6 classified according to Ward's method.

Table 3. Land cover differences characteristic in Groups 1–6 classified according to Ward's method.

	Group 1			Group 2			Group 3			Group 4			Group 5			Group 6		
	Mean	*Max*	*Min*	*Mean*	*Max*	*Min*	*Mean*	*Max*	*Min*	*Mean*	*Max*	*Min*	*Mean*	*Max*	*Min*	*Mean*	*Max*	*Min*
Water class	−3%	8%	−17%	3%	15%	−7%	−1%	3%	−8%	3%	14%	−2%	−17%	−8%	−24%	−5%	4%	−11%
Sand class	2%	22%	−18%	−13%	−5%	−26%	1%	10%	−5%	−5%	1%	−10%	5%	10%	−3%	−6%	3%	−13%
Grass class	−7%	2%	−25%	6%	16%	0%	0%	6%	−4%	−1%	6%	−6%	8%	13%	5%	2%	10%	−6%
Tree class	9%	24%	−4%	4%	12%	−3%	0%	6%	−5%	2%	9%	−9%	3%	8%	−1%	10%	24%	0%

Finally, the areas assigned to the given group and the structures of the changes in their land cover were verified. In 1938–1973, there were areas assigned to Group 5 where the changes reached a maximum value of 33%. The changes in the structure varied, primarily in terms of the loss of water, accompanied by an increase in areas occupied by beach and vegetation in the form of grass and trees, typical of high accumulation coasts.

In two groups (Groups 2 and 4), the area of land covered by water went up, which was linked to the erosion experienced by those terrains and the accompanying shrinkage of the beach. Group 1 comprised stabilized areas in which a loss of land covered by grass and water was observed, accompanied by a simultaneous incremental increase in the land occupied by trees and sand. Group 2 comprised coastal sections in which a rise in grass-covered land was seen, which may have been evidence of stabilized coastal conditions that benefited the growth of new grass. Increases in grass cover occurred at the expense of beach cover (sand class—areas free of vegetation). Group 3 included very stable coastal areas with low coastal dynamics. Group 4 comprised terrains where accumulation and erosion occurred and was followed by beach expansion and the dwindling of the vegetation cover, as well as terrains with increasing amounts of land covered by water. Group 6 contained territories in which the expansion of the tree cover occurred at the expense of diminishing amounts of land covered by water and beaches. These were the coastal sections along

The employed statistical analysis demonstrated that the examined fragment of the coast had terrains that were statistically similar to one another despite the distances between them. This finding was also confirmed by the statistical analysis of the differences in land cover changes. The observed area grouping regularity occurred at a distance of approximately 0.5 for both the analyses. A statistical analysis conducted for all the areas in the individual years and for the land cover differences further demonstrated that the grounds had changeable coast characteristics.

The six averaged land cover area groups shown in Figure 8 reflect different types of coast evolution found along the Polish dune coastline, with the spatiotemporal changes of the distinguished types shown in Figures 10 and 11.

5. Discussion

Aerial photographs are a source of valuable historical information on the vegetation cover and its status [18,19]. As resource managers' and scientists' demands for spatially explicit data continue to grow [18], the use of digitized or digital aerial photographs and the development of automatic analysis techniques can improve the accuracy, consistency, and efficiency of results [47]. Historic information from aerial photographs can also be helpful in monitoring land cover, land cover changes, and coastal dynamics.

Using aerial photos in automated image analysis brings about multiple challenges, including differences in contrast, spectral and spatial resolutions, the solar angle, and the time/year of acquisition. Nevertheless, historical photos in conjunction with plot-based records and appropriate ground truth information provide an important record of vegetation dynamics over time [20].

Investigating land cover and its spatial and temporal changes is essential in coastal sciences. Most recent work deals with their impact on economic development, investment, agricultural policy, and environmental protection policy [33,48–50]. The novelty of the proposed research is that it aims to use land cover changes as a basis for studying coastal dynamics, which allows for the determination of the long-term evolutionary trends of individual fragments of a dune coastline.

This approach differs from many other coastal analyses made worldwide based on simplified global shoreline movement patterns [10,51,52] or site-specific regional analyses of actual coastal changes, that are mostly studied using cross sections of the coast [3,53].

We should keep in mind that, regardless of the line or area indicator methodology, by capturing the historical shoreline position form maps and aerial photographs, one will have to deal with the uncertainty of delineating them. Based on RS data, we are able to mark only the temporal location of a water line. On the one hand, both types of analyses actually neglect the problem of water level change. On the other hand, using area indicators instead of line ones automatically provides additional information about the erosive or accumulative character of change. This appears to be a more reliable method to represent the mean shoreline position for the analyzed period.

Furthermore, the proposed method for determining land cover changes in basic fields containing both water and land surfaces represents an original approach. Although van der Meulen [54] has applied a similar method to the coast of the Netherlands, he only analyzed changes in land cover under the influence of an expected rise in sea level in cells that were 1000-m-long and 50-m-wide, and parallel to the shoreline.

Although the methodology for analyzing coastal dynamics based on basic fields proposed herein has its limitations, mostly associated with the short field, of 1 km in length, the observed changes with regard to the land covered by water, beach, and vegetation do reproduce coastal dynamics with acceptable accuracy and are comparable with earlier research. Thus, the approach where the structure of land cover is analyzed and subsequently used to conduct a statistical analysis of the changes appears to be an efficient technique and an alternative to methods only resorting to cross sections and line indicators [55].

Detailed results for land cover change dynamics confirm the general development tendency demonstrated by the shore as shown by Dudzińska-Nowak [30]. She analyzed the changes of a dune/cliff base occurring between 1938 and 1996 in Rewal Community (research areas no. 32-46, respectively). Based on the analysis, she identified areas subject to morphological erosion and accumulation processes. Her research helped to recognize the variations and dynamics of coast changes. The areas featuring strong accumulation and erosion between 1938 and 1951 turned out either to be stable in the subsequent 1951-73 period or changed their character to either erosive or accumulative. She showed that the period between 1973 and 1996 saw the further disappearance of shore sections of an accumulative nature in that region. Sections showing an accumulative trend in the previous period transformed into sections of stable shore, whereas sections of stable shore changed their character to erosive [30]. The results of our analysis confirm those tendencies, which proves that the implemented methodology for determining coastal dynamics by analyzing the land cover in basic fields can be compared to line-indicator methods.

Furthermore, the proposed combination of land cover changes derived from historical remote sensing data supported with a statistical analysis is in agreement with the dune coast evolution proposed as part of the general dune shore development model presented by Hesp [56] or Psuty [57]. According to those authors, the coastal foredune is the uppermost and inlandmost component of the sand-sharing system. It has accumulated sand in association with a range of pioneer vegetation types to create a positive landform perched above the dry sand beach. It is the most conservative portion of the coast, undergoing dimensional and temporal changes of a far lesser magnitude and frequency than the sand beach or the offshore zone. In that simple model, the coastal foredune exists on the boundary between the coastal processes on its seaward side and the continental processes occurring landwards. However, many coastal zones are not as simple as this profile, and there are multiple instances of variable dune configurations and areas immediately inland of the dune-beach profile that appear to be morphodynamically related to the processes active in the sand-sharing system [58].

According to our research, performed in the dune areas of the Pomeranian Bay's eastern coast and extending back to 1938, a succession of the vegetation cover was observed, chiefly with regard to the tree class. The fragment of the coast between Świnoujscie and Międzyzdroje (areas no. 1 to 15) boasted the most marked vegetation cover succession with simultaneous intensive accumulation processes. Foredunes may be classified into two types: incipient and established. Incipient foredunes are new, or developing, foredunes forming within pioneer plant communities. They may be formed by sand depositing within discrete or relatively discrete clumps of vegetation, or individual plants (types 1a and 1b of Hesp) [56,59]. Such types were identified in areas no. 1 to 15 between 1938 and 1973.

The beaches in the sandy parts of the Polish coast are of considerably varying widths and inclinations, determined by the dynamics in force in the given section of the coast [60]. These findings were confirmed by our results for several of the basic fields analyzed. In areas no. 16 to 23, slight changes to the land area occupied by water were observed, along with substantial proportions of vegetation and beach. The stretch to the east of Dziwnów and to the west of Rewal is dominated by land covered by water (areas 24 to 37). The changes occurring within this fragment of coast are represented by an increasing area carrying tree and shrub land cover classes, which have been successively invading the grass-covered area of land. In turn, a decline in the share of grass cover resulted from its inability to inhabit new terrain, with the beach area failing to expand in that particular section of the coast. In areas 38 to 51, land cover changes demonstrated intensive dynamics. In areas 45 and 50, a substantial growth of grass and tree cover was recorded. In 1938 and 1951, these sections of the coast carried initial vegetation that was characterized by irregular coverage (island coverage). In 1973, a dense structure was observed that

including a wide beach.

Foredunes are very dynamic forms which may rapidly convert into a different form likely to remain in the same location for years to come, until the erosional phase sets in [32,61]. Herein the dynamics of sand class building that forms is an indicator of long-term variability, and could be used as an alternative indicator of coastal dune changes.

The statistical analysis based on Ward's method demonstrated that in terms of land cover structure the studied areas could be classified into six groups according to the shore dynamics presented. Depending on the land cover structure, it is possible to predict the shore evolution stages. These results are consistent with a dune coast evolution model based on the differences in beach width, a strip of white dunes, and gray dunes overgrown with dense vegetation [62].

6. Conclusions

The methodology proposed, that accounts for long-term land cover changes as a record of coastal dynamics, is a novel research approach that has never been applied to studies of the Polish Baltic Sea coastal zone. Historical aerial photographs dating back to 1938, 1951, 1973, 1996, and 2017 were processed to conduct a detailed long-term analysis of land cover changes in sections of the dune coast stretching over a total of 51 km. This study was the first for the coastal zone of the Pomeranian Bay's eastern coast, demonstrating that the long-term evolution of dune shores could be analyzed based on changes in land cover using archival RS data.

The methodology developed herein helps discover new possibilities for defining coastal zone dynamics and can be used as an alternative solution to methods only resorting to cross sections and line indicators. The rates of change can be derived from the observed variability of areas of land covered by water or free of vegetation, while the share of the grass and tree classes constitutes a record of the long-term dynamics of the coastal zone. A multidimensional cluster analysis using Ward's method showed that there were six different groups of coast types differing in terms of the land cover structure that is likely to transform over time.

Supplementary Materials: The following are available online at https://www.mdpi.com/2072-4292/13/6/1068/s1, Table S1: Land cover area percentages in each year for areas no. 1–17 [%], Table S2: Land cover area percentages in each year for areas no. 18–36 [%], Table S3: Land cover area percentages in each year for areas no. 37–51 [%].

Author Contributions: Conceptualization, field experiments, and formal analysis: A.G.; methodology and data curation: A.G., P.T., P.C., and T.K.; writing original draft, visualization, and validation: A.G. and P.T.; supervision: A.G. All authors have read and agreed to the published version of the manuscript.

Funding: This research was co-financed within the framework of the program of the Ministry of Science and Higher Education under the name "Regional Excellence Initiative" during the period 2019–2022; project number 001/RID/2018/19; the amount of financing was PLN 10,684,000.00, and "The Oceanographic Data and Information System eCUDO.pl"—contract no. POPC.02.03.01-00-0062/18-00 was funded by the European Union through the Operational Programme Digital Poland 2014–2020 Fund.

Data Availability Statement: The data presented in this study are available on request from the corresponding author. The data are not publicly available due to privacy issues.

Conflicts of Interest: The authors declare no conflict of interest.

References

1. Deng, J.; Harff, J.; Zhang, W.; Schneider, R.; Dudzińska-Nowak, J.; Giza, A.; Terefenko, P.; Furmańczyk, K. The Dynamic Equilibrium Shore Model for the Reconstruction and Future Projection of Coastal Morphodynamics. In *Coastline Changes of the Baltic Sea from South to East*; Springer International Publishing: Berlin/Heidelberg, Germany, 2017; pp. 87–106.
2. Vousdoukas, M.I.; Voukouvalas, E.; Annunziato, A.; Giardino, A.; Feyen, L. Projections of extreme storm surge levels along Europe. *Clim. Dyn.* **2016**, *47*, 3171–3190. [CrossRef]
3. Terefenko, P.; Giza, A.; Paprotny, D.; Kubicki, A.; Winowski, M. Cliff retreat induced by series of storms at Międzyzdroje (Poland). *J. Coast. Res.* **2018**, *85*, 181–185. [CrossRef]
4. Tõnisson, H.; Suursaar, U.; Rivis, R.; Kont, A.; Orviku, K. Observations and analysis of coastal in the West Estonian Archipelago caused by storm Ulli (?Emil) in January 2012. *J. Coastal Res.* **2013**, *S165*, 832–837. [CrossRef]
5. Paprotny, D.; Terefenko, P. New estimates of potential impacts of sea level rise and coastal floods in Poland. *Nat. Hazards* **2017**, *85*, 1249–1277. [CrossRef]
6. Bugajny, N.; Furmańczyk, K. Comparison of Short-Term Changes Caused by Storms along Natural and Protected Sections of the Dziwnow Spit, Southern Baltic Coast. *J. Coastal Res.* **2017**, *33*, 775–785. [CrossRef]

7. Furmańczyk, K.; Andrzejewski, P.; Benedyczak, R.; Bugajny, N.; Cieszyński, Ł.; Dudzińska-Nowak, J.; Giza, A.; Paprotny, D.; Terefenko, P.; Zawiślak, T. Recording of selected effects and hazards caused by current and expected storm events in the Baltic Sea coastal zone. *J. Coastal Res.* **2014**, *70*, 338–342. [CrossRef]
8. Bugajny, N.; Furmańczyk, K. Short-term Volumetric Changes of Berm and Beachface during Storm Calming. *J. Coastal Res.* **2020**, *SI95*, 398–402. [CrossRef]
9. Terefenko, P.; Giza, A.; Paprotny, D.; Walczakiewicz, S. Characteristic of Winter Storm Xavier and Its Impacts on Coastal Morphology: Results of a Case Study on the Polish Coast. *J. Coastal Res.* **2020**, *SI95*, 684–688. [CrossRef]
10. Paprotny, D.; Terefenko, P.; Giza, A.; Czapliński, P.; Vousdoukas, M.I. Future losses of ecosystem services due to coastal erosion in Europe. *Sci. Total Environ.* **2021**, *760*, 144310. [CrossRef] [PubMed]
11. Hägerstrand, T. Samhälle och natur. *NordREFO* **1993**, *1*, 14–59.
12. Montereale-Gavazzi, G.; Roche, M.; Lurton, X.; Degrendele, K.; Terseleer, N.; Van Lancker, V. Seafloor change detection using multibeam echosounder backscatter: Case study on the Belgian part of the North Sea. *Mar. Geophys. Res.* **2018**, *39*, 229–247. [CrossRef]
13. Gaida, T.C.; van Dijk, T.A.G.P.; Snellen, M.; Vermaas, T.; Mesdag, C.; Simons, D.G. Monitoring underwater nourishments using multibeam bathymetric and backscatter time series. *Coast. Eng.* **2020**, *158*, 103666. [CrossRef]
14. Janowski, L.; Madricardo, F.; Fogarin, S.; Kruss, A.; Molinaroli, E.; Kubowicz-Grajewska, A.; Tegowski, J. Spatial and Temporal Changes of Tidal Inlet Using Object-Based Image Analysis of Multibeam Echosounder Measurements: A Case from the Lagoon of Venice, Italy. *Remote Sens.* **2020**, *12*, 2117. [CrossRef]
15. Morgan, J.L.; Gergel, S.E. Quantifying historic landscape heterogeneity from aerial photographs using object-based analysis. *Landsc. Ecol.* **2010**, *25*, 985–998. [CrossRef]
16. Morgan, J.L.; Gergel, S.E.; Coops, N.C. Aerial Photography: A Rapidly Evolving Tool for Ecological Management. *Bioscience* **2010**, *60*, 47–59. [CrossRef]
17. Tomscha, S.A.; Sutherland, I.J.; Renard, D.; Gergel, S.E.; Rhemtulla, J.M.; Bennett, E.M.; Daniels, L.D.; Eddy, I.M.S.; Clark, E.E. A guide to historical data sets for reconstructing ecosystem service change over time. *Bioscience* **2016**, *66*, 747–762. [CrossRef]
18. Cohen, W.B.; Kushla, J.D.; Ripple, W.J.; Garman, S.L. An introduction to digital methods in remote sensing of forested ecosystems: Focus on the Pacific Northwest. *Environ. Manag.* **1996**, *20*, 421–435. [CrossRef] [PubMed]
19. Fensham, R.J.; Fairfax, R.J. Aerial photography for assessing vegetation change: A review of applications and the relevance of findings for Australian vegetation history. *Aust. J. Bot.* **2002**, *50*, 415–429. [CrossRef]
20. Laliberte, A.S.; Rango, A.; Havstad, K.M.; Paris, J.F.; Beck, R.F.; McNeely, R.; Gonzalez, A.L. Object-oriented image analysis for mapping shrub encroachment from 1937 to 2003 in southern New Mexico. *Remote Sens. Environ.* **2004**, *93*, 198–210. [CrossRef]
21. Gibbens, R.P.; McNeely, R.P.; Havstad, K.M.; Beck, R.F.; Nolen, B. Vegetation changes in the Jornada Basin from 1858 to 1998. *J. Arid Environ.* **2005**, *61*, 651–668. [CrossRef]
22. Crowell, M.; Leatherman, S.P.; Buckley, M.K. Historical Shoreline Change: Error Analysis and Mapping Accuracy. *J. Coast. Res.* **1991**, *7*, 839–852.
23. Wziątek, D.Z.; Terefenko, P.; Kurylczyk, A. Multi-Temporal Cliff Erosion Analysis Using Airborne Laser Scanning Surveys. *Remote Sens.* **2019**, *11*, 2666. [CrossRef]
24. Terefenko, P.; Paprotny, D.; Giza, A.; Morales-Nápoles, O.; Kubick, A.; Walczakiewicz, S. Monitoring Cliff Erosion with LiDAR Surveys and Bayesian Network-based Data Analysis. *Remote Sens.* **2019**, *11*, 843. [CrossRef]
25. Deng, J.J.; Harff, J.; Giza, A.; Hartleib, J.; Dudzinska-Nowak, J.; Bobertz, B.; Furmanczyk, K.; Zolitz, R. Reconstruction of Coastline Changes by the Comparisons of Historical Maps at the Pomeranian Bay, Southern Baltic Sea. In *Coastline Changes of the Baltic Sea from South to East: Past and Future Projection*; Harff, J., Furmanczyk, K., Von Storch, H., Eds.; Coastal Research Library; Springer: Berlin/Heidelberg, Germany, 2017; Volume 19, pp. 271–287.
26. Lucas, C. Flora der Insel Wollin. *Verh. des Bot. Ver. Berl. Brandenbg.* **1860**, *2*, 25–68.
27. Short, A.D.; Hesp, P.A. Wave, beach and dune interactions in southeastern Australia. *Mar. Geol.* **1982**, *48*, 259–284. [CrossRef]
28. Carter, R. *Coastal Environments. An Introduction to the Physical, Ecological, and Cultural Systems of Coastlines*; Academic Press: Cambridge, MA, USA, 2013.

31. Grunewald, R. Assessment of Damages from Recreational Activities on Coastal Dunes of the Southern Baltic Sea. *J. Coast. Res.* **2006**, *5*, 1145–1157. [CrossRef]
32. Łabuz, T.A.; Grunewald, R. Studies on Vegetation Cover of the Youngest Dunes of the Świna Gate Barrier (Western Polish Coast). *J. Coast. Res.* **2007**, *1*, 160–172. [CrossRef]
33. Bielecka, E.; Jenerowicz, A.; Pokonieczny, K.; Borkowska, S. Land Cover Changes and Flows in the Polish Baltic Coastal Zone: A Qualitative and Quantitative Approach. *Remote Sens.* **2020**, *12*, 2088. [CrossRef]
34. Wolski, T.; Wiśniewski, B.; Giza, A.; Kowalewska-Kalkowska, H.; Boman, H.; Grabbi-Kaiv, S.; Hammarklint, T.; Holfort, J.; Lydeikaitė, Ž. Extreme sea levels at selected stations on the Baltic Sea coast. *Oceanologia* **2014**, *56*, 259–290. [CrossRef]

35. Różyński, G. Wave Climate in the Gulf of Gdańsk vs. Open Baltic Sea near Lubiatowo, Poland. *Arch. Hydroeng. Environ. Mech.* **2010**, *57*, 167–176.
36. Novak, K. Rectification of Digital Imagery. *Photogramm. Eng. Remote Sens.* **1992**, *58*, 339–344.
37. Hood, J.; Ladner, L.; Champion, R. Image Processing Techniques for Digital Orthophotoquad Production. *Photogramm. Eng. Remote Sens.* **1989**, *55*, 1323–1329.
38. Toutin, T. Geometric processing of remote sensing images: Models, algorithms and methods. *Int. J. Remote Sens.* **2004**, *25*, 1893–1924. [CrossRef]
39. Stone, K.H. A Guide to the Interpretation and Analysis of Aerial Photos. *Ann. Assoc. Am. Geogr.* **1964**, *54*, 318–328. [CrossRef]
40. Lillesand, T.; Kiefer, R.W. *Remote Sensing and Image Interpretation*, 4th ed.; John Wiley & Sons: New York, NY, USA, 2000.
41. Hudak, A.T.; Wessman, C.A. Textural Analysis of Historical Aerial Photography to Characterize Woody Plant Encroachment in South African Savanna. *Remote Sens. Environ.* **1998**, *66*, 317–330. [CrossRef]
42. Bähr, H.P.; Vögtle, T. *Digitale Bildverarbeitung. Anwendungen in Photogrammetrie, Fernerkundung und GIS*; Wichmann Verlag: Heidelberg, Germany, 2005.
43. Congalton, R. A Review of Assessing The Accuracy of Classifications of Remotely Sensed Data. *Remote Sens. Environ.* **1991**, *37*, 35–46. [CrossRef]
44. Stanisz, A. *Przystępny Kurs Statystyki z Zastosowaniem STATISTICA PL na Przykładach Medycyny*; StatSoft: Kraków, Poland, 2007.
45. Ward, J.K. Hierarchical groupings to optimize an objective function. *J. Am. Stat. Assoc.* **1963**, *58*, 236–244. [CrossRef]
46. Luszniewicz, A.; Słaby, T. *Statystyka z Pakietem Komputerowym STATISTICA PL. Teoria i Zastosowania*; Wydawnictwo C.H. Beck: Warsaw, Poland, 2003.
47. Harvey, K.; Hill, G.J.E. Vegetation mapping of a tropical freshwater swamp in the Northern Territory, Australia: A comparison of aerial photography, Landsat TM and SPOT satellite imagery. *Int. J. Remote Sens.* **2001**, *22*, 2911–2925. [CrossRef]
48. Lambin, E.F.; Geist, H.J.; Lepers, E. Dynamics of Land-Use and Land-Cover Change in Tropical Regions. *Annu. Rev. Environ. Resour.* **2003**, *28*, 205–241. [CrossRef]
49. Antso, K.; Palginõmm, V.; Szava-Kovats, R.; Kont, A. Dynamics of Coastal Land Use over the Last Century in Estonia. *J. Coast. Res.* **2011**, *SI64*, 1769–1773.
50. Chun, J.; Kim, C.-K.; Kang, W.; Park, H.; Kim, G.; Lee, W.-K. Sustainable Management of Carbon Sequestration Service in Areas with High Development Pressure: Considering Land Use Changes and Carbon Costs. *Sustainability* **2019**, *11*, 5116. [CrossRef]
51. Ibaceta, R.; Splinter, K.D.; Harley, M.D.; Turner, I.L. Enhanced coastal shoreline modelling using an Ensemble Kalman Filter to include non-stationarity in future wave climates. *Geophys. Res. Lett.* **2020**, *47*, e2020GL090724. [CrossRef]
52. Vousdoukas, M.I.; Ranasinghe, R.; Mentaschi, L.; Plomaritis, T.A.; Athanasiou, P.; Luijendijk, A.; Feyen, L. Sandy coastlines under threat of erosion. *Nat. Clim. Chang.* **2020**, *10*, 260–263. [CrossRef]
53. McCarroll, R.J.; Masselink, G.; Valiente, N.G.; Scott, T.; Wiggins, M.; Kirby, J.A.; Davidson, M. A novel rules-based shoreface translation model for predicting future coastal change: ShoreTrans. *Earth arXiv* **2020**. [CrossRef]
54. van der Meulen, F.; Witter, J.V.; Arens, S.M. The use of a GIS in assessing the impacts of sea level rise on nature conservation along the Dutch coast: 1990–2090. *Landsc. Ecol.* **1991**, *1*, 105–113. [CrossRef]
55. Keijsers, J.G.S.; De Groot, A.V.; Riksen, M.J.P.M. Modeling the biogeomorphic evolution of coastal dunes in response to climate change. *J. Geophys. Res. Earth Surf.* **2016**, *121*, 1161–1181. [CrossRef]
56. Hesp, P. Foredunes and blowouts: Initiation, geomorphology and dynamics. *Geomorphology* **2002**, *48*, 245–268. [CrossRef]
57. Psuty, N.P. The Coastal Foredune: A Morphological Basis for Regional Coastal Dune Development. In *Coastal Dunes*; Martínez, M.L., Psuty, N.P., Eds.; Ecological Studies; Springer: Berlin/Heidelberg, Germany, 2008; p. 171.
58. Martínez, M.L.; Psuty, N.P. *Coastal Dunes. Ecology and Conservation*; Springer: New York, NY, USA, 2004.
59. Hesp, P. A review of biological and geomorphological processes involved in the initiation and development of incipient foredunes. *Proc. R. Soc. Edinb. Sect. B Biol. Sci.* **1989**, *96*, 181–202. [CrossRef]
60. Musielak, S. Shoreline dynamics between Niechorze and Świnoujście. *J. Coast. Res.* **1995**, *22*, 288–291.
61. Łabuz, T.A. Polish coastal dunes: Affecting factors and morphology. *Landf. Anal.* **2013**, *22*, 33–59. [CrossRef]
62. Giza, A.; Furmańczyk, K. Model Rozwoju Profilu Brzegu Wydmowego. In *Geologia i Geomorfologia Pobrzeża i Południowego Bałtyku*; Florek, W., Ed.; Wydawnictwo Naukowe Pomorskiej Akademii Pedagogicznej w Słupsku: Słupsk, Poland, 2005; pp. 123–136.

Article

Mapping Coastal Dune Landscape through Spectral Rao's Q Temporal Diversity

Flavio Marzialetti [1], Mirko Di Febbraro [1], Marco Malavasi [2,*], Silvia Giulio [3], Alicia Teresa Rosario Acosta [3] and Maria Laura Carranza [1]

1 Envix-Lab, Department of Bioscience and Territory, University of Molise, Contrada Fonte Lappone, 86090 Pesche (Is), Italy; flavio.marzialetti@unimol.it (F.M.); mirko.difebbraro@unimol.it (M.D.F.); carranza@unimol.it (M.L.C.)

2 Department of Applied Geoinformatics and Spatial Planning, Faculty of Environmental Sciences, Czech University of Life Sciences Prague, Kamycka 129, 165 00 Prague, Czech Republic

3 Department of Sciences, University of Roma Tre, Viale G. Marconi 446, 00146 Rome, Italy; silvia.giulio@uniroma3.it (S.G.); aliciateresarosario.acosta@uniroma3.it (A.T.R.A.)

* Correspondence: malavasi@fzp.czu.cz

Received: 27 June 2020; Accepted: 17 July 2020; Published: 18 July 2020

Abstract: Coastal dunes are found at the boundary between continents and seas representing unique transitional mosaics hosting highly dynamic habitats undergoing substantial seasonal changes. Here, we implemented a land cover classification approach specifically designed for coastal landscapes accounting for the within-year temporal variability of the main components of the coastal mosaic: vegetation, bare surfaces and water surfaces. Based on monthly Sentinel-2 satellite images of the year 2019, we used hierarchical clustering and a Random Forest model to produce an unsupervised land cover map of coastal dunes in a representative site of the Adriatic coast (central Italy). As classification variables, we used the within-year diversity computed through Rao's Q index, along with three spectral indices describing the main components of the coastal mosaic (i.e., Modified Soil-adjusted Vegetation Index 2—MSAVI2, Normalized Difference Water Index 2—NDWI2 and Brightness Index 2—BI2). We identified seven land cover classes with high levels of accuracy, highlighting different covariates as the most important in differentiating them. The proposed framework proved effective in mapping a highly seasonal and heterogeneous landscape such as that of coastal dunes, highlighting Rao's Q index as a sound base for natural cover monitoring and mapping. The applicability of the proposed framework on updated satellite images emphasizes the procedure as a reliable and replicable tool for coastal ecosystems monitoring.

Keywords: coastal habitats; ecosystem monitoring; land cover mapping; random forest algorithm; Sentinel-2; modified soil-adjusted vegetation index 2–MSAVI2; normalized difference water index 2–NDWI2; brightness index 2–BI2

(i.e., climate and land-cover change, invasive species and habitat loss) are promoting changes in global biodiversity at an unprecedented rate in human history [1–3]. The effects of such changes are particularly severe on coastal dune landscapes [4,5], despite the fact that they host a highly specialized biodiversity [6] and provide essential benefits to society [7,8].

Coastal dunes are found at the boundary between continents and seas, representing unique transitional mosaics hosting highly dynamic habitats undergoing frequent and substantial changes in physical extent and environmental conditions. Along with this typical transitional condition, seasonality also plays a pivotal role in such ever-changing nature, affecting both abiotic (e.g., magnitude and intensity of weather and sea conditions) and biotic (e.g., phenology) factors, whose interaction finally gives rise to an extremely dynamic and complex mosaic of psammophilous plant communities, bare surfaces and water surfaces [9,10]. Moreover, being highly endangered, coastal dunes are of conservation concern worldwide and, as such, they need updated monitoring and mapping protocols [11,12]. Such protocols, traditionally based on field biodiversity data and habitat photointerpretation of aerial imagery [13,14], have improved in the last decades with the support of remotely sensed images and biomass spectral information (e.g., [15]). Nonetheless, in order to better discriminate between different types of standing biomass, remote sensing approaches for habitat mapping and monitoring in coastal areas are conventionally performed through classification of images captured at the peak of the plant growing season (see [16]). That said, developing an efficient mapping protocol that accounts for the dynamic nature of coastal systems, thus capturing all of the main components of coastal dune mosaics together with their seasonal variation (i.e., vegetation, bare surfaces and water surfaces) [17], still represents a challenge for delivering more accurate and viable maps.

At present, the availability of remotely sensed data, at various spatial, spectral and temporal resolutions, offers a great potential to carry out accurate and cost-effective studies accounting for ecosystem phenology and seasonality [18–20], thus reinforcing this research field, which traditionally relies on ground-based observations [21,22]. The analysis of remotely sensed phenological trends may effectively support land cover classification and mapping [23,24]. In addition, this potential is steadily growing with the continuous improvement of the temporal resolution of orbiting satellites [25,26].

The remotely sensed analysis of seasonality, traditionally based on spectral bands [27,28], has been improved over time by including the time-series analysis of spectral indices [20]. Several metrics have been proposed to quantify environmental seasonality and vegetation phenology based on remotely sensed data [29]. Such metrics commonly focus on describing the cyclic behavior of spectral indices, intended as the combination of spectral reflectance from two or more bands indicating the relative abundance of features of interest, and on identifying key transition dates (e.g., start, peak, end, and length of seasonal periods) [30]. Additionally, other indices, considering the whole array of available bands and used in the past for describing spectral diversity across space, have been recently extended to quantify variations in diversity across time [31,32]. Among them, the Rao's Q index, borrowed from community and landscape ecology [33–35], is able to take into account both the proportion of cells assuming different spectral values and their spectral distance [36]. Consequently, extending Rao's Q rationale to the temporal dimension (e.g., comparing spectral values and distances of a given location in a multi-temporal stack), makes it a good candidate to accurately synthesize ecosystems' seasonality. The calculation of Rao's Q on a temporal stack generates a new layer of temporal diversity, with high or low values indicating seasonal or stable habitats, respectively. Several diversity layers summarizing temporal stacks (bands or spectral indices) can be combined instead of the rough spectral values traditionally used on multi-temporal classification [16,36]. As far as we know, few studies, if any, extending Rao's Q to the spectral temporal diversity for land cover mapping, currently exist in transitional systems.

In this context, the present work sets out to provide and test a land cover classification approach specifically designed for coastal landscapes based on the within-year temporal diversity, computed through Rao's Q index. In particular, the within-year diversity will be calculated upon the seasonal behavior of three spectral indices properly describing the main components of the coastal mosaic (i.e., vegetation, bare surfaces and water surfaces). Such behavior will be then classified to provide a map of natural and seminatural land cover types. While accounting for the dynamic/temporal dimension highly characterizing coastal landscape, we expect that such an approach will provide a high level of classification accuracy.

2. Materials and Methods

2.1. Study Area

The study was carried out on the Adriatic coast of Central Italy (Molise region, Figure 1). This area, of approximately 9000 km^2, is mainly composed of sandy beaches, a few river mouths and channels, and one rocky promontory. Dunes, occupying a narrow strip parallel to the seashore, are low and relatively recent (formed in the Holocene period) [37,38].

Along a sea-to-inland gradient, the typical vegetation zonation ranges from embryonic dunes in the seashore, followed by mobile dunes with perennial herbaceous vegetation, fixed dunes covered by evergreen shrub and small sclerophyllous trees and, in the inner sectors, by wooded dunes covered by coniferous forests (Figure 1) [39,40]. The Molise coast hosts several ecosystems of conservation concern in Europe (European Directive 92/43/EEC) [41]. For this reason, such an area is largely included in the European Natura2000 system [41] and is part of the European LTER monitoring network (Long Term Ecological Research network) [40,42].

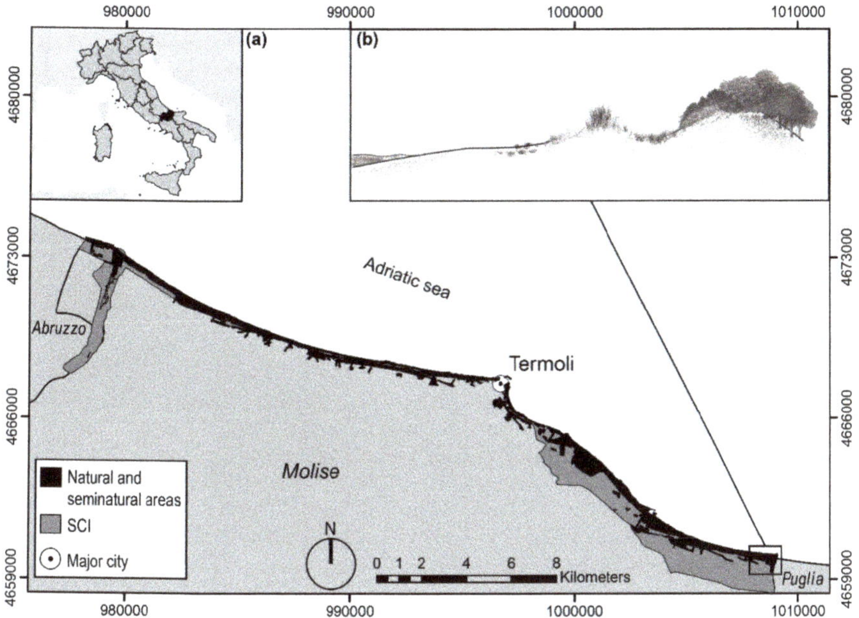

Figure 1. (**a**) In black, the study area including the coastal dunes of the Molise Region (Italy). Most of the analyzed coastal sectors are included in Sites of European Conservation Concern (SCI, European Directive 92/43/EEC); Foce Trigno-Marina di Petacciato [illegible]

2.2. Methodology

We followed an unsupervised approach and classified coastal dune natural and semi-natural land cover types through a hierarchical cluster analysis. The classes identified in the clustering phase were then used as the response variable in a Random Forest (RF) model (an accurate learning method for discriminating differences among classes [43,44]), in order to quantify their accuracy and to predict their occurrence in the study area. In particular, the procedure was organized in the following steps:

(1) Sentinel-2 imagery selection; (2) spectral indices calculation and temporal variability computation; (3) variables selection and classification; (4) accuracy assessment (Figure 2).

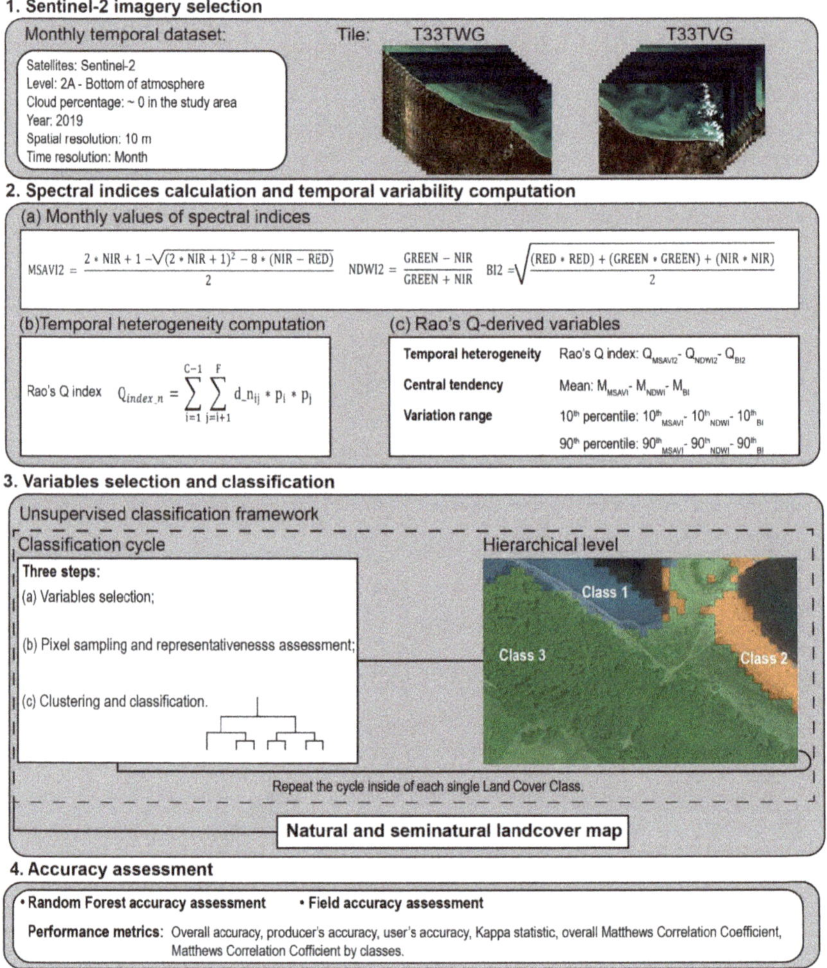

$$Q_{index\_n} = \sum_{i=1}^{C-1} \sum_{j=i+1}^{F} d\_n_{ij} * p_i * p_j$$

Figure 2. Workflow synthesizing the full mapping procedure of coastal dune Semi-natural and natural cover types with temporal MSAVI2, NDWI2, BI2 series and Random Forest classification approach.

2.2.1. Sentinel-2 Imagery Selection

We used Sentinel-2 mission images as a sound support for land monitoring [45], with good spatial (10, 20 or 60 m), temporal (revisit time of 2–3 days at mid-latitudes) and spectral (13 bands ranging from 400 nm to visible to 2400 nm) resolutions [25]. Multispectral images of Sentinel-2 satellites were downloaded from Copernicus Open Access Hub (https://scihub.copernicus.eu/) at the level of bottom of atmosphere (Level-2A), and used to build a monthly temporal dataset for the year 2019. As the study area falls in two tiles (T33TVG, T33TWG), we selected 24 images (12 for tile) of each month with low cloud coverage (<5%, Supplementary Material). All images were atmospherically corrected by ESA through the "sen2cor" processor algorithm (Figure 2, box 1) [46]. We specifically used Green band, with wavelength 560 ± 36 nm; Red band, with wavelength 665 ± 31 nm, and NIR band, with wavelength 833 ± 106 nm, with 10 m of resolution.

2.2.2. Spectral Indices Calculation and Temporal Variability Computation

For each image, we calculated three spectral indices: Modified Soil-Adjusted Vegetation Index 2 (MSAVI2), Normalized Difference Water Index 2 (NDWI2) and Brightness Index 2 (BI2), as they are good indicators of vegetation biomass, the presence of water surfaces, and bare surfaces, respectively (Figure 2 box 2 a, Table 1).

MSAVI2 is a vegetation index that quantifies the photosynthetic biomass based on Red and NIR bands (Table 1). MSAVI2 has been developed for mapping landscapes characterized by high percentages of bare surfaces [47,48] and its spatial behavior is a good support for land classification [49]. MSAVI2 ranges from −1 (absence of biomass vegetation) to 1 (maximum of biomass vegetation) with higher values indicating higher percentages of photosynthetic biomass [48].

NDWI2 is a water index that is useful to identify water surfaces, exploiting the Green and NIR bands (Table 1). NDWI2 is a remotely sensed index that is particularly efficient for identifying water surfaces and for mapping water-land transitions [50,51]. NIR band and Green band present opposite reflectance values behaviors and NDWI2 values range from −1 to 1, where values greater than 0 indicate water surfaces [52].

The BI2 index is sensitive to soil brightness, hence it quantifies the bare surfaces through the square root of brightness of each pixel (Table 1) [53]. BI2 accurately discriminates the bare surfaces from vegetation in heterogeneous environments [54,55]. The minimum value of BI2 is 0 and indicates the absence of bare surfaces, as growing positive values correspond to increasing percentages of bare surfaces.

We used ESA's Sentinel-2 toolbox—ESA Sentinel Application Platform 7.0 (SNAP) for index calculation.

Table 1. Spectral indices selected for analyzing the temporal diversity, serving as proxies of seasonality in vegetation biomass, presence of water surfaces, and bare surfaces.

Acronym	Name	Formula	Index of	References
MSAVI2	Modified Soil-Adjusted Vegetation Index 2	$MSAVI2 = \frac{2*NIR+1-\sqrt{(2*NIR+1)^2-8*(NIR-RED)}}{2}$	photosynthetic biomass	[48]
NDWI2	Normalized Difference Water Index 2	$NDWI2 = \frac{GREEN-NIR}{GREEN+NIR}$	presence of water surfaces	[51]
BI2	Brightness Index 2	$BI2 = \sqrt{\frac{(RED*RED)+(GREEN*GREEN)+(NIR*NIR)}{3}}$	presence of bare surfaces	[52]

For each spectral index, we built an annual stack containing the 12 month values (i.e., $MSAVI2_{2019}$, $NDWI2_{2019}$, $BI2_{2019}$), standardized between 0 and 1 to make them comparable [56]. To summarize the within-year heterogeneity of ecological conditions (e.g., biomass, water surfaces and bare surfaces yearly variation), we calculated the temporal Rao's Q index for each annual stack (Figure 2 box 2 b). The Rao's Q index has recently been borrowed from functional ecology and successfully applied in remote sensing contexts as an [illegible]

The Rao's Q diversity index was applied on each annual stack (i.e., $MSAVI2_{2019}$, $NDWI2_{2019}$, and $BI2_{2019}$) according to the following formula (Equation (1)) [57,58]:

$$Q_{index\_n} = \sum_{i=1}^{C-1} \sum_{j=i+1}^{F} d\_n_{ij} * p_i * p_j \quad (1)$$

where:

Q_{index_n} = Rao's Q quantifying the within-year variability of a spectral index ($MSAVI2_{2019}$, $NDWI2_{2019}$ or $BI2_{2019}$) for the pixel n

p = relative abundance of the index value assumed by the pixel n within a temporal stack

i = month i

j = month j

d_n_{ij} = distance between the month i-th and j-th index value of pixel n ($d_{ij} = d_{ji}$ and $d_{ii} = 0$).

As similarly done in spectral diversity applications, Rao's Q adapted to detect temporal diversity quantifies the expected dissimilarity between two combinations of pixel values randomly selected within a pixel temporal stack (Figure 3). Pixels representing highly seasonal cover types (e.g., deciduous or annual formations) are characterized by a pronounced within-year variability of biomass values and bare surfaces cover. Therefore, such pixels should assume high temporal Rao's Q values. On the contrary, pixels representing temporally stable coastal areas (e.g., open sand, pine wood) portraying weak or absent seasonal variations, should score low Rao's Q values. The temporal Rao's Q for each spectral index (Q_{MSAVI2}, Q_{NDWI2}, Q_{BI2}) was computed through R package 'spacetimerao' 0.1 [59].

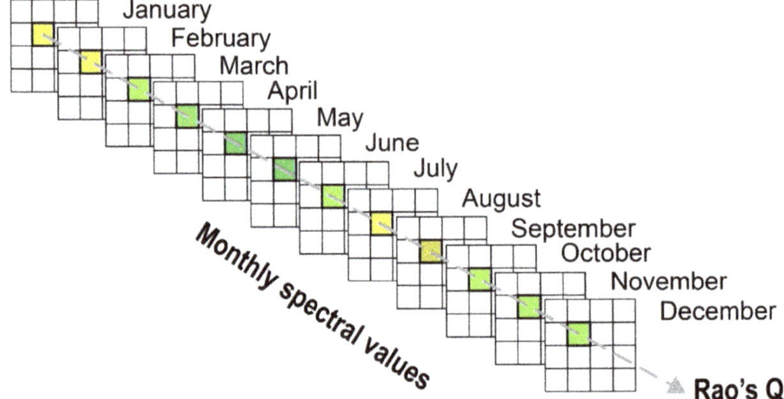

Figure 3. Schematic representation of temporal Rao's Q diversity calculation implemented on year stacks of spectral indices MSAVI2, NDWI2 and BI2 to summarize the seasonal behavior of vegetation biomass, water and bare soil surfaces on coastal dunes.

In order to summarize the range of annual variation for each index, we also calculated their mean values (M_{BI2}, M_{MSAVI2}, M_{NDWI2}), along with their 10th and 90th percentiles (10^{th}_{BI2}, 10^{th}_{MSAVI2}, 10^{th}_{NDWI2}, 90^{th}_{BI2}, 90^{th}_{MSAVI2}, 90^{th}_{NDWI2}, Figure 2 box 2 c).

2.2.3. Variables Selection and Classification

The classification phase was implemented through consecutive cycles, each structured into three steps (Figure 2 box 3). Each step consists of a sequence of procedures: (a) variables selection, (b) pixel sampling and representativeness assessment, (c) clustering and classification. As covariates for classification, we used 12 variables, which describe the temporal heterogeneity (Q_{MSAVI2}, Q_{NDWI2}, Q_{BI2}), the central tendency (M_{MSAVI2}, M_{NDWI2}, M_{BI2}), and the variation range (10^{th}_{MSAVI2}, 10^{th}_{NDWI2}, 10^{th}_{BI2}, 90^{th}_{MSAVI2}, 90^{th}_{NDWI2}, 90^{th}_{BI2}) of the three spectral indices (i.e., MSAVI2, NDWI2, BI2).

(a) Variables Selection

During each cycle, the 12 covariates were checked for their multicollinearity (Graham 2003) using the Variance Inflation Factor (VIF, Figure 2 box 3 a). VIF quantifies the multiple correlation of a variable with respect to all the other variables through linear regressions [60]. As VIF values greater than 5 indicate multicollinearity problems [61], we retained, for the analysis, the variables with VIF < 5 [62].

(b) Pixel Sampling and Representativeness Assessment

To reduce the computational costs, we ran the classification process on a subset of pixels randomly extracted from the study area. In particular, classification was carried out on a 20%-pixel sample and the results were extrapolated throughout the entire study area using an RF algorithm. The degree of extrapolation on values of covariates lying outside the RF calibration range was assessed through the Multivariate Environmental Similarity Surface, (MESS, Figure 2 box 3 b) [63]. MESS quantifies the similarity of all the pixels in the study area with respect to the covariates of the extracted 20% sample used for classification. Low MESS values indicate a high similarity between covariate values of calibration and prediction pixels, suggesting a high representativeness of the former. In each classification cycle, we indicated the percentage of highly dissimilar pixels (MESS value < 0).

(c) Clustering and Classification

The random sample of pixels was classified through a hierarchical cluster analysis using the Ward's minimum variance method [64], which minimizes the cluster's internal variability using the sum-of-squares [65,66]. We used as distance the Bray–Curtis dissimilarity metric [67,68] measured as follows (Equation (2))

$$BC_{pq} = \frac{\sum_{i=1}^{n} \left| x_{pi} - x_{qi} \right|}{\sum_{i=1}^{n} \left(x_{pi} - x_{qi} \right)} \quad (2)$$

where BC_{pq} is the Bray–Curtis dissimilarity between pixels (p, q), x_{pi} are the values of the n selected remotely sensed variables (e.g., Q_{MSAVI2}, M_{NDWI2}, $90^{th}{}_{B12}$) on pixel p and x_{qi} the variables value on pixel q. BC_{pq} ranges from 0 when the pixels are identical, to 1 when the pixels are completely different [69].

On each cycle, we identified the optimal number of classes (form from 2 to 10 classes; Supplementary Material) by calculating three indices (Silhouette index [70]; Calinski–Harabasz index [71] and Davies–Bouldin index [72]) on the just built hierarchical cluster. The selected indices summarize two cluster characteristics: the compactness of classes (e.g., how closely pixels are grouped inside a class), and the separation between classes (e.g., how the classes are different from each other).

The classes identified in the clustering phase were used as response variables in a RF model (R package 'caret' 6.0-85; Supplementary Material) [73–75] in order to evaluate their accuracy, to extrapolate the classification throughout all the pixels of the study area, and to evaluate the variables' contribution in defining such classes. To optimize RF parameters, we set a high number of uncorrelated decision trees (*Ntree* = 1000) [76–78], while testing different combinations of the number of variables randomly selected at each node (*Mtry* parameter) and split rules [79–82], then choosing the combination that yielded the highest Kappa statistic value. Specifically, we tested *Mtry* values ranging from 2 to the total number of variables in the cycle, and the Gini index and Extra-Trees algorithm as possible split rules (Supplementary Material).

Once the optimal RF model was identified, it was used to predict the membership of all the study area pixels to one of the classes identified in the clustering phase. Moreover, we estimated the relative variables' [illegible]

[illegible] were then interpreted in terms of coastal cover types through an expert-based approach based on field detection and visual interpretation of high resolution aerial images (~1 m).

2.2.4. Accuracy Assessment

The RF predictive accuracy was assessed by an internal 10-fold cross validation and an independent validation based on 300 random checkpoints. We selected 300 points as to assure a minimum standard number of 20 control points for each land cover class [84]. In the independent accuracy assessment, we compared the assignment of the checkpoints according to the land cover classes obtained by satellite classification with their description obtained from field observations and the visual inspection of Google Earth images. Because of the differences in spatial resolution between Sentinel-2 (10 × 10 m) and Google Earth images (~1 m), we focused our visual inspection on circular buffer areas of 5-m radius (10 m′ diameter) around each check point [85–88]. In both accuracy assessments, we calculated the same performance metrics. We built a confusion matrix and calculated the percentages of overall accuracy, producer's accuracy, user's accuracy and Kappa statistic [84]. Moreover, we calculated the Matthews Correlation Coefficient, both overall and for each land cover class [89]. This coefficient is widely used to evaluate the accuracy of classifications [86,90], as it proved reliable to evaluate the accuracy of classification when the classes had different sizes; its range of −1 to 1 and values close to 1 represent a perfect accuracy [87,91].

3. Results

3.1. Unsupervised Classification

We obtained seven land cover classes after three classification cycles, which are organized in two hierarchical levels, with marked differences in temporal diversity values and in the range of the spectral indices (Figures 4 and 5). During the first classification cycle, we identified the first hierarchical level including three classes (see Table 2, Figure S1): Water (W), Sand (S), and Vegetation (V, Figure 4, Table 2). The second and third classification cycles established the second hierarchical level in which sand and vegetation clusters were split into two and four classes, respectively (see Table 2). Particularly, the second RF cycle applied to Sand class (S) identified two categories (Table 2, Figure S2): Water Edge (WE) and Open Sand (OS; Figures 4 and 5). The WE class represents the sea–land transition including the inter-tidal area, while OS includes dry sand dunes partially covered by sparse annual vegetation. In the third cycle, RF split the class Vegetation (V) into four categories (Table 2, Figure S3): Mobile Dune Herbaceous Vegetation (MDHV), Fixed Dune Herbaceous Vegetation with Sparse Shrub (FHVSS), Evergreen Woody Vegetation (EWV), and Deciduous and Humid Herbaceous Vegetation (DHHV, Figures 4 and 5). Since we focused on terrestrial cover types, we did not explore the presence of sub-classes inside the Water class (Table 3). For each cycle, RF showed extremely low MESS values, suggesting no extrapolation effect on models' predictions throughout the study area. These outcomes evidenced that the selected pixels for each classification cycle are representative of the study area.

Table 2. Description of classification cycles (1°, 2°,3° cycles) in terms of the optimal number of classes (Op. num. classes) according to Silhouette (S), Calinski–Harabasz (CH) and Davies–Bouldin (DB) indexes; the MESS values (%); the selected variables according to VIF selection, the variables' importance in the definition of classes and the obtained cover classes (Classes). Q_{NDWI2}: temporal Rao of NDWI2, Q_{BI2}: temporal Rao of BI2, Q_{MSAVI2}: temporal Rao of MSAVI2, $10^{th}{}_{NDWI2}$: 10th percentile of NDWI2, $10^{th}{}_{BI2}$: 10th percentile of BI2, $90^{th}{}_{BI2}$: 90th percentile of BI2, $90^{th}{}_{MSAVI2}$: 90th percentile of MSAVI2, M_{MSAVI2}: mean of MSAVI2.

Cycle	1° Cycle			2° Cycle			3° Cycle		
Op. num. classes	S 3	CH 3	DB 3	S 2	CH 2	DB 3	S 4	CH 4	DB 3
MESS (%)		0.0005			0.0020			0.0003	
Variables	Variable selection	VIF	Variable imp. (%)	Variable selection	VIF	Variable imp. (%)	Variable selection	VIF	Variable imp. (%)
	$10^{th}{}_{NDWI2}$	2.30	63	Q_{MSAVI2}	1.73	63	$10^{th}{}_{BI2}$	2.89	33
	$10^{th}{}_{BI2}$	2.09	22	Q_{BI2}	1.39	26	Q_{MSAVI2}	1.32	32
	Q_{NDWI2}	3.03	11	$90^{th}{}_{BI2}$	1.78	11	Q_{BI2}	2.40	26
	[illegible]	1.78	1	$90^{th}{}_{MSAVI2}$	1.16	0	Q_{NDWI2}	1.92	9
	Q_{MSAVI2}	1.34	0				[illegible]	1.17	0
Classes		Water (W) Sand (S) Vegetation (V)			Water Edge (WE) Open Sand (OS)			Mobile Dune Herbaceous V. (MDHV) Fixed Dune Herbaceous V. with Shrubs (FDHVS) Evergreen Woody V. (EWV) Deciduous and Humid Herbaceous V. (DHHV)	

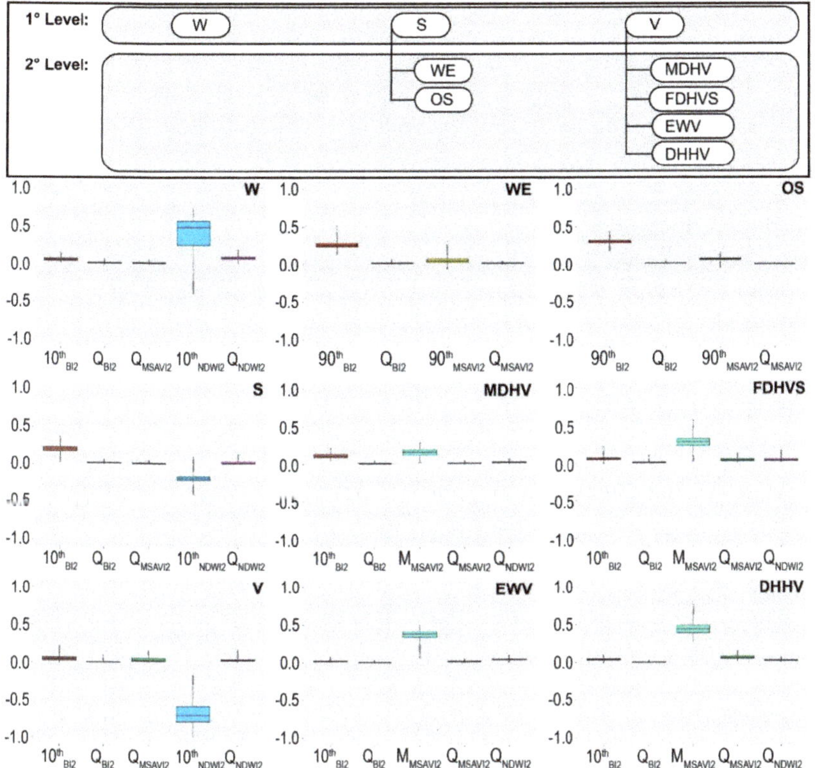

Figure 4. Scheme of the obtained natural and semi-natural cover classes (in the top) along with the respective boxplots showing variables' values. Reported variables are those selected by VIF analysis and used for classification on each cycle. Classes are: Water (W), Sand (S), Vegetation (V), Water Edge (WE), Open Sand (OS), Mobile Dune Herbaceous Vegetation (MDHV), Evergreen Woody Vegetation (EWV), Deciduous and Humid Herbaceous Vegetation (DHHV), Fixed Dune Herbaceous Vegetation with Sparse Shrub (FHVSS). Variables are Q_{NDWI2}: temporal Rao of NDWI2, Q_{BI2}: temporal Rao of BI2, Q_{MSAVI2}: temporal Rao of MSAVI2, 10^{th}_{NDWI2}: 10th percentile of NDWI2, 10^{th}_{BI2}: 10th percentile of BI2, 90^{th}_{BI2}: 90th percentile of BI2, 90^{th}_{MSAVI2}: 90th percentile of MSAVI2, M_{MSAVI2}: mean of MSAVI2.

3.2. Accuracy Assessment

In all the three cycles, the RF results achieved very high accuracy values according to both cross-validation and field assessments (Table 3).

For the first cycle, the overall accuracy, Kappa statistic and MCC indicate an almost perfect

cycle reported cross-validation performances of 99% for overall accuracies, 96% for Kappa statistic, and 0.98 for Matthews Correlation Coefficient, while the performance assessed through the field assessment is slightly lower (overall accuracy: 87%, Kappa statistic: 72%, Matthews Correlation Coefficient: 0.66, Table 3). All performance metrics by land cover classes calculated through cross-validation indicate an almost perfect agreement, whereas the comparison of the natural and semi-natural land cover map between the field and visual inspection shows a decrement of accuracy (Table S4). Indeed, the Water Edge class evidences an almost perfect agreement only for the Producer's accuracy, while the agreement of User's accuracy and Matthews

Correlation Coefficient shows a substantial agreement (Producer's accuracy: 96.67, User's accuracy: 72.22, Matthews Correlation Coefficient: 0.66, Table S4). The Open Sand class displays a higher agreement than the previous class, and, only for the Matthews Correlation Coefficient, the agreement drops to under 0.70 (Table S4). Finally, the performance of the third classification cycle has very high results under cross-validation assessment, with all metrics above 95% (overall accuracy: 97%, Kappa statistic: 96%, Matthews Correlation Coefficient: 0.96), while in the field accuracy assessment, the overall metrics show a substantial agreement (Table 3). Similar to the previous cycles, all land cover classes were accurately predicted, as showed by cross-validation results (Table S5).

Table 3. The percentages of overall accuracy assessment values for three classification cycles, in particular Overall accuracy (O ACC), Kappa statistic (K), overall Matthews Correlation Coefficient (O MCC). In the Random Forest accuracy assessment are indicated the mean values and standard deviations for the overall performance metrics.

		1° Classification Cycle			
		Random Forest accuracy assessment			
O ACC (%)	99.61 ± 0.14	K (%)	99.31 ± 0.25	O MCC	0.993 ± 0.002
		Field accuracy assessment			
O ACC (%)	96	K (%)	92.47	O MCC	0.925
		2° Classification Cycle			
		Random Forest accuracy assessment			
O ACC (%)	99.03 ± 0.05	K (%)	97.95 ± 1.12	O MCC	0.980 ± 0.011
		Field accuracy assessment			
O ACC (%)	87.50	K (%)	72.09	O MCC	0.656
		3° Classification Cycle			
		Random Forest accuracy assessment			
O ACC (%)	97.38 ± 0.38	K (%)	96.49 ± 0.52	O MCC	0.965 ± 0.005
		Field accuracy assessment			
O ACC (%)	82.80	K (%)	76.50	O MCC	0.735

In the field accuracy assessment, the user's accuracy of all classes shows substantially good performances, with the metrics values resulting higher than 75%, especially in the Mobile Dune Herbaceous Vegetation class (Table S5). Equally, in the producer's accuracy, all classes show a substantial agreement, and the values are higher than 70%; indeed, the values range from the minimum of substantial agreement (73%) in Deciduous and Humid Herbaceous Vegetation to the maximum of almost perfect agreement (93%) in Fixed Dune Herbaceous Vegetation with Shrubs (Table S5).

3.3. *Variables Importance*

Among the most important variables in the first RF cycle, the 10th percentile of NDWI2 ($10^{th}{}_{NDWI2}$) showed a 63% importance, followed by the 10th percentile of BI2 ($10^{th}{}_{BI2}$) with 22%, the temporal Rao of NDWI2 (Q_{NDWI2}, 11%), and the temporal Rao of BI2 (Q_{BI2},4%). Water class (W) is characterized by high values of $10^{th}{}_{NDWI2}$ and Q_{NDWI2} (Figure 4) and includes the sea, rivers, channels and wetland. In this class, the $10^{th}{}_{BI2}$ shows overall low values and variability, while Q_{BI2} and Q_{MSAVI2} are close to zero. These features are also evident when inspecting the annual trend of each of the three spectral indices (MSAVI2, NDWI2 and BI2; Figure 4). For the Water class, NDWI2 values are above 0.5 in all months. The BI2 and the NDWI2 values are always lower than 0.25 and 0, respectively.

The most important variables in the second RF cycle are temporal Rao of MSAVI2 (63%, Q_{MSAVI2}) followed by temporal Rao of BI2 (26%, Q_{BI2}) and the 90th percentile of BI2 (11%, $90^{th}{}_{BI2}$, Table 4). Water Edge and Open Sand showed similar trends in annual spectral indices (Figures 4 and 5), only differing in the upper values of brightness ($90^{th}{}_{BI2}$), which is higher in Open Sand. Open Sand shows higher BI2 values in summer (Figure 4), and lower variability in the upper values of biomass ($90^{th}{}_{MSAVI2}$, Figure 4).

The most important variables in the third RF cycle are the lower values of bare surfaces (10^{th}_{BI2}, 33%) and the temporal variability of biomass (Q_{MSAVI2}, 32%) and of bare surfaces (Q_{BI2}, 26%, Table 2). Mobile Dune Herbaceous Vegetation (MDHV, Figures 4 and 5) is characterized by high values of bare surfaces (high 10^{th}_{BI2} and low Q_{BI2}) and moderate values of biomass (low M_{MSAVI2} and low Q_{MSAVI2}). The MSAVI2 index profile depicts a mosaic of bare sand with sparse open perennial and annual vegetation that corresponds to coastal herbaceous vegetation referable to mobile dunes (Figure 5). Fixed Dune Herbaceous Vegetation with Shrubs (FDHVS, Figures 4 and 5) is characterized by a strong seasonality (intermediate M_{MSAVI2} and very high Q_{MSAVI2}), intermediate biomass values and moderate presence of bare surfaces (moderate 10^{th}_{BI2} and Q_{BI2}) and water (moderate Q_{NDWI2}). The FDHVS corresponds to herbaceous vegetation with shrubs growing on fixed dunes and dune slacks, more inland with respect to the previous class of ruderal vegetation. Evergreen Woody Vegetation (EWV, Figures 4 and 5) is characterized by high biomass values, a low seasonality (intermediate M_{MSAVI2} and very low Q_{MSAVI2}) and almost no bare surfaces (very low 10^{th}_{BI2} and Q_{BI2}). EWV includes the Mediterranean maquis, pine woods and shrubs and woody evergreen formations. The Deciduous and Humid Herbaceous Vegetation class (DHHV, Figures 4 and 5) is characterized by high biomass values and pronounced phenology (very high M_{MSAVI2} and very high Q_{MSAVI2}), as well as moderate bare surfaces occurrence (moderate 10^{th}_{BI2} and Q_{BI2}). The DHHV includes herbaceous and woody vegetation of humid areas close to river mouths, riparian vegetation and residual patches of lowland woods.

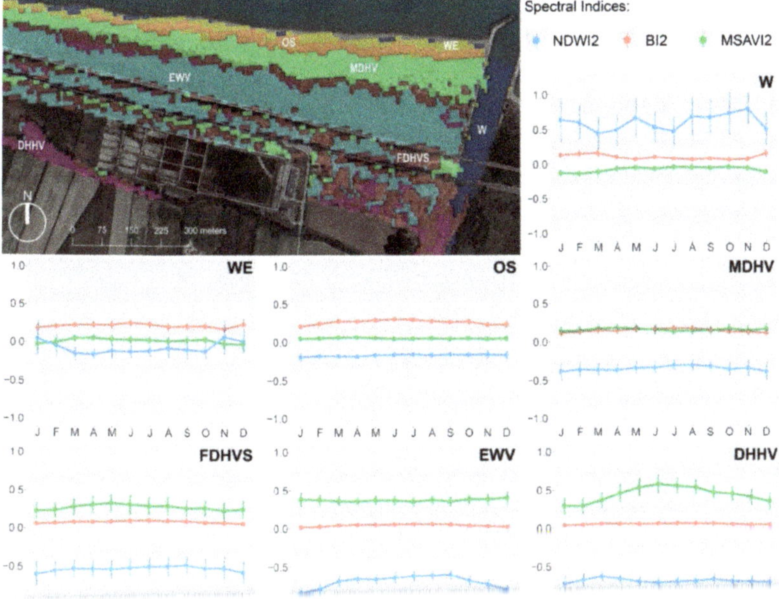

Water (W), Sand (S), Vegetation (V), Water Edge (WE), Open Sand (OS), Mobile Dune Herbaceous Vegetation (MDHV), Fixed Dune Herbaceous Vegetation with Shrub (FDHVS), Evergreen Woody Vegetation (EWV), Deciduous and Humid Herbaceous Vegetation (DHHV).

4. Discussion

In this study, we used information from annual fluctuations of vegetation biomass, water surface, and bare soil combined into the temporal Rao's Q index, in an effort to identify and map seven land cover classes that clearly depicted the sequence of coastal dune natural and semi-natural ecosystems occurring along the sea–inland gradient.

Our modelling approach showed high levels of classification accuracy, substantially confirming the usefulness of time-series analysis for semi-natural and natural land cover mapping [16,92,93]. Moreover, we pointed out the high potential of integrating such spectral indices as MSAVI2, NDWI2 and BI2 to describe land cover seasonality in such heterogeneous environments as those of coastal dunes.

The classification framework that made it possible to identify and map seven coastal semi-natural cover types was organized in two hierarchical levels. For each set of cover types, different variables emerged as the most important for classification. In the first RF cycle implemented on the overall coastal landscape, water surface was split from the other cover classes because of the typical behavior of water-related spectral indices. Specifically, the minimum of NDWI2 ($10^{th}{}_{NDWI2}$) and its temporal diversity (Q_{NDWI2}) assumed particularly high values and emerged as the most important variables in the first RF cycle. As previously observed, the spectral responses of water and dryland are very different [94,95]. Such differences clearly emerged at coarser classification levels on the transition area we analyzed here. However, the percentage of bare surfaces also appeared as an important variable in the first cycle, revealing the presence of open sand and sparse vegetation in the coastal dune mosaic. In the second and third cycles performed on non-water areas, the temporal diversity of biomass (Q_{MSAVI2}) showed an important role in land cover classification, confirming the importance of phenology in coastal landscape mapping [16]. In particular, the second cycle, which focused on open sand and sparse vegetation areas, reported the phenology (Q_{MSAVI2}) along with the annual fluctuations on bare soil (Q_{BI2}) as important variables discriminating bare areas from complex mosaics of pioneer vegetation and sand [96,97]. The third RF cycle focused mainly on vegetated areas hosting herbaceous and woody land cover classes, reporting the temporal variability of vegetation biomass (Q_{MSAVI2}) and the low percentages of bare soil ($10th_{BI2}$) as the most important predictors for classification. In addition, low temporal variation on bare soil cover (low Q_{BI2}) helped to discriminate seasonal deciduous formations from the contiguous evergreen vegetation [38,96].

The classification of Rao's Q temporal diversity measured on spectral indices made it possible to identify and map the mosaic characterizing the different sectors of coastal dune zonation with high accuracy levels. However, such transition classes as Water Edge and Fixed Dune Herbaceous Vegetation with Shrubs evidenced the lowest values, perhaps due to the spatial dimension of Sentinel-2 images (10 m), which is too coarse to describe this intricate transition mosaic.

The seven identified classes clearly depicted the sequence of coastal dune natural and semi natural cover classes occurring from the sea to the inland, each characterized by a specific seasonal pattern. Indeed, classes ranged from the Water class including the sea close to the seashore, to Deciduous formations and Humid Herbaceous Vegetation growing in the inner dune slacks [16,41,92]. As for the Water Edge class, it represented the inter-tidal area, with higher presence of water during the winter months possibly caused by storms. The Open Sand class included the dry sand beach with the sparse annual vegetation on the drift line, while the embryonic shifting dune formations, such as the Dense Herbaceous Vegetation type, represented the mobile dunes with *Ammophila arenaria* (see also [16]). The Fixed Dune Herbaceous Vegetation with Shrubs class included a fine mosaic of herbaceous and shrub vegetation occurring on transition areas, as well as fuzzy edges between herbaceous and woody vegetation on dune sectors towards the sea and shrubs, and ruderal formations on inland disturbed areas [16,96]. As for the Evergreen Woody Vegetation class, we reported low temporal diversity values, both in vegetation biomass (Q_{MSAVI2}) and bare surfaces (Q_{BI2}), confirming the already known absence of seasonality of this cover class [16,98]. The Deciduous and Humid Herbaceous Vegetation class enclosed riparian deciduous woody vegetation and vegetation of humid environments. This class was characterized by a similar seasonality of vegetation biomass and presented two peaks in spring and summer that alternate with lower values in autumn and winter [99,100].

The results we obtained in this study clearly identified remotely sensed seasonality and Rao's Q temporal diversity as an effective source of information for landscape mapping in coastal areas.

Using Sentinel-2 images with accurate spatial, temporal and spectral resolutions [25], we derived sound temporal heterogeneity variables for land cover mapping summarizing spectral indices previously implemented separately (e.g., using MSAVI2 [48], or NDWI2 [52], or BI2 [53]).

Land cover classification based on multi-temporal remotely sensed data analysis has improved here for landscape mapping on Mediterranean coasts by relying on a set spectral indices rarely analyzed together, as well as by the introduction of Rao's Q for summarizing their temporal variability. Furthermore, the implemented RF classification organized in sequential cycles ensured accurate and ecologically coherent results. For instance, water presence and seasonality evidenced by the yearly behavior of NDWI2 were important variables in the first cycle of classification implemented on overall coastal area extent, which confirmed its potential for discriminating water from emerged areas [94,95] and suggested its role when mapping transition systems between terrestrial and marine realms. Similarly, our results confirmed that biomass and bare soil indexes and their temporal variability are important variables for mapping terrestrial cover classes [16,28,92,101]. Based on our results, it possible to affirm that for remotely sensed monitoring and mapping, the inclusion of variables depicting bare soil, water and biomass seasonality are highly advisable.

5. Conclusions

The proposed procedure for classifying and mapping natural and semi-natural land cover effectively summarized the dynamic nature of coastal systems, capturing all of the main components of coastal dune mosaics together with their seasonal variation (i.e., vegetation, bare surfaces and water surfaces) by combining spectral indices that are rarely used together (MSAVI2, NDWI2, or BI2).

The information provided by Rao's Q offered a sound base for natural and semi-natural land cover monitoring and mapping. In particular, the here proposed implementation of Rao's Q temporal heterogeneity allowed the accurate depiction of the seasonality of the different cover types conforming the coastal dune mosaic along the sea–inland gradient (e.g., biomass phenology, water seasonality, yearly variation on bare surfaces). Furthermore, the applicability of the proposed framework on the available updated sentinel images emphasized the procedure as a promising tool for cover monitoring and reporting. Even more interestingly, the integration of other remotely sensed data with higher spatial resolutions derived by satellite, such as Planet images, UAV, or LiDAR data, may further improve the classification of coastal zonation.

From an applied perspective, the natural and semi-natural land cover map provided in this study yields relevant knowledge for coastal monitoring and management; therefore, we hope new studies exploring increasingly larger areas will be analyzed to further test the proposed classification and, at the same time, to provide homogeneous information for coasts in the Mediterranean.

Supplementary Materials: The following are available online at http://www.mdpi.com/2072-4292/12/14/2315/s1, Table S1: Sentinel-2 dataset indicating for each image the month, day, platform (Sentinel-2A or Sentinel-2B) and the hour of acquisition, the cloud percentage, and the relative tile (T33TVG for Molise north, T33TWF for Molise south). Table S2: The optimal number of classes for each cycle was identified by Silhouette, Calinski–Harabasz, and Davies–Bouldin indices calculated on a specific hierarchical cluster produced through Ward's minimum variance. These three indices identify the optimal number of classes by comparing the intra-class compactness (degree of [illegible]

[illegible] Silhouette, Calinski–Harabasz, and Davies–Bouldin indices used for identifying the optimal number of classes in the 2° cycle of Random Forest. Two (Silhouette and Calinski–Harabasz) of the three indices have established two as the optimal number of classes. Figure S3: Graphical representation of Silhouette, Calinski–Harabasz, and Davies–Bouldin indices used for identifying the optimal number of classes in the 3° cycle of Random Forest. Two (Silhouette and Calinski–Harabasz) of the three indices suggested four as the optimal number of classes. Table S3: The percentages of accuracy assessment values by land cover classes for first classification cycle: Water (W), Sand (S), Vegetation (V). In particular, User's accuracy (U ACC), Producer's accuracy (P ACC), and Matthews Correlation Coefficient (MCC). In the Random Forest accuracy assessment are indicated the mean values and standard deviation for the performance metrics. Table S4: The percentages of accuracy assessment values by land cover classes for second classification cycle: Water Edge (WE), Open Sand (OS). In particular, User's accuracy (U ACC), Producer's accuracy (P ACC), and Matthews Correlation Coefficient (MCC). In the Random Forest accuracy assessment are indicated the mean values and standard deviation for the performance metrics. Table S5: The percentages of accuracy assessment values by land cover classes for the third classification cycle: Dense Herbaceous Vegetation (DHV), Evergreen Woody Vegetation (EWV), Fixed Dune Herbaceous Vegetation with Shrubs (FDHVS), Deciduous and Humid Herbaceous Vegetation (DHHV). In particular, User's accuracy (U ACC), Producer's accuracy (P ACC), and Matthews Correlation Coefficient (MCC). In the Random Forest accuracy assessment are indicated the mean values and standard deviation for the performance metrics.

Author Contributions: All authors contributed substantially to the work: F.M., M.M. and M.L.C. conceived and designed the study; F.M. collected the data; F.M., M.D.F., M.L.C. analyzed the data; F.M., M.M., S.G., M.L.C. led the writing of the manuscript; A.T.R.A. and M.L.C. supervised the research. All authors contributed critically to the drafts and gave final approval for publication. All authors have read and agreed to the published version of the manuscript.

Funding: This research received no external funding.

Acknowledgments: This study was carried out with the partial support of PON-AIM (Italian program for research and innovation 2014-2020—AIM1897595-2) in collaboration with the bilateral program Italy–Israel DERESEMII (Developing state-of-the-art remote sensing tools for monitoring the impact of invasive plant species in coastal ecosystems in Israel and Italy) and INTERREG Italy-Croazia CASCADE (CoAStal and marine waters integrated monitoring systems for ecosystems proteCtion AnD management - Project ID 10255941). The authors acknowledge the Principal Investigator(s) of the Sentinel-2 mission for providing datasets in the archive and the developers of SNAP software used for data analysis. Sentinel-2 images are freely downloadable from the Copernicus Open Access Hub https://scihub.copernicus.eu/). Furthermore, the authors acknowledge the entire team of EnvixLab, in particular Ludovico Frate, Gabriella Sferra, Marco Varricchione, Luca Francesco Russo, for the help given. The authors also thank the editor and the anonymous reviewers for their comments that helped us to improve the manuscript.

Conflicts of Interest: The authors declare no conflict of interest.

References

1. Allan, E.; Manning, P.; Alt, F.; Binkenstein, J.; Blaser, S.; Blüthgen, N.; Böhm, S.; Grassein, F.; Hölzel, N.; Klaus, V.H.; et al. Land use intensification alters ecosystem multifunctionality via loss of biodiversity and changes to functional composition. *Ecol. Lett.* **2015**, *18*, 834–843. [CrossRef] [PubMed]
2. Mantyka-Pringle, C.S.; Visconti, P.; Di Marco, M.; Martin, T.G.; Rondinini, C.; Rhodes, J.R. Climate change modifies risk of global biodiversity loss due to land-cover change. *Biol. Conserv.* **2015**, *187*, 103–111. [CrossRef]
3. Oliver, T.H.; Heard, M.S.; Isaac, N.J.B.; Roy, D.B.; Procter, D.; Eigenbrod, F.; Freckleton, R.; Hector, A.; Orme, C.D.L.; Petchey, O.L.; et al. Biodiversity and resilience of ecosystem functions. *Trends Ecol. Evol.* **2015**, *30*, 673–684. [CrossRef] [PubMed]
4. Brown, A.C.; McLachlan, A. Sandy shore ecosystems and threats facing them: Some predictions for the year 2025. *Environ. Conserv.* **2002**, *29*, 62–77. [CrossRef]
5. Defeo, O.; McLachlan, A.; Schoeman, D.S.; Schlacher, T.A.; Dugan, J.; Jones, A.; Lastra, M.; Scapini, F. Threats to sandy beach ecosystems: A review. *Estuar. Coast. Shelf Sci.* **2009**, *81*, 1–12. [CrossRef]
6. Martínez, M.L.; Psuty, N.P. *Coastal Dunes. Ecology and Conservation*, 2nd ed.; Springer: Berlin/Heidelberg, Germany, 2008; pp. 1–387.
7. Martínez, M.L.; Intralawan, A.; Vázquez, G.; Pérez-Maqueo, O.; Sutton, P.; Landgrave, R. The coasts of our world: Ecological, economic and social importance. *Ecol. Econ.* **2007**, *63*, 254–272. [CrossRef]
8. Drius, M.; Jones, L.; Marzialetti, F.; de Francesco, M.C.; Stanisci, A.; Carranza, M.L. Not just a sandy beach. The multi-service value of Mediterranean coastal dunes. *Sci. Total Environ.* **2019**, *668*, 1139–1155. [CrossRef]
9. Hesp, P.A. Ecological processes and plant adaptations on coastal dunes. *J. Arid Environ.* **1991**, *21*, 165–191. [CrossRef]
10. Acosta, A.T.R.; Blasi, C.; Carranza, M.L.; Ricotta, C.; Stanisci, A. Quantifying ecological mosaic connectivity and hemeroby with a new topoecological index. *Phytocoenologia* **2003**, *33*, 623–631. [CrossRef]
11. Henle, K.; Bauch, B.; Auliya, M.; Külvik, M.; Pe'er, G.; Schmeller, D.S.; Framstad, E. Priorities for biodiversity monitoring in Europe: A review of supranational policies and a novel scheme for integrative prioritization. *Ecol. Indic.* **2013**, *33*, 5–18. [CrossRef]

12. Janssen, J.A.M.; Rodwell, J.S.; García Criado, M.; Gubbay, S.; Haynes, T.; Nieto, A.; Sanders, N.; Landucci, F.; Loidi, J.; Ssymank, A.; et al. *European Red List of Habitats. Part 2. Terrestrial and Freshwaters Habitats*; European Union: Luxemburg, 2016; pp. 1–38.
13. Acosta, A.T.R.; Carranza, M.L.; Izzi, C.F. Combining land cover mapping of coastal dunes with vegetation analysis. *Appl. Veg. Sci.* **2005**, *8*, 133–138. [CrossRef]
14. Malavasi, M.; Santoro, R.; Cutini, M.; Acosta, A.T.R.; Carranza, M.L. What has happened to coastal dunes in the last half century? A multitemporal coastal landscape analysis in Central Italy. *Landsc. Urban Plan.* **2013**, *119*, 54–63. [CrossRef]
15. Brownett, J.M.; Mills, R.S. The development and application of remote sensing to monitor sand dune habitats. *J. Coast. Conserv.* **2017**, *21*, 643–655. [CrossRef]
16. Marzialetti, F.; Giulio, S.; Malavasi, M.; Sperandii, M.G.; Acosta, A.T.R.; Carranza, M.L. Capturing coastal dune natural vegetation types using a phenology-based mapping approach: The potential of Sentinel-2. *Remote Sens.* **2019**, *11*, 1506. [CrossRef]
17. Doody, J.P. *Sand Dune Conservation Management and Restoration*, 1st ed.; Springer: Dordrecht, The Netherlands, 2013; pp. 9–20.
18. Williams, C.M.; Ragland, G.J.; Betini, G.; Buckley, L.B.; Cheviron, Z.A.; Donohue, K.; Hereford, J.; Humphries, M.M.; Lisovski, S.; Marshall, K.E.; et al. Understanding evolutionary impacts of seasonality: An introduction to the symposium. *Integr. Comp. Biol.* **2017**, *57*, 921–933. [CrossRef] [PubMed]
19. Tonkin, J.D.; Bogan, M.T.; Bonada, N.; Rios-Touma, B.; Lytle, D.A. Seasonality and predictability shape temporal species diversity. *Ecology* **2017**, *98*, 1201–1216. [CrossRef]
20. Gómez, C.; White, J.C.; Wulder, M.A. Optical remotely sensed time series data for land cover classification: A review. *ISPRS J. Photogramm. Remote Sens.* **2016**, *116*, 55–72. [CrossRef]
21. Senf, C.; Leitão, P.J.; Pflugmacher, D.; van der Linden, S.; Hostert, P. Mapping land cover in comlex Mediterranean landscapes using Landsat: Improved classification accuracies from integrating multi-seasonal and synthetic imagery. *Remote Sens. Environ.* **2015**, *156*, 527–536. [CrossRef]
22. Luo, J.; Li, X.; Ma, R.; Li, F.; Duan, H.; Hu, W.; Qin, B.; Huang, W. Applying remote sensing techniques to monitoring seasonal and interannual changes of acquatic vegetation in Taihu Lake, China. *Ecol. Indic.* **2016**, *60*, 503–513. [CrossRef]
23. Mannel, S.; Price, M. Comparing classification results of multi-seasonal TM against AVIRIS imagery-seasonality more important then number of bands. *Photogramm. Fernerkund. Geoinf.* **2012**, *5*, 603–612. [CrossRef]
24. Knight, J.F.; Luneta, R.S.; Ediriwickrema, J.; Khorran, S. Regional scale land cover characterization using MODIS-NDVI 250 m multi-temporal imagery: A phenology-based approach. *GIScience Remote Sens.* **2006**, *43*, 1–23. [CrossRef]
25. Drusch, M.; Del Bello, U.; Carlier, S.; Colin, O.; Fernandez, V.; Gascon, F.; Hoersch, B.; Isola, C.; Labertini, P.; Martimort, P.; et al. Sentinel-2: ESA's optical high-resolution mission for GMES operational services. *Remote Sens. Environ.* **2012**, *120*, 25–36. [CrossRef]
26. Wulder, M.A.; Hilker, T.; White, J.C.; Coops, N.C.; Masek, J.G.; Pflugmacher, D.; Crevier, Y. Virtual constellations for global terrestrial monitoring. *Remote Sens. Environ.* **2015**, *170*, 65–76. [CrossRef]
27. Hill, R.A.; Wilson, A.K.; George, M.; Hinsley, S.A. Mapping tree species in temperate deciduous woodland using time-series multi-spectral data. *Appl. Veg. Sci.* **2010**, *13*, 86–99. [CrossRef]
28. Rapinel, S.; Mony, C.; [illegible]
29. [illegible] New York, NY, USA, 2009; pp. 1–275.
30. Donohue, R.J.; McVicar, T.R.; Roderick, M.L. Climate-related trends in Australian vegetation cover as inferred from satellite observations, 1981–2006. *Glob. Chang. Biol.* **2009**, *15*, 1025–1039. [CrossRef]
31. Rocchini, D.; Marcantonio, M.; Da Re, D.; Chirici, G.; Galluzzi, M.; Lenoir, J.; Ricotta, C.; Torresani, M.; Ziv, G. Time-lapsing biodiversity: And open source method for measuring diversity changes by remote sensing. *Remote Sens. Environ.* **2019**, *231*, 111192. [CrossRef]

32. Torresani, M.; Rocchini, D.; Sonnenschein, R.; Zebisch, M.; Marcantonio, M.; Ricotta, C.; Tonon, G. Estimating tree species diversity from space in an alpine conifer forest: The Rao's Q diversity index meets the spectral variation hypothesis. *Ecol. Inform.* **2019**, *52*, 26–34. [CrossRef]
33. Ricotta, C.; Carranza, M.L.; Avena, G.; Blasi, C. Quantitative comparison of the diversity of landscapes with actual vs. potential natural vegetation. *Appl. Veg. Sci.* **2000**, *3*, 157–162. [CrossRef]
34. Ricotta, C.; Moretti, M. CWM and Rao's quadratic diversity: A unified framework for functional ecology. *Oecologia* **2011**, *167*, 181–188. [CrossRef]
35. Ricotta, C.; Carranza, M.L. Measuring scale-dependent landscape structure with Rao's quadratic diversity. *ISPRS Int. J. Geo-Inf.* **2013**, *2*, 405–412. [CrossRef]
36. Rocchini, D.; Marcantonio, M.; Ricotta, C. Measuring Rao's Q diversity index from remote sensing: An open source solution. *Ecol. Indic.* **2017**, *72*, 234–238. [CrossRef]
37. Acosta, A.T.R.; Carranza, M.L.; Izzi, C.F. Are there habitats that contribute best to plant species diversity in coastal dunes? *Biodivers. Conserv.* **2009**, *18*, 1087–1098. [CrossRef]
38. Carranza, M.L.; Acosta, A.T.R.; Stanisci, A.; Pirone, G.; Ciaschetti, G. Ecosystem classification for EU habitat distribution assessment in sandy coastal environments: An application in central Italy. *Environ. Monit. Assess.* **2008**, *140*, 99–107. [CrossRef]
39. Stanisci, A.; Acosta, A.T.R.; Carranza, M.L.; Feola, S.; Giuliano, M. Gli habitat di interesse comunitario sul litorale molisano e il loro valore naturalistico su base floristica. *Fitosociologia* **2007**, *44*, 171–175.
40. Drius, M.; Malavasi, M.; Acosta, A.T.R.; Ricotta, C.; Carranza, M.L. Boundary-based analysis for the assessment of coastal dune landscape integrity over time. *Appl. Geogr.* **2013**, *45*, 41–48. [CrossRef]
41. Stanisci, A.; Acosta, A.T.R.; Carranza, M.L.; de Chiro, M.; Del Vecchio, S.; Di Martino, L.; Frattaroli, A.R.; Fusco, S.; Izzi, C.F.; Pirone, G.; et al. EU habitats monitoring along the coastal dunes of the LTER sites of Abruzzo and Molise (Italy). *Plant Sociol.* **2014**, *51*, 51–56. [CrossRef]
42. Prisco, I.; Stanisci, A.; Acosta, A.T.R. Mediterranean dunes on the go: Evidence from a short term study on coastal herbaceous vegetation. *Estuar. Coast. Shelf Sci.* **2016**, *182*, 40–46. [CrossRef]
43. Gislason, P.O.; Benediktsson, J.A.; Sveinsson, J.R. Random forests for land cover classification. *Pattern Recognit. Lett.* **2006**, *27*, 294–300. [CrossRef]
44. Pal, M. Random forest classifier for remote sensing classification. *Int. J. Remote Sens.* **2005**, *26*, 217–222. [CrossRef]
45. Berger, M.; Moreno, J.; Johannessen, J.A.; Levelt, P.F.; Hanssen, R.F. ESA's sentinel missions in support of Earth system science. *Remote Sens. Environ.* **2012**, *120*, 84–90. [CrossRef]
46. Main-Knorn, M.; Pflug, B.; Louis, J.; Debaecker, V.; Müller-Wilm, U.; Gascon, F. Sen2Cor for Sentinel-2. In *Image and Signal Processing for Remote Sensing XXIII, Proceedings of Spie Remote Sensing, Warsaw, Poland, 21–24 September*; Bruzzone, L., Bovolo, F., Eds.; SPIE: Bellingham, WA, USA, 2017.
47. Rondeaux, G.; Steven, M.; Baret, F. Optimization of soil-adjusted vegetation indices. *Remote Sens. Environ.* **1996**, *55*, 95–107. [CrossRef]
48. Qi, J.; Chehbouni, A.; Huete, A.R.; Kerr, Y.H.; Sorooshian, S. A modified soil adjusted vegetation index. *Remote Sens. Environ.* **1994**, *48*, 118–126. [CrossRef]
49. Wu, Z.; Velasco, M.; McVay, J.; Middleton, B.; Vogel, J.; Dye, D. MODIS derived vegetation index for drought detection on the San Carlos Apache reservation. *Int. J. Adv. Remote Sens. GIS* **2016**, *5*, 1524–1538. [CrossRef]
50. Adam, E.; Mutanga, O.; Rugege, D. Multispectral and hyperspectral remote sensing for identification and mapping of wetland vegetation: A review. *Wetl. Ecol. Manag.* **2010**, *18*, 281–296. [CrossRef]
51. Maglione, P.; Parente, C.; Vallario, A. Coastline extraction using high resolution WorldView-2 satellite imagery. *Eur. J. Remote Sens.* **2014**, *47*, 685–699. [CrossRef]
52. McFeeters, S.K. The use of the Normalized Difference Water Index (NDWI) in the delineation of open water features. *Int. J. Remote Sens.* **1996**, *17*, 1425–1432. [CrossRef]
53. Escadafal, R. Remote sensing of soil color: Principles and applications. *Remote Sens. Rev.* **1989**, *7*, 261–279. [CrossRef]
54. Todd, S.W.; Hoffer, R.M.; Milchunas, D.G. Biomass estimation on grazed and ungrazed rangelands using spectral indices. *Int. J. Remote Sens.* **1998**, *19*, 427–438. [CrossRef]
55. Timm, B.C.; McGarigal, K. Fine-scale remotely-sensed cover mapping of coastal dune and salt marsh ecosystems at Cape Cod National Seashore using Random Forests. *Remote Sens. Environ.* **2012**, *127*, 106–117. [CrossRef]

56. Milligan, G.W.; Cooper, M.C. A study of standardization of variables in cluster analysis. *J. Classif.* **1988**, *5*, 181–204. [CrossRef]
57. Rocchini, D.; Balkenhol, N.; Carter, G.M.; Foody, G.M.; Gillespie, T.W.; He, K.S.; Kark, S.; Levin, N.; Lucas, K.; Luoto, M.; et al. Remotely sensed spectral heterogeneity as a proxy of species diversity: Recent advances and open challenges. *Ecol. Inform.* **2010**, *5*, 318–319. [CrossRef]
58. Rao, C.R. Diversity and dissimilarity coefficients: A unified approach. *Theor. Popul. Biol.* **1982**, *21*, 24–43. [CrossRef]
59. Bascietto, M. Compute Rao's Diversity Index on Numeric Vectors and Matrices, R Package Version 0.1.0. 2018. Available online: https://github.com/mbask/spacetimerao/ (accessed on 28 September 2019).
60. Zuur, A.F.; Ieno, E.N.; Elphick, C.S. A protocol for data exploration to avoid common statistical problems. *Methods Ecol. Evol.* **2010**, *1*, 3–14. [CrossRef]
61. Chatterjee, S.; Hadi, A.S. *Regression Analysis by Example*, 5th ed.; John Wiley & Sons, Inc.: Hoboken, NJ, USA, 2012; pp. 248–251.
62. Montgomery, D.C.; Peck, E.A.; Vining, G. *Introduction to Linear Regression Analysis*, 5th ed.; John Wiley & Sons Inc.: Hoboken, NJ, USA, 2012; pp. 1–645.
63. Elith, J.; Kearney, M.; Philips, S. The art of modelling range-shifting species. *Methods Ecol. Evol.* **2010**, *1*, 330–342. [CrossRef]
64. Ward, J.H. Hierarchical grouping to optimize and objective function. *J. Am. Stat. Assoc.* **1963**, *58*, 236–244. [CrossRef]
65. Legendre, P.; Legendre, L. *Numerical Ecology*, 3rd ed.; Elsevier: Oxford, UK, 2012; pp. 360–367.
66. Murtagh, F.; Legendre, P. Ward's hierarchical agglomerative clustering method: Which algorithms implement ward's criterion? *J. Classif.* **2014**, *31*, 274–295. [CrossRef]
67. Bray, J.R.; Curtis, J.T. An ordination of the upland forest communities of Southern Wisconsin. *Ecol. Monogr.* **1957**, *27*, 325–349. [CrossRef]
68. Goslee, S.C. Correlation analysis of dissimilarity matrices. *Plant Ecol.* **2010**, *206*, 279–286. [CrossRef]
69. Ricotta, C.; Podani, J. On some properties of the Bray-Curtis dissimilarity and their ecological meaning. *Ecol. Complex.* **2017**, *31*, 201–205. [CrossRef]
70. Kaufman, L.; Rousseeuw, P.J. *Finding Groups in Data*, 10th ed.; John Wiley & Sons: New York, NY, USA, 2005; pp. 1–342.
71. Calinski, T.; Harabasz, J. A dendrite method for cluster analysis. *Commun. Stat. Theory Methods* **1974**, *3*, 1–27. [CrossRef]
72. Davies, D.; Bouldin, D. A cluster separation measure. *IEEE Trans. Pattern Anal. Mach. Intell.* **1979**, *1*, 224–227. [CrossRef]
73. Breiman, L. Random Forest. *Mach. Learn.* **2001**, *45*, 5–32. [CrossRef]
74. Breiman, L.; Friedman, J.H.; Olshen, R.A.; Stone, C.J. *Classification and Regression Trees*, 3rd ed.; Chapman & Hall: Boca Ration, FL, USA, 1984; pp. 1–358.
75. Kuhn, M.; Wing, J.; Weston, S.; Williams, A.; Keefer, C.; Engelhardt, A.; Cooper, T.; Mayer, Z.; Kenkel, B.; Benesty, M. Classification and Regression Training, R Package Version 6.0-85. 2020. Available online: https://github.com/topepo/caret/ (accessed on 28 September 2019).
76. Kattenborn, T.; Lopatin, J.; Förster, M.; Braun, A.C.; Fassnacht, F.E. UAV data as alternative to field sampling to map woody invasive species based on combined Sentinel-1 and Sentinel-2 data. *Remote Sens. Environ.* **2019**, *227*, 61–73. [CrossRef]

[illegible] classifications (RandomForest). *Remote Sens. Environ.* **2006**, *100*, 356–362. [CrossRef]

79. Noi, P.T.; Kappas, M. Comparison of Random Forest, k-Nearest Neighbor and Support Vector Machine classifiers for land cover classification using Sentinel-2 imagery. *Sensors* **2018**, *18*, 18. [CrossRef]
80. Belgiu, M.; Drăgut, L. Random forest in remote sensing: A review of applications and future directions. *ISPRS J. Photogramm. Remote Sens.* **2016**, *114*, 24–31. [CrossRef]
81. Gastwirth, J.L. The estimation of the Lorenz curve and Gini index. *Rev. Econ. Stat.* **1972**, *54*, 306–316. [CrossRef]
82. Geurts, P.; Ernest, D.; Wehenkel, L. Extremely randomized trees. *Mach. Learn.* **2006**, *63*, 3–42. [CrossRef]

83. Cutler, D.R.; Edwards, T.C.; Beard, K.H.; Cutler, A.; Hess, K.T.; Gibson, J.; Lawler, J.J. Random forests for classification in ecology. *Ecology* **2007**, *88*, 2783–2792. [CrossRef] [PubMed]
84. Congalton, R.G.; Green, K. *Assessing the Accuracy of Remotely Sensed Data. Principles and Practices*, 2nd ed.; CRC Press and Taylor & Francis Group: Boca Raton, FL, USA, 2009; pp. 1–183.
85. Dorais, A.; Cardille, J. Strategies for incorporating high-resolution Google Earth databases to guide and validate classifications: Understanding deforestation in Borneo. *Remote Sens.* **2011**, *3*, 1157–1176. [CrossRef]
86. Mohammed, N.Z.; Landry, R., Jr. Assessing horizontal positional accuracy of Google Earth imagery in the city of Montreal, Canada. *Geod. Cartogr.* **2016**, *43*, 55–65. [CrossRef]
87. Tilahun, A.; Teferie, B. Accuracy assessment of land use land cover classification using Google Earth. *Am. J. Environ. Prot.* **2015**, *4*, 193–198. [CrossRef]
88. Rwanga, S.S.; Ndambuki, J.M. Accuracy assessement of land use/land cover classification using remote sensing and GIS. *Int. J. Geosci.* **2017**, *8*, 611–622. [CrossRef]
89. Matthews, B.W. Comparison of the predicted and observed secondary structure of T4 phage lysozyme. *Biochim. Biophys. Acta (BBA) Protein Struct.* **1975**, *405*, 442–451. [CrossRef]
90. Boughorbel, S.; Jarray, F.; El-Anbari, M. Optimal classifier for imbalanced data using Matthews Correlation Coefficient metric. *PLoS ONE* **2017**, *12*, e0177678. [CrossRef]
91. Bekkar, M.; Djemaa, H.K.; Alitouche, T.A. Evaluation measures for models assessment over imbalanced data set. *J. Inf. Eng. Appl.* **2013**, *3*, 27–39.
92. Pesaresi, S.; Mancini, A.; Quattrini, G.; Casavechia, S. Mapping mediterranean forest plant associations and habitats with functional principal component analysis using Landsat 8 NDVI time series. *Remote Sens.* **2020**, *12*, 1132. [CrossRef]
93. Sun, C.; Fagherazzi, S.; Liu, Y. Classification mapping of salt marsh vegetation by flexible monthly NDVI time-series using Landsat imagery. *Estuar. Coast. Shelf Sci.* **2018**, *213*, 61–80. [CrossRef]
94. Chatziantoniou, A.; Petropoulos, G.P.; Psomiadis, E. Co-orbital Sentinel 1 and 2 for LULC mapping with emphasis on wetlands in a Mediterranean setting based on machine learning. *Remote Sens.* **2017**, *9*, 1259. [CrossRef]
95. Prajesh, P.J.; Kannan, B.; Pazhanivelan, S.; Kumaraperumal, R.; Ragunath, K.P. Monitoring and mapping of seasonal vegetation trend in Tamil Nadu using NDVI and NDWI imagery. *J. Appl. Nat. Sci.* **2019**, *11*, 54–61. [CrossRef]
96. Bazzichetto, M.; Malavasi, M.; Acosta, A.T.R.; Carranza, M.L. How does dune morphology shape coastal EC habitats occurrence? A remote sensing approach using airborne LiDAR on the Mediterranean coast. *Ecol. Indic.* **2016**, *71*, 618–626. [CrossRef]
97. Sciandrello, S.; Tomaselli, G.; Minissale, P. The role of natural vegetation in the analysis of the spatio-temporal changes of coastal dune system: A case study in Sicily. *J. Coast. Conserv.* **2015**, *19*, 199–212. [CrossRef]
98. Bonari, G.; Acosta, A.T.R.; Angiolini, C. Mediterranean coastal pine forest stands: Understory distinctiveness or not? *For. Ecol. Manag.* **2017**, *391*, 19–28. [CrossRef]
99. Zoffoli, M.L.; Kandus, P.; Madanes, N.; Calvo, D.H. Seasonal and interannual analysis of wetlands in South America using NOAA-AVHRR NDVI time series: The case of the Parana delta region. *Landsc. Ecol.* **2008**, *23*, 833–848. [CrossRef]
100. Fu, B.; Burgher, I. Riparian vegetation NDVI dynamics and its relationships with climate, surface water and groundwater. *J. Arid Environ.* **2015**, *113*, 59–68. [CrossRef]
101. Grabska, E.; Hostert, P.; Pflugmacher, D.; Ostapowicz, K. Forest stand species mapping using the Sentinel-2 time series. *Remote Sens.* **2019**, *11*, 1197. [CrossRef]

Article

Measuring Surface Moisture on a Sandy Beach based on Corrected Intensity Data of a Mobile Terrestrial LiDAR

Junling Jin [1,*], Lars De Sloover [1], Jeffrey Verbeurgt [1], Cornelis Stal [1,2], Greet Deruyter [3], Anne-Lise Montreuil [4], Philippe De Maeyer [1] and Alain De Wulf [1]

1 Department of Geography, Ghent University, Krijgslaan 281 S8, 9000 Ghent, Belgium; Lars.DeSloover@ugent.be (L.D.S.); Jeffrey.Verbeurgt@ugent.be (J.V.); Cornelis.Stal@ugent.be (C.S.); Philippe.DeMaeyer@ugent.be (P.D.M.); Alain.DeWulf@ugent.be (A.D.W.)
2 Department of Real-estate and Applied Geomatics, University College Ghent, Valentin Vaerwyckweg 1, 9000 Ghent, Belgium
3 Department of Civil Engineering, Ghent University, Technologiepark 904, 9052 Ghent, Belgium; Greet.Deruyter@ugent.be
4 Hydrology and Hydraulic Engineering, Vrije Universiteit Brussel, Pleinlaan 2, 1050 Elsene, Belgium; Anne-Lise.Montreuil@vub.be
* Correspondence: Junling.Jin@ugent.be; Tel.: +32-92644625

Received: 25 November 2019; Accepted: 6 January 2020; Published: 8 January 2020

Abstract: Surface moisture plays a key role in limiting the aeolian transport on sandy beaches. However, the existing measurement techniques cannot adequately characterize the spatial and temporal distribution of the beach surface moisture. In this study, a mobile terrestrial LiDAR (MTL) is demonstrated as a promising method to detect the beach surface moisture using a phase-based Z&F/Leica HDS6100 laser scanner mounted on an all-terrain vehicle. Firstly, two sets of indoor calibration experiments were conducted so as to comprehensively investigate the effect of distance, incidence angle and sand moisture contents on the backscattered intensity by means of sand samples with an average grain diameter of 0.12 mm. A moisture estimation model was developed which eliminated the effects of the incidence angle and distance (it only relates to the target surface reflectance). The experimental results reveal both the distance and incidence angle influencing the backscattered intensity of the sand samples. The standard error of the moisture model amounts to 2.0% moisture, which is considerably lower than the results of the photographic method. Moreover, a field measurement was conducted using the MTL system on a sandy beach in Belgium. The accuracy and robustness of the beach surface moisture derived from the MTL data was evaluated. The results show that the MTL is a highly suitable technique to accurately and robustly measure the surface moisture variations on a sandy beach with an ultra-high spatial resolution (centimeter-level) in a short time span (12 × 200 m per minute).

1. Introduction

The measurement of surface moisture on a beach is a fundamental component of field studies that seek to model the aeolian transport from the beach (which contributes to dune growth and recovery after erosion from storm-wave processes) [1–9], or investigate the distribution of the beach groundwater [10–13]. However, the surface moisture is determined by complex hydraulics of tidal and wave action, groundwater and capillary flow, and evaporation and precipitation. As a result, the distribution of surface moisture can vary greatly over space and time [14]. This requires measuring techniques to adequately detect the surface moisture on a substantial beach section and over timescales of seconds to months [15]. The techniques which have been used for measuring the beach surface moisture could generally be classified into three approaches: (a) soil moisture probes, (b) sample gravimetric method, (c) the optical remote sensing methods. The soil moisture probes are mainly applied to detect the near-surface moisture through inserting probes in the sand rather than the actual top surface, which is most important for the aeolian sand transport [1–4,10,16–19]. Although the probes could be modified to reduce the sampling depth, their measurement results overestimate the actual surface moisture by an average of 2.5% and 4.4% gravimetric moisture contents for the 1.5 cm and 6 cm sampling depths respectively, according to [1]. By means of the sample gravimetric method, one needs to collect sand samples from the beach surface (i.e., 5 mm thickness) for a laboratory analysis. After drying the samples, one determines the gravimetric moisture contents by calculating the difference between the wet and dry samples [1,5,8,20]. While it is time-consuming to sample the moisture contents across large areas with a detailed resolution, this method is more accurate than the soil moisture probes. Secondly, the ability to repetitively sample at the exact same location is compromised due to the surface destruction of sampling sites, which restrict the utility of this approach for many applications (e.g., an analysis of temporal variation in the surface moisture).

Instead of using destructive/disturbing and time-consuming methods, the optical remote sensing method holds great promise with a faster and repeatable detection of the real surface moisture [1]. This approach is based on the principle that wet sand darkens upon wetting because of a reduced reflectance and there are at least two different theoretical hypotheses explaining this phenomenon. The first hypothesis is the reasoning that the total internal reflection within water films surrounding the sand grains decreases the sand reflectance [14,21,22]. Another explanation is that the relative index of the refraction between water and sand is lower than the one between air and sand. As with the case of internal reflection, this increases the interaction of light with sand and results in more light absorption by the sand [23,24]. Philpot [25] attributed the darkening of wetted soil to multiple mechanisms and believed this two fundamental explanations are both important. Several studies employed the optical remote sensing techniques so as to calculate the beach surface moisture by relating the beach surface brightness derived from digital cameras to the surface moisture contents [9,15,26–29]. However, this technique only works during daylight hours and requires careful control for changing illumination. In addition, the accuracy level of the approach is comparatively small, with the lowest standard error of 3–4% moisture [15,30]. According to [23,25], the reliance on the visible wavelengths is the main contributing factor to the low accuracy in these cases. Once the pore water surrounding the sand grains is sufficient to cover the sand grains, the increasing moisture contents no longer have a significant impact on the surface reflectance. The infrared wavelengths are considered to have better characteristics than the visible wavelengths for determining the moisture contents of the beach surface due to a stronger absorption of the light energy by water. Using infrared spectroscopy, a portable narrow band radiometer (λ = 1940 nm) [30] and a spectroradiometer (λ = 970 nm) [1] were tested for measuring the beach surface moisture, the standard error of the narrow band radiometer averaged to about 1% moisture, which is comparable to the gravimetric moisture contents determined from the 1.5 mm deep surface scrapes. While the narrow band radiometer slightly outperforms the spectroradiometer [30], the two instruments are still time-consuming to deploy on a large spatial scale.

Terrestrial laser scanning (TLS) is a more convenient remote sensing technique for measuring beach surface moisture, which records both the target position and backscattered intensity of the beach surface. Because the TLS is an active sensor, it can scan the beach repetitively without correction for changes in illumination [7,31–34]. Based on the LiDAR equation, the backscattered intensity is a function of the beach surface properties and scanning geometry, which can be calibrated to the beach surface moisture when the other surface properties remain constant (e.g., the mineral composition, grain size distribution, packing density and surface roughness) [14,33,35–39]. The potential of terrestrial laser scanning was first demonstrated for measuring the beach surface moisture by [7,31,32], where a Leica Scanstation 2 with a wavelength of 532 nm was used to quantify the relation between the beach surface moisture contents and the TLS backscattered intensity. The experimental results revealed that the latter could discriminate within 1–2% gravimetric moisture contents in the moisture below 7% [32]. The same principle was tested by [33] and [34] with a RIEGL VZ-400 on a sandy beach, where the RIEGL VZ-400 is a laser scanner with a higher moisture sensitivity utilizing a 1550 nm wavelength. A robust negative relation between the beach surface reflectance and surface moisture was found for the full range of possible surface moisture contents (0–25%) with a standard error of 2.7% moisture. However, because of the existence of intensity reducing in the near-distance, only the intensity values within 20 to 60 m were used. In addition, the intensity values were only corrected by a factor of $1/R^2$ and the impact of incidence angles was ignored [33]. For the sandy beach of (very) fine sand (grain diameter < 0.25 mm), the roughness level of the beach surface is comparatively low and thus the effects of the incidence angles (especially when > 60°) on the backscattered intensity should also be taken into account.

Here, we examine the suitability of a mobile terrestrial LiDAR (MTL) to measure the sandy beach surface moisture, where a red laser scanner Z&F/Leica HDS6100 (λ = 650–690 nm) was mounted on an all-terrain vehicle. Although the most notable absorption peaks of light in water occur at the infrared wavelengths (e.g., 760, 970, 1200, 1470 and 1940 nm), there are still two small absorption peaks visible between the wavelengths 650 nm and 690 nm [14]. In this study, two sets of indoor calibration experiments were conducted in order to investigate the effect of the distance, incidence angles and sand moisture contents on the backscattered intensity in detail using the Z&F/Leica HDS6100, where the moisture contents of sand samples vary from 0% to 25%, the incidence angles from 0° to 80° and the scanning distance from 1 to 20 m. Afterwards, a moisture estimation model was developed so as to correct the influence of the incidence angles and distance on the backscattered intensity and to relate the corrected intensity values to the sand moisture contents. At last, a field measurement was carried out using the MTL system on a Belgian beach with very fine sand (and the accuracy and robustness of the derived beach surface moisture from the MTL data was evaluated). We also explored how the derived surface moisture from the MTL data could be utilized to compute the variation on the beach surface moisture with a high spatial and temporal resolution. We start with a theoretical introduction on the manner in which the distance, incidence angles and sand moisture contents' impact on the backscattered intensity should be quantified, followed by the indoor calibration experiments and data processing.

2. Theory

2.1. Surface Moisture-Reflectance Models

The surface reflectance of a sandy beach depends predominantly on its surface moisture contents, assuming that other surface properties such as the mineral composition, grain size distribution, packing density and surface roughness are relatively constant. Hence, it is possible to quantify the relation between the beach surface moisture and its reflectance [14,33,35–40]. The interaction between the incident light and wetted sand is a complex optical phenomenon which encompasses reflection, refraction and diffraction. It is determined by the light's wavelength [illegible]

sand grains and absorption by the water (path R_2).

Philpot [25] proposed an exponential model to express the effect of various moisture conditions on the soil surface reflectance (Equation (1)), which is also applicable to describe the relation between the sand surface moisture and its reflectance. Several previous studies also used exponential models [14,15,23,42,43] but the meaning of coefficients and variables are not exactly the same. The exponential model as proposed by Philpot is given by

$$\rho = f_w \rho_w + (1 - f_w)\rho_s e^{\alpha d} \tag{1}$$

where ρ, ρ_w and ρ_s denote the total reflectance, the Fresnel reflectance from the water surface and the reflectance from sand grains, respectively. f_w is a fraction which describes the contribution of the water surface reflection to the total reflectance. d is the optical path length of light through the pore water which describes the contribution of the water absorption to the reduction of reflectance, and α (a negative value) represents the wavelength dependent absorption coefficient for liquid water. In the special case of absence of surface water, f_w equals zero and a simplified reflectance expression would be

$$\rho = \rho_s e^{\alpha d} \tag{2}$$

Here, only the reflection from sand grains and pore water absorption have an impact on the final reflectance. According to the study findings of [14,25], the optical path length d increases almost linearly upon the soil wetting. We try to describe d with $= bM$, b and M demonstrating the slope coefficient and the moisture contents in sand respectively. When the optical path length $d = 0$, the reflectance ρ is equal to the ρ_s. Furthermore, based on the LiDAR equation [44,45], the backscattered intensity is proportional to the spectral reflectance of the target surface. We try to describe the backscattered intensity I with $= k\rho$, where k represents the slope coefficient. Thus, the Equation (2) could be modified as

$$I = k\rho_s e^{\alpha b M} \tag{3}$$

setting $\delta_0 = k\rho_s$ and $c = \alpha b$, the Equation (3) could be expressed as

$$I = \delta_0 e^{cM} \tag{4}$$

where parameters δ_0 and c are two constants for a fixed laser wavelength and sand type and their values can be obtained by means of a regression analysis.

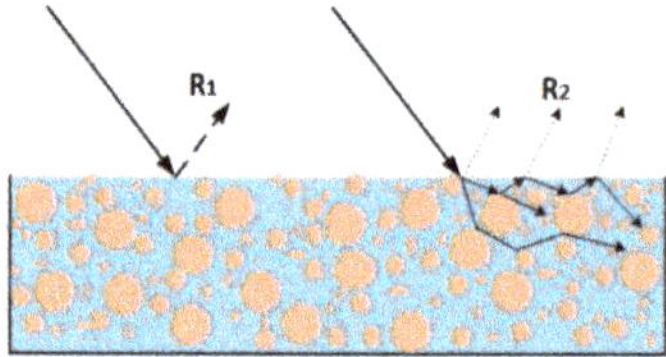

Figure 1. Illustration of the interaction between the incidence light and wetted sand. The path R_1 denotes the light directly reflected from the surface water film of saturated sand. Path R_2 demonstrates the remaining light that is transmitted through the water surface and subject to reflection from sand grains and absorption by water.

2.2. Correction of the Backscattered Intensity

In this study, a simplified LiDAR formula is adopted to explain the effect of the distance, incidence angle and surface reflectance on the backscattered intensity. As given by

$$I = \frac{\rho \cos\theta}{R^2} C \tag{5}$$

where I shows the backscattered intensity, ρ stands for the reflectance of the target's surface, θ is the incidence angle and R denotes the range. The parameter C describes the system and the atmospheric factor. Numerous studies [33,39,45–48] revealed that the original backscattered intensity is inapplicable to discriminate the target surface properties directly. The radiometric correction is necessary to eliminate the impact of the distance and incidence angle on the intensity data and to convert the raw intensity into a corrected value that is proportional or equal to the target's reflectance [39,45,46,49–61]. However, regarding the distance effect, due to the possible existence of automatic reducers for the near-distance backscattered signals and amplifiers for weak backscattered signals [46,62], only the part of the intensity data within a specific distance follows the theoretical LiDAR formula [39,48,61,63] and can be corrected effectively through $1/R^2$ [33,34]. On the other hand, the effect of the incidence angle is mainly related to the target surface properties and surface irregularities [35,39,64,65] and the cosine law is the most common method to rectify the effect of the incidence angles on the backscattered intensity [45]. Tan [47] proposed an empirical method in order to rectify the impact of the incidence angle using four standard Lambertian targets. However, considering the possible differences between the standard Lambertian targets and the real sand surface, the empirical method should be investigated by means of real sand samples at various incidence angles. However, few studies [35,39,65] investigated the effect of the incidence angle (from 0° to 40°) on the TLS intensity using sandblasting sand samples of two different grain sizes (0.1–0.6 mm and 0.5–1.2 mm). Nield [32] and Smit [33] believe that the influence of the incidence angle could be ignored when detecting the sandy beach surface moisture using TLS. It is noteworthy that the roughness level of the beach surface is comparatively lower in this study due to the smaller grain size, which means that the effect of the incidence angle on the backscattered intensity increases. Moreover, the height of the MTL scanner used in this study is merely about 1.75 m, where the incidence angle exceeds 81° at a corresponding scanning distance of 12 m. Thus, the impact of incidence angles should also be taken into account. In this study, an empirical correction method is adopted using sand collected from the selected research area. Based on the LiDAR formula, the effects of the distance and incidence angle on the intensity are independent from each other and could be corrected separately by a regression analysis [46,47,58–60,65]. The original backscattered intensity I can be expressed as [46,47,58–60]

$$I = F_1(\rho) \cdot F_2(\cos\theta) \cdot F_3(R) \tag{6}$$

where F_1, F_2 and F_3 represent a function of the target reflectance ρ, the cosine of the incidence angle $\cos\theta$ and the distance from the laser scanner R respectively and $0 \leq \cos\theta \leq 1$, $R_{min} \leq R \leq R_{max}$ (R_{min} and R_{max} denote the minimum and maximum distance that the TLS could detect respectively). According to the Weierstrass approximation theorem, a continuous function on a closed interval can be approximated by a polynomial series [47]. Therefore, the functions F_2 and F_3 can be expressed by a polynomial. Combining Equations (4) and (6) generates:

$$I = \delta_0 e^{cM} \cdot \sum_{i=0}^{N_2} \left[\beta_i (\cos\theta)^i\right] \cdot \sum_{i=0}^{N_3} \left(\gamma_i R^i\right) \tag{7}$$

with

$$\begin{cases} F_1(\rho) = \delta_0 e^{cM} \\ F_2(\cos\theta) = \sum_{i=0}^{N_2} \left[\beta_i (\cos\theta)^i\right] \\ F_3(R) = \sum_{i=0}^{N_3} \left(\gamma_i R^i\right) \end{cases} \tag{8}$$

where N_2 and N_3 denote the degree of the polynomials and β_i and γ_i are the polynomial coefficients. In previous studies, the intensity correction normally converts the original intensity into a corrected [illegible]

[illegible] M can be described by

$$M = \frac{1}{c} ln \left(\frac{I}{\delta_0 \left(\sum_{i=0}^{N_2} \left[\beta_i (\cos\theta)^i\right] \cdot \sum_{i=0}^{N_3} \left(\gamma_i R^i\right) \right)} \right) \tag{9}$$

For the determination of the parameters of Equation (8), a number of indoor calibration experiments were conducted using the sand samples. The principle is similar to the methods in [46,47,59] but in the calibration experiments commercial target panels of known reflectance were adopted rather than natural targets. In order to estimate the parameters of function $F_1(\rho)$, the incidence angle and scanning distance of the target samples should be unchanged, while the moisture contents of the samples vary in the indoor experiments. In these conditions, $F_2(\cos\theta)$ and $F_3(R)$ are considered as two constants, setting $c_1 = F_2(\cos\theta)\cdot F_3(R)$ and the intensity is merely related to the moisture contents M expressed as

$$I = c_1\delta_0 e^{cM} \tag{10}$$

Through a least-squares fit, the values of $c_1\delta_0$ and c can be estimated. To reduce random errors, several sets of $c_1\delta_0$ and c should be considered, defining a total average of c as the final parameter of the function $F_1(\rho)$. Similarly, to estimate the parameters of the polynomial $F_2(\cos\theta)$, the target samples should be scanned at various incidence angles, keeping the moisture and distance constant. In these conditions, $F_1(\rho)$ and $F_3(R)$ are two constants, setting $c_2 = F_1(\rho)\cdot F_3(R)$. The intensity level I can be expressed as

$$I = c_2\sum_{i=0}^{N_2}\left[\beta_i(\cos\theta)^i\right] \tag{11}$$

the value $c_2\beta_i$ could be obtained by a least-squares fit. The value of N_2 is determined to maintain a balance of the simplicity and accuracy of the model. Setting the coefficient of the highest degree $\beta_{N_2} = 1$ resulted in $c_2 = c_2\beta_{N_2}$ and $\beta_i = c_2\beta_i/c_2$. The average of β_i is calculated as the final polynomial coefficients of $F_2(\cos\theta)$ based on several sets of incidence angle experiments. Similarly, setting $c_3 = F_1(\rho)\cdot F_2(\cos\theta)$, the intensity value I can be expressed as Equation (12). The parameters of the polynomial $F_3(R)$ can be estimated based on the distance control experiments where the target samples are scanned at a variable distance but with a constant moisture level and incidence angle.

$$I = c_3\sum_{i=0}^{N_3}\left(\gamma_i R^i\right) \tag{12}$$

After estimating the parameters $F_1(\rho)$, $F_2(\cos\theta)$ and $F_3(R)$, the intensity value I can be expressed as

$$I = Ke^{cM}\cdot\sum_{i=0}^{N_2}\left[\beta_i(\cos\theta)^i\right]\cdot\sum_{i=0}^{N_3}\left(\gamma_i R^i\right) \tag{13}$$

The value of K can be obtained by substituting the intensity data obtained from the calibration experiments and the known values of the incidence angles, distance and moisture contents into the Equation (13) and then determining the average of K. The final function of the target moisture contents M can be expressed as

$$M = \frac{1}{c}ln\left(\frac{I}{K\left(\sum_{i=0}^{N_2}\left[\beta_i(\cos\theta)^i\right]\cdot\sum_{i=0}^{N_3}(\gamma_i R^i)\right)}\right) \tag{14}$$

It is noteworthy that Equation (14) not only derives the target moisture contents but it also eliminates the effects of the incidence angles and distance at the same time. In previous studies [7,31–34], the intensity correction and target moisture derivation were mostly conducted in two steps, which can contribute to additional errors. In terms of the calculation of the incidence angle, we conduct a plane-fitting with the adjacent points surrounding the target points and then obtain the normal vector of the fitted planes. The equation of the incidence angle is expressed as

$$\cos\theta = \left|\frac{v_{ps}\cdot v_n}{R}\right| \tag{15}$$

where v_{ps} and v_n denote the vectors from the target points to the scanner centre and the normal vectors of the fitted planes respectively. R represents the distance between the target points and the scanner centre. The length of the normal vector v_n is equal to 1.

3. Indoor Calibration Experiments

3.1. Terrestrial Laser Scanner

A phase-based laser scanner Z&F/Leica HDS6100 was used in the indoor calibration experiments, with the main parameters listed in Table 1. It utilizes a red laser with a sampling frequency of up to 508,000 points/second and a position accuracy of 5 mm within a 1–25 m range. Its scan density is up to 1.6 × 1.6 mm at a 10 m distance and 4.0 × 4.0 mm at 25 m for the *'ultra-high'* density mode; 6.3 × 6.3 mm at 10 m and 15.9 × 15.9 mm at 25 m for the *'high density'* mode. In this study, we adopt the *ultra-high* density mode for the indoor calibration experiments and the *high density* mode for the field beach measurements. The point cloud data are recorded in the form of x, y and z coordinates and the intensity of the return signal. Raw intensity values are stored as dimensionless numbers to denote the target reference level and do not have any physical meaning. It should be noted that the Z&F/Leica HDS6100 can theoretically measure distances up to 50 m at 18% albedo but in actual scans, the maximum measured distance measures considerably less than 50 m, especially on a wet beach.

Table 1. Main parameters of the Z&F/Leica HDS6100.

Attribute	Value
Wavelength	650–690 nm
Maximum range	50 m at 18% albedo
Field of View	360° × 310°
Laser beam divergence	0.22 mrad
Position accuracy	5 mm (1–25 m), 9 mm (25–50 m)

3.2. Incidence Angle Experiments

Several studies have investigated the relation between the surface reflectance and moisture contents in sand or soil over the full range of the gravimetric moisture contents (0–25%) [14,17,19,29]. However, few studies [31,35,61] investigated the effect of various incidence angles (0–40°) on the sand surface reflectance at a fixed TLS scanning distance, because it is hard to avoid disturbing the surface of the sand samples due to movement of dry sand and flowing of water inside the sand samples. In this study, we placed the sand samples horizontally by means of the specific structure of an indoor staircase. As shown in Figure 2a, the laser scanner was placed on the stair landing of a staircase. Eleven black platforms were horizontally fixed at the stairs, using triangle brackets which had the same distance (5 m) from the scanner centre but various incidence angles ranging from 0 to 80° in steps of 10°, supplemented with 2 platforms at 65 and 75°. The black platforms avoid the disturbing influence of the environmental lightning on the backscattered intensity signal, the positions and levelness were checked by a pre-scanning. The sand used in these experiments were collected from the study area of Groenendijk beach (Koksijde, Belgium) and was oven dried at 105 °C for 24 hours before executing the experiments. The sand grain size was analyzed by a sieving device which sorts the sand particles according to their size. Once the sand is sieved, it is weighed and the ratio of the total weight is [illegible]

[illegible] with a certain weight of pure water and then carefully put in a 5 mm depth plastic container that was laid on each of the eleven horizontal platforms. As shown in Figure 3, in the experiment of the incidence angles, the gravimetric moisture contents of the samples includes 0%, 1%, 2%, 3%, 4%, 5% and then, in steps of 2%, up from a 7% to a 25% moisture. In order to test the reflection of the surface water film of saturated sand, two additional sand samples (the sample of 26% moisture and the sample with a 2 mm water layer) were scanned too. The samples were weighed using an electronic scale with a milligram precision. According to the studies of [35,53,64], the roughness and irregularity of the natural target surface plays a key role in the incidence angle effect. In this study, in order to approximate the real irregularity level of a sandy beach surface, a scraper was used to flat the sand surface by means of gently removing the overflow sand from the plastic containers. A total of 18 scans were executed.

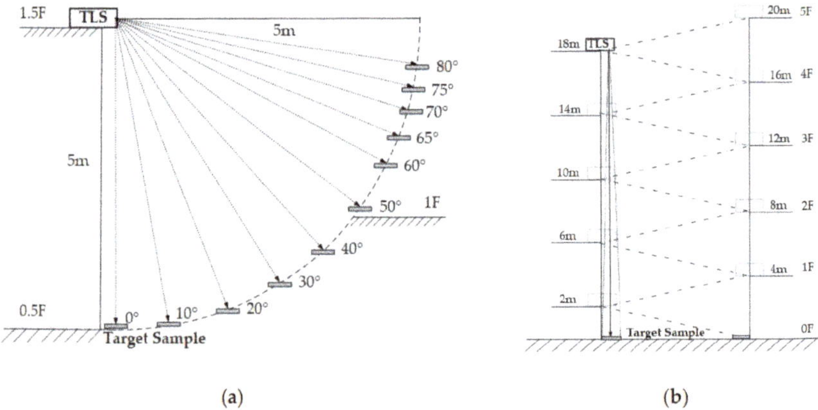

Figure 2. (**a**) The sketch of incidence angle experiments: the laser scanner is placed on the stair landing of a staircase and the sand samples are laid on eleven horizontal platforms. These eleven platforms have a fixed distance (5 m) from the scanner centre but different incidence angles varying from 0° to 80°. (**b**) The sketch of the distance experiments: the distance from the laser scanner to the target samples varied in steps of 1 m from 1 to 4 m and then in steps of 2 to 20 m by setting up the scanner at different floors. The sand samples were placed on the ground floor.

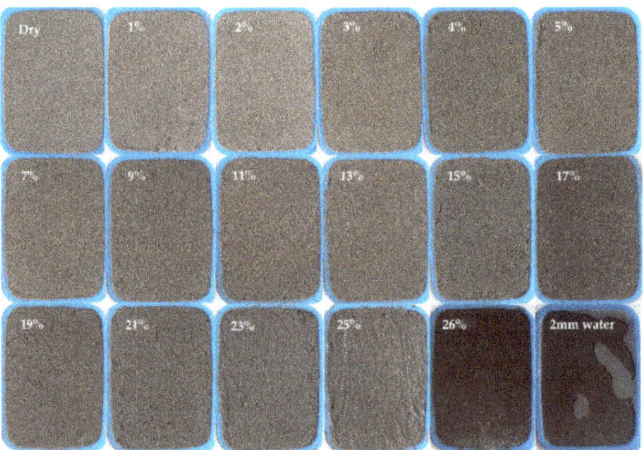

Figure 3. The close-up images of the sand samples with different moisture contents. The size of each plastic container has a 5 mm depth × a 10 cm width × a 20 cm length.

3.3. Distance Experiments

Numerous studies [39,46,47,53,65] investigated the effect of the TLS scanning distance on the target surface intensity using standard reflectance panels. However, to our knowledge, no study had investigated the distance effect on the backscattered intensity by means of sand samples of various moisture contents. In this study, a distance experiment was conducted using the sand samples whose moisture contents ranged from 0% to 25% moisture

As shown in Figure 2b, the target sand samples were placed on the ground level. Through setting up the scanner at different floors, the distance from the scanner centre to the target samples varied in 1 m increments from 1 to 4 m and then in 2 m increments from 6 to 20 m, the incidence angle was kept constant at a normal incidence. Each time the scanner setup was established at a new height, a pre-scanning was conducted to check the distance to the target samples, then the heights of the target samples were fine-tuned to meet the designed scanning distances. Based on the prior experiments of beach scanning using Z&F/Leica HDS6100, the effective maximum scanning range on the wet beach amounted to approximately 12 m with a scanner height of 1.75 m. Considering that the goal of the calibration experiments is the application of the MTL on the beach, longer-distance experiments were not conducted. The preparation process of the sand samples is similar to the experiment of the incidence angles. At each scanner position, 16 sand samples with different moisture contents were scanned in turn and then the scanner moved to the next floor. A total of 192 scans were executed.

4. Results

4.1. Experimental Data Analysis

After executing the indoor calibration experiments, the point clouds were manually sampled over the centre area (5 × 5 cm) of each plastic container using CloudCompare v2.11, the average intensity of each sample was calculated for the subsequent analysis. The effect of the incidence angle on the intensity is presented in Figure 4. One could notice that the overall decreasing trend of the original intensity upon the increasing incidence angle is similar for different moisture contents, with, as expected, the dry sand sample showing the highest intensities. The value of the original intensity decreases slightly with the increasing incidence angles from 0° to 30°, which is similar to the results in [65] and then declines gradually until 80°. Comparing with the experiments of [35,39,65], the incidence angle effect on the original intensity of TLS seems to be more significant in this study. A possible reason is the grain size of the sand used in this study, which is comparatively smaller. Consequently, the incidence angle effect should be eliminated for a further exploitation of the intensity data. On the other hand, the overall trend of the original intensity data upon the increasing gravimetric moisture contents is almost the same for different incidence angle from 0° to 80° and the impact of the sand moisture contents on the backscattered intensity seems to follow the exponential model Equation (4), which is similar to the findings of the studies [14,33,53]. The intensity values initially decrease dramatically from 0% to 3%. This means that a little water inside the sand samples has a huge effect on the backscattered intensity of the TLS. Beyond a 3% moisture, the intensity value decreases more gradually from 3% to 26%. Therefore, we could draw the conclusion that the intensity data of the Z&F/Leica HDS6100 have potential to discriminate the sand moisture contents for a full range of 0% to 25 % when scanning at non-normal incidence angles. It is noteworthy that the intensity evolution in the incidence angles near 80° fluctuate due to the decreased point density.

The effect of the distance on the intensity is presented in Figure 5. Initially, the original intensity increases drastically and reaches the intensity peak at about 3 m. Afterwards, the intensity decreases gradually until 20 m. Therefore, the intensity trend upon the increasing scanning distance [illegible]

[illegible] of the beach surface moisture. In this study, the intensity data from the distance below 3 m, which do not follow the LiDAR formula, were also drawn.

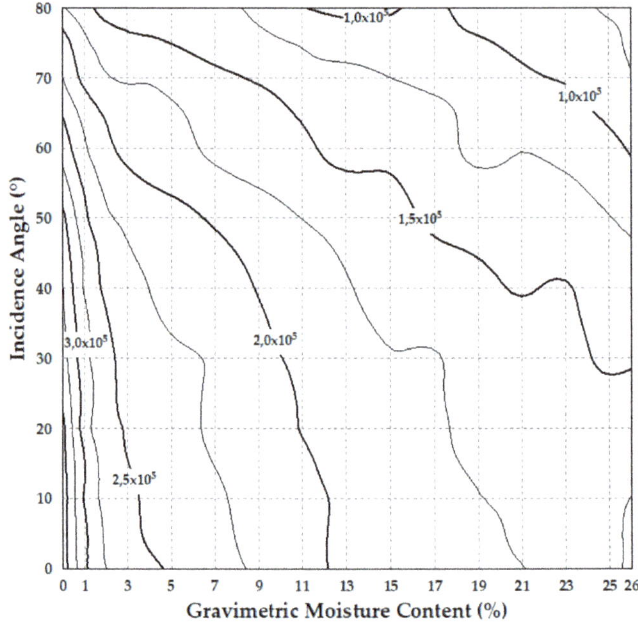

Figure 4. Contour lines of the Z&F/Leica HDS6100 original intensity versus the incidence angle (0–80°) and the gravimetric moisture contents (0–26%) when scanning sand samples from a fixed 5 m distance.

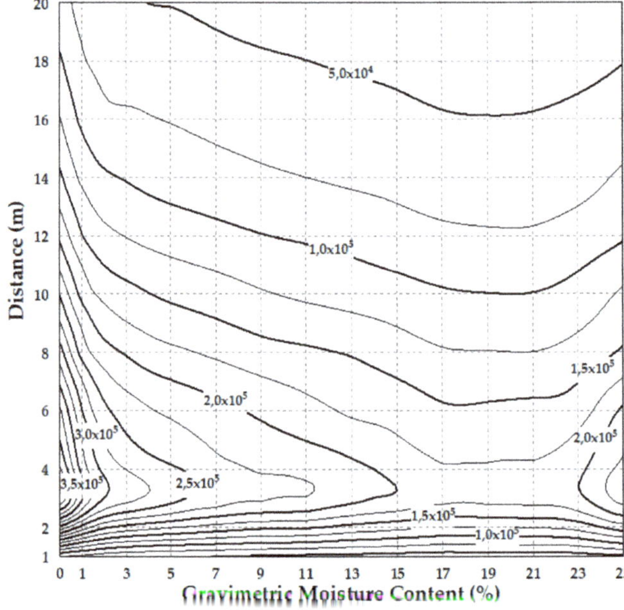

Figure 5. Contour lines of the Z&F/Leica HDS6100 original intensity versus the distance (1–20 m) and the gravimetric moisture contents (0–25%) when scanning sand samples orthogonally.

The overall trend (of the original intensity data upon the increasing sand moisture) is similar at different scanning distance from 1 to 20 m. Comparing with the results of the incidence angle experiments in Figure 4, an obvious difference includes the fact that the intensity values have been increasing gradually from 21% to 25%. The reflection of the surface water film coating upon the sand grains is considered to contribute to this phenomenon. More obvious evidence is the original backscattered intensity of the sand sample with a 2 mm water layer (in the additional incidence angles experiments, not presented in Figure 6), which peaked at 249×10^4, which is factor 7 above the intensity value of dry sand at a normal incidence angle. This means that the effect of the sand moisture contents on the backscattered intensity of the TLS does not completely follow the exponential model of the Equation (4) when scanning orthogonally.

4.2. Moisture Estimation

The surface moisture of the sand samples was estimated according to Equation (14). Before the parameter estimation, the intensity value of the dry sand scanned from 5 m and 70° had been normalized to 1. Considering the height of the MTL scanner used in the beach measurement, which is merely about 1.75 m where the effective scanning range measures from about 2 to 12 m with incidence angles varying from about 30 to 80°. Only this part of the experimental data was adopted to fit the best moisture model parameters for the next usage of the MTL data in this study. On the other hand, based on the analysis in Section 4.1, the intensity data of a 23% and a 25% moisture (when scanning at normal incidence) obviously do not follow the exponential model and were therefore not adopted to estimate the parameters. Based on the method introduced in Section 2, a first-degree polynomial and a fifth-degree polynomial were adopted to fit the $F_2(\cos\theta)$ and $F_3(R)$, respectively, after testing different degrees of polynomials. Afterwards, the parameters of $F_1(\rho)$, $F_2(\cos\theta)$ and $F_3(R)$ were estimated by the least-squares adjustment with the mean correlation-coefficient square (R^2) 0.90 ± 0.05, 0.98 ± 0.01 and 0.99 ± 0.003 respectively. Finally, the mean values of these parameters were calculated (see Table 2) and after obtaining the average of K, the surface moisture of the sand samples could be calculated based on Equation (14). If the derived moisture values were negative, we set them to zero. The process of the moisture estimation was executed in MATLAB 2014a.

Table 2. Mean values of the parameters in Equation (14).

F_1		c					
		−3.23 ± 0.28					
F_2	N_2	β_0	β_1				
	1	0.75 ± 0.13	1				
F_3	N_3	γ_0	γ_1	γ_2	γ_3	γ_4	γ_5
	5	−10,398.95 ± 1509.62	13,064.05 ± 2391.26	−3990.40 ± 540.70	564.62 ± 47.56	−38.29 ± 1.48	1
K		**1.65×10^{-4}**					

Figure 6 and Table 3 present the derived sand moisture from the intensity data of the [illegible]

[illegible] incidence angle of 80°, where the fluctuation of the derived moisture values is high due to the decreased point density. The maximum standard deviation is 1.8% (moisture), meaning that the impact of incidence angles on the moisture measurements of sand samples seem to have been eliminated. There are better correction results for the data of the distance experiments as shown in Figure 7 and Table 4. The standard deviations of the derived sand moisture are even less than 1.0% except for the sample of 25% moisture. Therefore, we can draw the conclusion that the derived moisture contents of sand samples no longer depend on the scanning incidence angles and distance after correction and that they are therefore solely associated with the intrinsic reflectance of the sand samples.

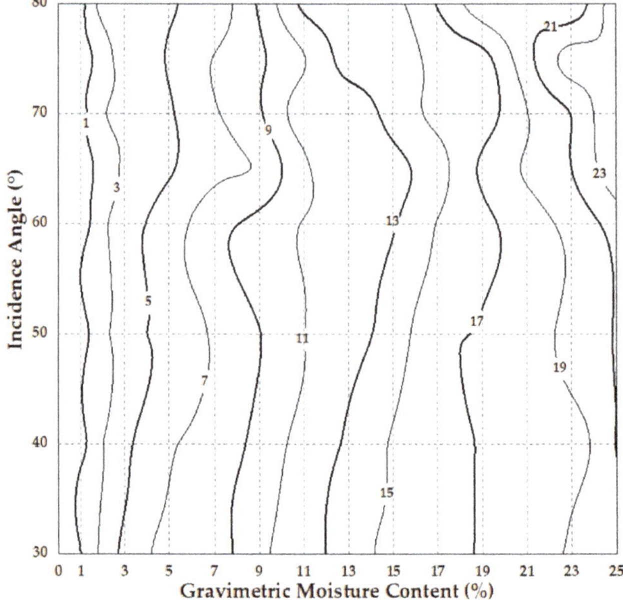

Figure 6. Contour lines of the derived moisture versus the incidence angle (30–80°) and the gravimetric moisture contents (0–25%) based on the incidence angle experiments.

Table 3. Derived samples' moisture contents based on the incidence angle experiments.

Moisture (%)	**Incidence Angle**								**Mean**	**Std.Dev.**	**Std.Error**
	30°	**40°**	**50°**	**60°**	**65°**	**70°**	**75°**	**80°**			
0	0	0	0	0	0	0	0	0	0	0	**0**
1	0.8	0.2	0	0	0	0.3	0	0.1	0.2	0.3	**0.8**
2	3.7	3.1	2.7	2.6	1.8	2.8	2.0	4.3	2.9	0.8	**0.9**
3	5.5	4.5	3.6	4.2	3.3	3.8	3.9	4.7	4.2	0.7	**1.2**
4	6.9	5.8	5.1	5.0	4.1	3.4	4.0	4.4	4.9	1.1	**1.0**
5	7.4	6.8	5.3	5.9	4.7	4.7	5.3	4.9	5.6	1.0	**0.8**
7	8.1	7.7	7.3	8.4	6.6	6.8	7.1	5.8	7.2	0.8	**0.7**
9	10.6	9.7	8.9	9.6	7.2	8.7	8.7	9.3	9.1	1.0	**0.7**
11	12.1	11.8	11.0	11.1	10.7	12.2	10.8	13.5	11.6	1.0	**0.8**
13	14.0	13.3	12.2	11.3	12.1	11.9	13.7	13.9	12.8	1.0	**0.9**
15	15.4	15.2	13.9	12.8	12.4	13.7	13.9	14.4	14.0	1.1	**1.2**
17	15.3	16.2	16.6	15.1	14.1	15.6	15.6	17.1	15.7	0.9	**1.3**
19	17.4	17.2	17.1	16.4	17.4	15.9	16.2	20.2	17.2	1.3	**2.1**
21	18.6	18.7	18.4	18.0	19.2	18.9	20.3	20.6	19.1	0.9	**1.9**
23	19.2	17.9	19.4	19.6	20.9	20.9	24.0	19.4	20.2	1.8	**3.1**
25	20.8	21.2	21.2	21.9	24.5	25.0	23.2	24.4	22.8	1.7	**2.2**
									Total Std. Error:		**1.2**

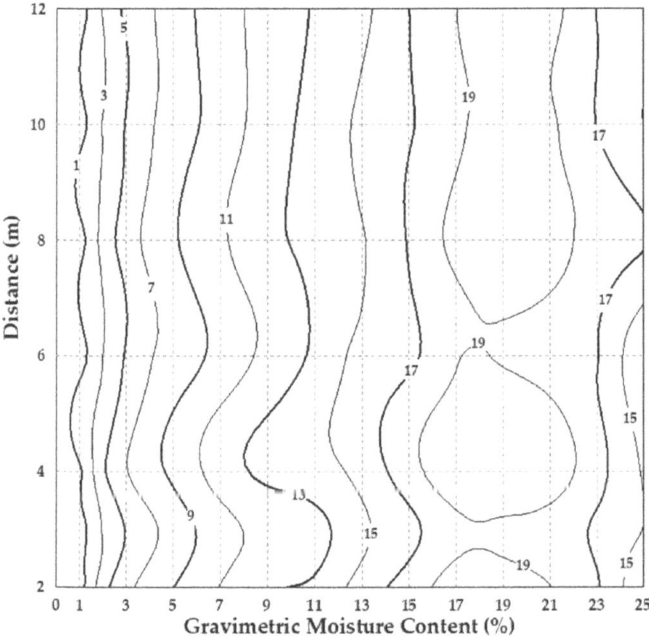

Figure 7. Contour lines of the derived moisture versus the distance (2–12 m) and the gravimetric moisture contents (0–25%) based on the distance experiments.

Table 4. Derived samples' moisture contents based on the data of the distance experiments.

Moisture (%)	Distance							Mean	Std.Dev.	Std.Error
	2m	3m	4m	6m	8m	10m	12m			
0	0	0	0	0	0	0	0	0	0	**0**
1	0	0	0.4	0	0	0	0	0.1	0.2	**0.9**
2	4.4	3.3	4.9	3.2	4.0	3.3	3.4	3.8	0.7	**1.8**
3	6.5	5.0	6.9	5.2	5.9	5.1	5.5	5.7	0.7	**2.7**
4	7.8	6.5	8.3	6.9	7.6	6.8	6.7	7.2	0.7	**3.2**
5	9.0	7.8	9.4	7.6	8.8	7.9	8.0	8.4	0.7	**3.4**
7	11.1	10.1	12.0	9.6	10.7	9.8	10.0	10.5	0.9	**3.5**
9	12.9	11.5	13.5	11.5	12.5	12.0	11.9	12.3	0.8	**3.3**
11	13.3	12.2	14.2	13.3	13.5	13.7	13.1	13.3	0.6	**2.3**
13	16.0	14.6	15.7	15.7	14.8	15.4	14.5	15.3	0.6	**2.3**
15	17.9	16.3	18.3	16.4	17.2	16.7	17.0	17.1	0.7	**2.1**
17	20.0	18.6	20.4	18.9	19.6	18.7	19.0	19.3	0.7	**2.3**
19	19.6	18.7	20.3	18.8	20.0	19.4	19.9	19.5	0.6	**0.7**
[illegible]										**2.9**

It has to be mentioned that the derived sand moisture from the samples of 23% and 25% moisture are less than the real moisture values, especially those based on the data of the distance experiments (where the scans were conducted orthogonally). Their standard errors measure up to 5.8% and 10.0% moisture, respectively. The remarkable increase of the backscattered intensity, due to the reflection of the significant surface water film, contributes to the errors. In other moisture levels (0–21%), including the range of a 0% to 8% moisture where the threshold lies for the aeolian transport [33], the derived moisture values show good concordance with the maximum standard error of 2.1% for the non-normal incidence scanning and 3.5% for the orthogonal scanning. The total average standard error of all computed moisture contents is only around 2.0%. The average error is 1.2% moisture for the non-normal angle experiments in Table 3 and 2.9% for the distance experiments in Table 4. Both results are considerably better than the results of the photographic methods [1,30] and the formerly mentioned TLS method with a wavelength of 532 nm [31,32]. In the actual measurements of the TLS or MTL, the incidence angle of scanning is non-normal; therefore, we could theoretically compute quite accurate moisture levels using this method. Regarding the sand samples with moisture contents ranging from 19% to 25% (cfr. Table 3), the moisture values could be further corrected by adding a 2% moisture to the initially calculated value.

5. Application of MTL

5.1. Study Site

A field measurement using the MTL was conducted at the North Sea beach of Groenendijk (Koksijde, Belgium) as shown in Figure 8a. It is a gently sloping (1–1.5%) and ultra-dissipative natural beach with a width of about 500 m at low tide. There are three main beach surface morphologies except the inundated bar-troughs on the measurement beach (Figure 8b): flat surface, flat surface with razor shell dumps and flat surface with ripples. The Belgian coast is situated in a macro-tidal regime ranging from 3.5 m at neap tide to 5 m at spring tide [66]. The orthometric height values in Belgium are generally taken towards the TAW reference level (Tweede Algemene Waterpassing), which is an equipotential gravity surface of approximately 2.3 m under the conventional geoid EGM96. The coordinate reference system used in this study is called the Belgian Lambert 72, an orthogonal cone projection with two intersecting parallels at 49d50′ and 51d10′.

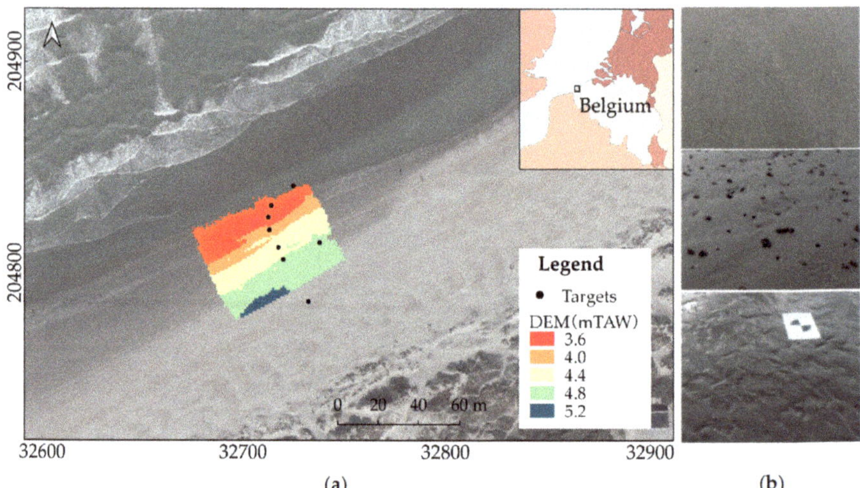

Figure 8. (**a**) Study site at the beach of Groenendijk and DEM of a 40 × 50 m area scanned three times. The black points denote the positioning targets. The photos in (**b**) represent three main morphologies on the measurement beach: flat surface, flat surface with razor shell dumps and flat surface with ripples.

5.2. MTL System and Beach Measurements

MTL is a complex real-time, multi-tasking and multi-sensor system and in this study, all the sensors were mounted on an all-terrain vehicle (Kymco) (Figure 9). The main devices include the same Z&F/Leica HDS6100 laser scanner used in the calibration experiments, an inertial measurement unit (IMU), two GNSS antennas with Real-Time Kinematic (RTK) precision and a rugged PC with hydrographic data acquisition software (QINSy). By the aid of RTK, the GNSS could provide positioning accuracy at centimeter level. However, it is virtually impossible to maintain the GNSS signal throughout the entire mobile survey because of the multipath effects and periods of GNSS outage [67], and as a result, the combination of GNSS with IMU was adopted in this study. It is noteworthy that the calibration must be executed to find the mounting angle errors (roll, pitch and heading) of the MTL system before the measurements. The time delay errors were solved by the PPS device and a latency check was done by surveying a line with two different speeds over a slope (based on the principle that the two slopes should be on top of each other when there is no latency) [68].

The field measurement using MTL was carried out on 14 November 2018 and the weather was cloudy. The surveying area measured about 200 × 200 m from the dyke to the intertidal zone in which a small area of 40 × 50 m near the high-tidal line was repeatedly scanned three times at 12:30, 14:00 and 15:30, respectively, between the low (11:31) and high tide (17:19). The DEM of the repeated scanning area was shown in Figure 8a. On the other hand, eight representative samples of beach sand were collected by a scraper from the sites next to the targets (Figure 8) after the first scanning. The sand samples taken demonstrated a 5 mm thickness over an area of 0.4 × 0.4 m. Their gravimetric moisture contents were calculated by double weighing, before and after the oven drying.

(a) (b)

Figure 9. (**a**) Kymco mobile platform equipped with the following sensors: a Z&F/Leica HDS6100 laser scanner, an IMU and two RTK-GNSS antennas and (**b**) MTL measurement on the beach where the effective scanning range on the wet beach is about 12 m with a scanner height of 1.75 m.

5.3. Beach Surface Moisture

Before the calculation of the beach surface moisture, a pre-processing of the point cloud data was performed by CloudCompare v2.11 and MATLAB 2014a. The non-ground outliers were manually

[illegible] incidence angle). In this study, a known dry sand area near the dunes was selected as the reference area where surface sand grains could move easily by breeze blowing. The point cloud at a 5 m distance and 70° incidence angle was extracted from this reference region and the mean intensity (215,386) of these points was calculated as the final denominator of normalization. In fact, the mean intensity was almost equal to the intensity value (224,820) of the indoor experiments.

Based on the parameter values in Table 3, the beach surface moisture was calculated using Equation (14). The negative derived values were set to zero denoting dry sand and the values larger than 26% were set to 26% denoting saturated sand or water surface. Afterwards, the point cloud data were transferred to a raster (DEM) with a 10 × 10 cm cell size. As shown in Figure 10a, before correction, the original intensity data heavily depend on the scanning incidence angles and distance. In contrast, the beach surface moisture in Figure 10b is independent of the scanning geometry. The details of the beach terrain also become clearer, such as the ripples in Line 1 and 2 and the tire tracks in Line 3 and 4. In order to further assess the accuracy of the derived beach surface moisture, the average moisture values at eight sampling sites were computed from the point cloud data over a 0.4 × 0.4 m area and compared with the real moisture contents using the sample gravimetric method. Besides, the average moisture values over a larger area of 1 × 1 m at eight sampling sites were also calculated for comparison. As shown in Table 5, the moisture contents of the sampling sites is representative and the surface of the sampling area is relative flat (except for the inundated trough area of S1 and S7). Overall, the derived surface moisture is relatively accurate with a maximum difference of a 2.7% moisture in S3 compared to the real moisture contents. It is noteworthy that in S3 and S4, the differences between the real and the derived moisture are relative large but at the same time, the standard deviation of the surface moisture in these areas is also larger than others. Moreover, a lot of shell debris is present at the sampling site of S3 (as shown in the close-up image of S3 in Table 5). For the sampling site of S1 and S7, several points were obtained from the non-water areas.

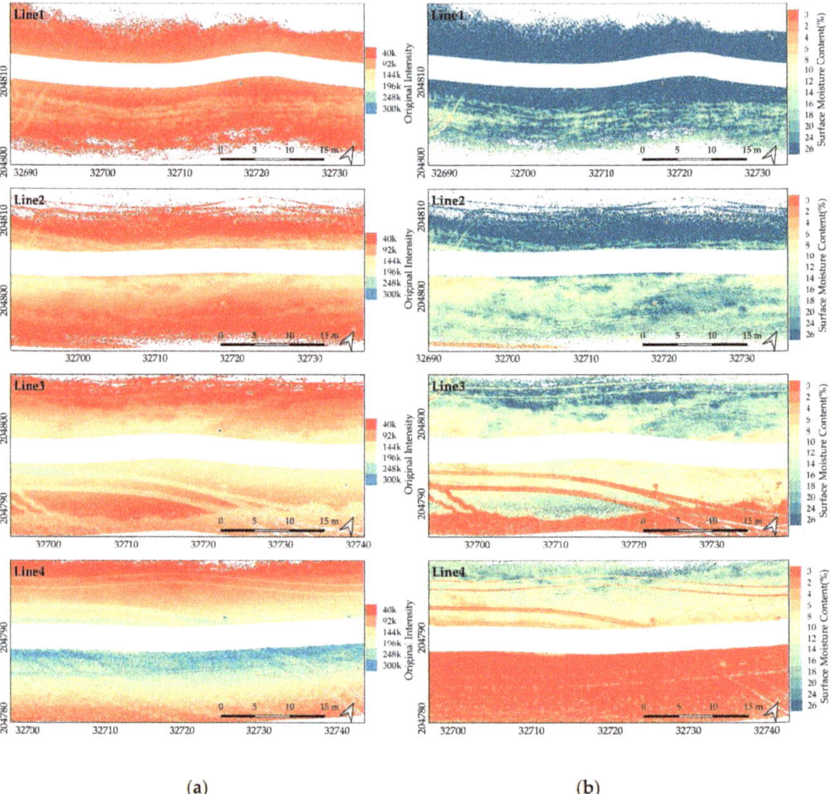

Figure 10. (**a**) Beach surface with the original intensity of four survey lines before correction and (**b**) beach surface moisture content of the same four survey lines after correction.

Table 5. The difference between the real sample moisture contents and the derived moisture content based on eight sand samples collected from the target beach.

Moisture (%)		S1	S2	S3	S4	S5	S6	S7	S8
Real moisture		Trough	26.9	20.1	13.5	8.1	5.9	Trough	1.2
0.4 × 0.4 m	Derived moisture (Std.Dev.)	No points	26.0 ± 0.2	22.8 ± 4.2	15.2 ± 5.2	9.4 ± 1.8	7.3 ± 3.4	25.4 ± 1.9	0.1 ± 1.4
	Difference		**−0.9**	**2.7**	**1.7**	**1.3**	**1.4**	**None**	**−1.1**
1 × 1 m	Derived moisture (Std.Dev.)	22.8 ± 3.9	25.7 ± 2.4	22.0 ± 5.0	16.7 ± 4.3	9.6 ± 2.1	7.9 ± 3.6	25.5 ± 1.7	0.1 ± 1.0
	Difference	**None**	**−1.2**	**1.9**	**3.2**	**1.5**	**2.0**	**None**	**−1.1**

The robustness of the overlapping strips is also important for the MTL measurements. Figure 11a depicts the absolute values of the surface moisture differences in the overlapping strip of adjacent survey lines for the first scan. The moisture raster maps were subtracted using the GIS raster calculator of QGIS. As shown in Figure 11a, the difference of almost overlapping areas amounts to less than a 2% moisture. Relative larger differences exist along the tracks and the edge of scanning strips, such as in the overlapping area of Line 2–Line 3. The possible reason is the distortion of the laser footprint at a great scanning distance and incidence angles. Another explanation could be the scanning areas of Line 2 and 3 at the transition zone from trough area to flat beach, in which the beach surface moisture varies greatly as shown in Line 1 of Figure 10b and the existence of tided debris also influences the accuracy of the derived surface moisture. In Figure 11b, there are relatively greater standard deviations of the surface moisture content noticeable along the tire tracks and in the transition zone from the trough area to the flat beach.

5.4. Beach Surface Moisture Variation

Figure 12a shows the rasterized surface moisture maps on three different moments in time, 12:30, 14:00 and 15:30, on 14 November 2018 during the rising tide. The red colours indicate dry sand, which are visible on the back beach in front of the dunes. The surface moisture varies spatially from 0% to 25%. Towards the sea, the beach surface moistures increase, depicted with green, yellow, blue to dark blue colours. The tire tracks have a lower moisture than their surroundings. Only few points are noticeable in the trough during the first scan but over time, more data points become visible. Figure 12b visualizes the surface moisture differences between the three time moments which give a clear illustration of the manner in which the beach surface moisture varies over time. Over three hours, the latter shows great variation, especially in the area between the dry sand and trough. The tire tracks demonstrate the greatest moisture variation in which the majority of the moisture variation is exceeds the 8% moisture.

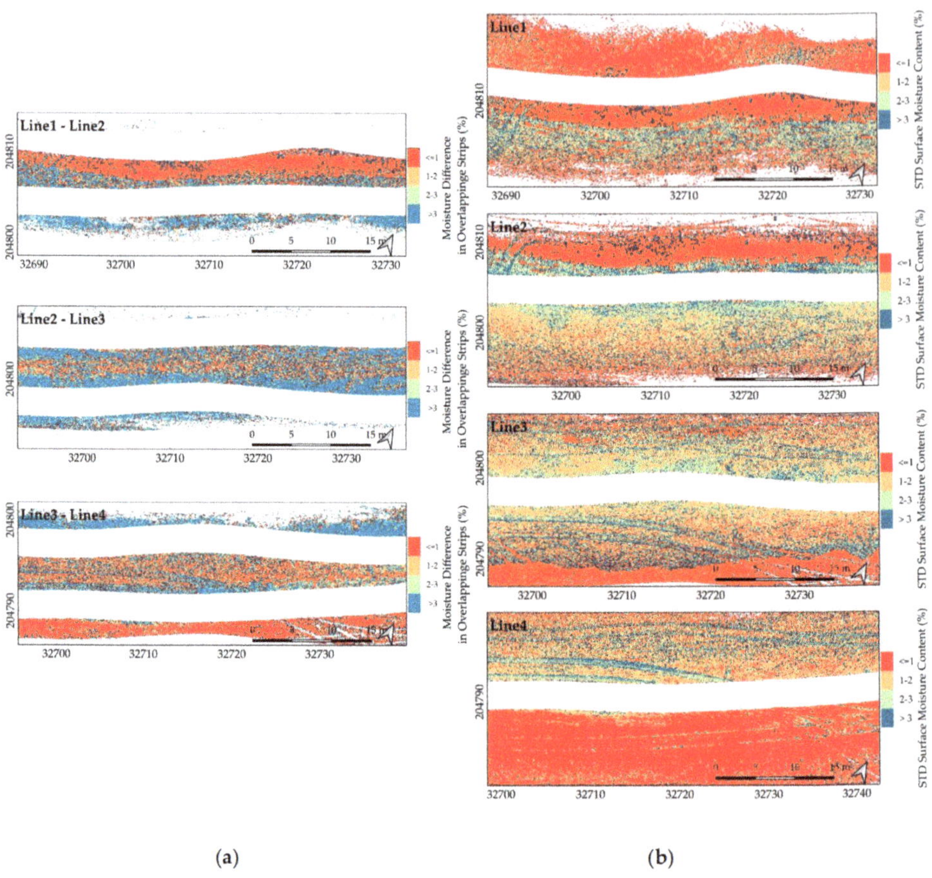

Figure 11. (**a**) Absolute values of the surface moisture difference in the overlapping strip of adjacent survey lines and (**b**) standard deviation of the derived moisture contents with grid cells of 10 × 10 cm.

Additionally, a moisture map of a 200 × 200 m area is shown in Figure 13, which integrates 24 survey lines of MTL. In the left corner, a simultaneously acquired orthophoto of the study site is given (which was obtained through UAV photogrammetry). The black dotted box indicates the repeated scanning area in Figure 12 and the points S1–S8 localize the eight sampling sites. By comparing with the orthophoto, the derived moisture map could accurately describe the distribution of the dry sand area (red colour) and trough (dark blue colour or no data points). The division between the areas with a different surface moisture is very clear. It is noteworthy that it took in total only 40 min to measure the 200 × 200 m area and each survey line (12 × 200 m) lasted about one minute, which means that the MTL could map the moisture variations of the mesoscale beach surface in a timescale of minutes.

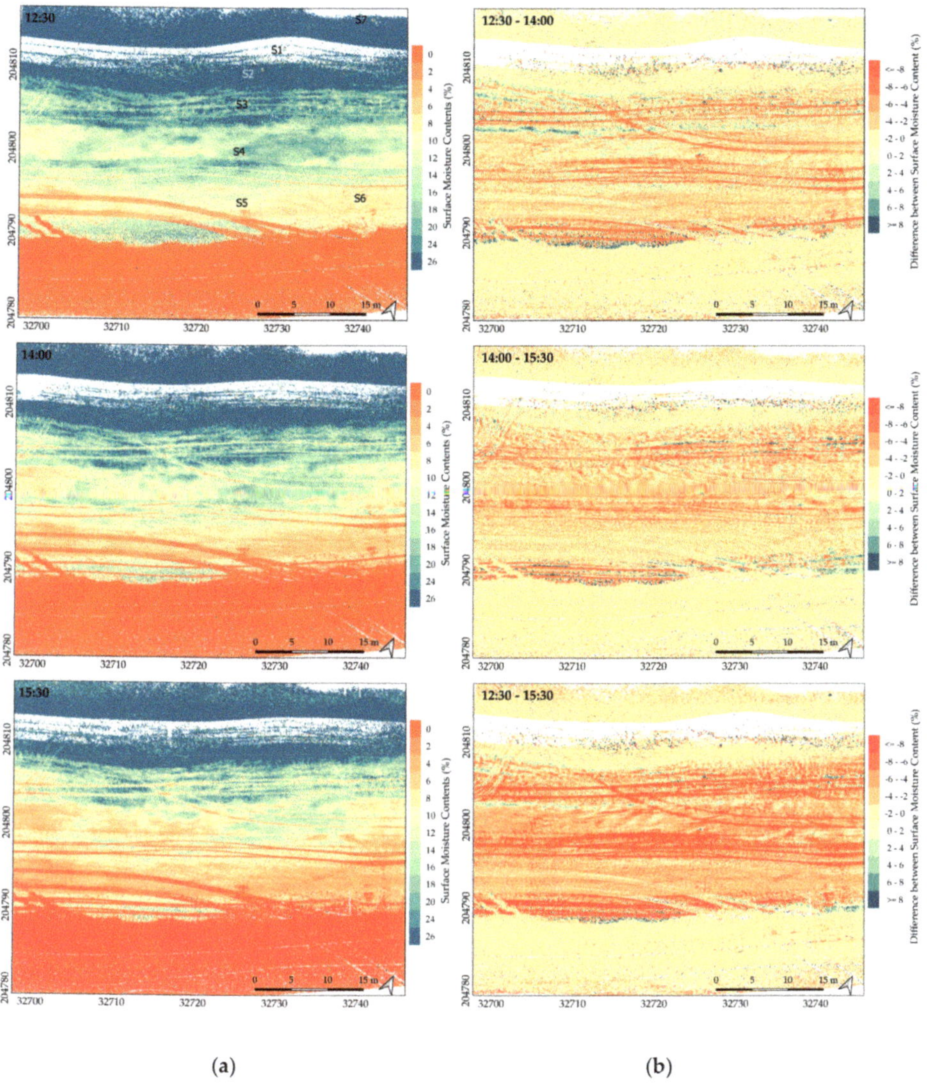

(**a**) (**b**)

Figure 12. (**a**) The beach surface moisture at 12:30, 14:00 and 15:30, respectively and (**b**) the moisture differences between these three time moments. The moisture map of 12:30 also presents the position of

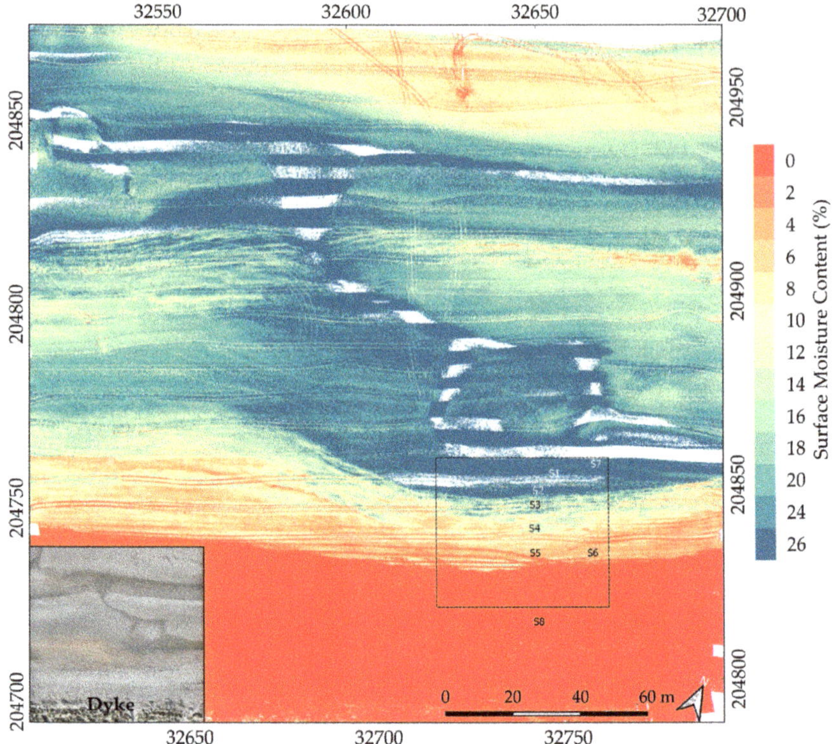

Figure 13. The beach surface moisture map of 200 × 200 m on the beach of Groenendijk, Belgium based on 24 survey lines of MTL. In the left corner, a simultaneously acquired orthophoto of the study site is given where we clearly see the distribution of the trough. The black dotted box indicates the repeated scanning area and the points S1–S8 localize the eight sampling sites.

6. Conclusions

This study comprehensively investigates the effect of the sand moisture contents on the backscattered intensity with moisture contents of sand samples from 0% to 25%, scanning incidence angles from 0 to 80° and measuring distances from 1 to 20 m using a Z&F/Leica HDS6100 laser scanner. Based on the experimental calibration data, a moisture estimation model was developed which eliminates the impact of the incidence angle and scanning distance and only reflects the surface reflectance. Moreover, to our knowledge, this is the first time that an MTL system was used to measure the beach surface moisture on a sandy beach. The main conclusions of this study include:

- Both the scanning incidence angle and scanning distance have an influence on the backscattered intensity of the flat sand samples (the average sand grain diameter is 0.12 mm), in which the intensity values decrease gradually with incidence angles between 30 to 80°. When scanning orthogonally, the surface water film coating sand grains could contribute to a remarkable increase of the backscattered intensity (which is factor 7 above the intensity value of dry sand).
- The original intensity values decrease strongly from dry sand to a 5% moisture. This means that even a little moisture in sandy areas has a huge effect on the backscattered intensity of TLS.
- The proposed moisture model could effectively eliminate the effect of the scanning geometry on the backscattered intensity of TLS. The accuracy of the derived surface moisture amounts to a total standard error of a 2.0% moisture. Regarding the non-normal scanning, the method could

accurately derive the full range of the sand moisture contents from 0% to 25% with a standard error of a 1.2% moisture. The result is considerably better than the previous TLS methods (using a Leica Scanstation 2 or a RIEGL VZ-400).

- Based on the proposed moisture model, the MTL system (using a Z&F/Leica HDS6100 laser scanner) is a promising technique to accurately and robustly measure the surface moisture on a sandy beach with an ultra-high spatial resolution (centimeter level) in a short time span (12 × 200 m per minute).
- On the other hand, some improvements should be considered in future studies. For example, the calibration experiments for varying distances could be conducted with a non-normal incidence angle. In addition, more sand samples could be collected from the target beach, in Groenendijk-Belgium, by a multi-person cooperation and the scanning efficiency of MTL could be increased by augmenting the laser scanner height or using a long-range LiDAR.

Author Contributions: J.J. provided the overall conception of this research, planned the indoor experiments and the MTL beach measurements, carried out the indoor experiments and most of the data analyses and wrote the majority of the manuscript. L.D.S. and J.V. executed the MTL beach measurements, performed the data pre-processing, helped with the production of maps and writing the manuscript. C.S. and A.-L.M. helped with providing the overall conception of this research and commented on the manuscript. G.D., P.D.M. and A.D.W. provided guidance, proposed the overall research field, provided the MTL system and commented on the manuscript. All authors have read and agreed to the published version of the manuscript.

Funding: This research was funded by the Chinese Scholarship Council, grant number 201406410069. The MTL measurements were performed within the CREST Project (www.crest-project.be), funded by the Strategic Basic Research (SBO) programme of the 'Instituut voor Innovatie door Wetenschap en Technologie' (IWT), grant number 150028.

Acknowledgments: The authors wish to thank Paul Schaepelynck, Bart De Wit and Samuel Van Ackere (Ghent University) for their logistic and technical assistance while performing the MTL measurements. The MTL measurements would not have been possible without the cooperation of Wijnand Vanhille (MOW-ATO) and the people of Surf Club Windekind (Oostduinkerke). We also thank Sabine Cnudde for the thorough proofreading of the manuscript.

Conflicts of Interest: The authors declare no conflict of interest.

References

1. Edwards, B.L.; Schmutz, P.P.; Namikas, S.L. Comparison of surface moisture measurements with depth-integrated moisture measurements on a fine-grained beach. *J. Coast. Res.* **2013**, *29*, 1284–1291.
2. Bauer, B.; Davidson-Arnott, R.; Hesp, P.; Namikas, S.; Ollerhead, J.; Walker, I. Aeolian sediment transport on a beach: Surface moisture, wind fetch, and mean transport. *Geomorphology* **2009**, *105*, 106–116. [CrossRef]
3. Davidson-Arnott, R.G.; Yang, Y.; Ollerhead, J.; Hesp, P.A.; Walker, I.J. The effects of surface moisture on aeolian sediment transport threshold and mass flux on a beach. *Earth Surf. Process. Landf. J. Br. Geomorphol. Res. Group* **2008**, *33*, 55–74. [CrossRef]
4. Oblinger, A.; Anthony, E.J. Surface moisture variations on a multibarred macrotidal beach: Implications for aeolian sand transport. *J. Coast. Res.* **2008**, *24*, 1194–1199. [CrossRef]
5. Wiggs, G.; Baird, A.; Atherton, R. The dynamic effects of moisture on the entrainment and transport of sand [illegible]

[illegible] **2016**, *41*, 738–753. [CrossRef]
7. Nield, J.M.; Wiggs, G.F.; Squirrell, R.S. Aeolian sand strip mobility and protodune development on a drying beach: Examining surface moisture and surface roughness patterns measured by terrestrial laser scanning. *Earth Surf. Process. Landf.* **2011**, *36*, 513–522. [CrossRef]
8. Davidson-Arnott, R.G.; MacQuarrie, K.; Aagaard, T. The effect of wind gusts, moisture content and fetch length on sand transport on a beach. *Geomorphology* **2005**, *68*, 115–129. [CrossRef]

9. McKenna Neuman, C.; Langston, G. Measurement of water content as a control of particle entrainment by wind. *Earth Surf. Process. Landf. J. Br. Geomorphol. Res. Group* **2006**, *31*, 303–317. [CrossRef]
10. Namikas, S.; Edwards, B.; Bitton, M.; Booth, J.; Zhu, Y. Temporal and spatial variabilities in the surface moisture content of a fine-grained beach. *Geomorphology* **2010**, *114*, 303–310. [CrossRef]
11. Atherton, R.J.; Baird, A.J.; Wiggs, G.F. Inter-tidal dynamics of surface moisture content on a meso-tidal beach. *J. Coast. Res.* **2001**, *17*, 482–489.
12. Smit, Y.; Donker, J.J.; Ruessink, G. Spatiotemporal surface moisture variations on a barred beach and their relationship with groundwater fluctuations. *Hydrology* **2019**, *6*, 8. [CrossRef]
13. Schmutz, P.P.; Namikas, S.L. Measurement and modeling of moisture content above an oscillating water table: Implications for beach surface moisture dynamics. *Earth Surf. Process. Landf.* **2013**, *38*, 1317–1325. [CrossRef]
14. Nolet, C.; Poortinga, A.; Roosjen, P.; Bartholomeus, H.; Ruessink, G. Measuring and modeling the effect of surface moisture on the spectral reflectance of coastal beach sand. *PLoS ONE* **2014**, *9*, e112151. [CrossRef] [PubMed]
15. Darke, I.; Davidson-Arnott, R.; Ollerhead, J. Measurement of beach surface moisture using surface brightness. *J. Coast. Res.* **2009**, *25*, 248–256. [CrossRef]
16. Yang, Y.; Davidson-Arnott, R.G. Rapid measurement of surface moisture content on a beach. *J. Coast. Res.* **2005**, *21*, 447–452. [CrossRef]
17. Tsegaye, T.D.; Tadesse, W.; Coleman, T.L.; Jackson, T.J.; Tewolde, H. Calibration and modification of impedance probe for near surface soil moisture measurements. *Can. J. Soil Sci.* **2004**, *84*, 237–243. [CrossRef]
18. Edwards, B.L.; Namikas, S.L. Small-scale variability in surface moisture on a fine-grained beach: Implications for modeling aeolian transport. *Earth Surf. Process. Landf.* **2009**, *34*, 1333–1338. [CrossRef]
19. Schmutz, P.P.; Namikas, S.L. Utility of the Delta-T Theta Probe for obtaining surface moisture measurements from beaches. *J. Coast. Res.* **2011**, *27*, 478–484.
20. Sherman, D.J.; Jackson, D.W.; Namikas, S.L.; Wang, J. Wind-blown sand on beaches: An evaluation of models. *Geomorphology* **1998**, *22*, 113–133. [CrossRef]
21. Williams, R.; Brasington, J.; Vericat, D.; Hicks, D. Hyperscale terrain modelling of braided rivers: Fusing mobile terrestrial laser scanning and optical bathymetric mapping. *Earth Surf. Process. Landf.* **2014**, *39*, 167–183. [CrossRef]
22. Ångström, A. The albedo of various surfaces of ground. *Geogr. Ann.* **1925**, *7*, 323–342.
23. Lobell, D.B.; Asner, G.P. Moisture effects on soil reflectance. *Soil Sci. Soc. Am. J.* **2002**, *66*, 722–727. [CrossRef]
24. Twomey, S.A.; Bohren, C.F.; Mergenthaler, J.L. Reflectance and albedo differences between wet and dry surfaces. *Appl. Opt.* **1986**, *25*, 431–437. [CrossRef] [PubMed]
25. Philpot, W. Spectral reflectance of wetted soils. *Proc. ASD IEEE GRS* **2010**, *2*. [CrossRef]
26. Darke, I.; Neuman, C.M. Field study of beach water content as a guide to wind erosion potential. *J. Coast. Res.* **2008**, *24*, 1200–1208. [CrossRef]
27. Delgado-Fernandez, I.; Davidson-Arnott, R.; Ollerhead, J. Application of a remote sensing technique to the study of coastal dunes. *J. Coast. Res.* **2009**, *25*, 1160–1167. [CrossRef]
28. Delgado-Fernandez, I. Meso-scale modelling of aeolian sediment input to coastal dunes. *Geomorphology* **2011**, *130*, 230–243. [CrossRef]
29. Delgado-Fernandez, I.R. Davidson-Arnott. Meso-scale aeolian sediment input to coastal dunes: The nature of aeolian transport events. *Geomorphology* **2011**, *126*, 217–232. [CrossRef]
30. Edwards, B.L.; Namikas, S.L.; D'sa, E.J. Simple infrared techniques for measuring beach surface moisture. *Earth Surf. Process. Landf.* **2013**, *38*, 192–197. [CrossRef]
31. Nield, J.M.; Wiggs, G.F. The application of terrestrial laser scanning to aeolian saltation cloud measurement and its response to changing surface moisture. *Earth Surf. Process. Landf.* **2011**, *36*, 273–278. [CrossRef]
32. Nield, J.M.; King, J.; Jacobs, B. Detecting surface moisture in aeolian environments using terrestrial laser scanning. *Aeolian Res.* **2014**, *12*, 9–17. [CrossRef]
33. Smit, Y.; Ruessink, G.; Brakenhoff, L.B.; Donker, J.J. Measuring spatial and temporal variation in surface moisture on a coastal beach with a near-infrared terrestrial laser scanner. *Aeolian Res.* **2018**, *31*, 19–27. [CrossRef]

34. Ruessink, G.; Brakenhoff, L.; van Maarseveen, M. Measurement of surface moisture using infra-red terrestrial laser scanning. In Proceedings of the EGU General Assembly 2014, Vienna, Austria, 27 April–2 May 2014; Volume 16. EGU2014–2797.
35. Kukko, A.; Kaasalainen, S.; Litkey, P. Effect of incidence angle on laser scanner intensity and surface data. *Appl. Opt.* **2008**, *47*, 986–992. [CrossRef]
36. Franceschi, M.; Teza, G.; Preto, N.; Pesci, A.; Galgaro, A.; Girardi, S. Discrimination between marls and limestones using intensity data from terrestrial laser scanner. *ISPRS J. Photogramm. Remote Sens.* **2009**, *64*, 522–528. [CrossRef]
37. Kaasalainen, S.; Kukko, A.; Lindroos, T.; Litkey, P.; Kaartinen, H.; Hyyppa, J.; Ahokas, E. Brightness measurements and calibration with airborne and terrestrial laser scanners. *IEEE Trans. Geosci. Remote Sens.* **2008**, *46*, 528–534. [CrossRef]
38. González-Jorge, H.; Gonzalez-Aguilera, D.; Rodriguez-Gonzalvez, P.; Arias, P. Monitoring biological crusts in civil engineering structures using intensity data from terrestrial laser scanners. *Constr. Build. Mater.* **2012**, *31*, 119–128. [CrossRef]
39. Kaasalainen, S.; Jaakkola, A.; Kaasalainen, M.; Krooks, A.; Kukko, A. Analysis of incidence angle and distance effects on terrestrial laser scanner intensity: Search for correction methods. *Remote Sens.* **2011**, *3*, 2207–2221. [CrossRef]
40. Kaasalainen, S.; Niittymaki, H.; Krooks, A.; Koch, K.; Kaartinen, H.; Vain, A.; Hyyppa, H. Effect of target moisture on laser scanner intensity. *IEEE Trans. Geosci. Remote Sens.* **2010**, *48*, 2128–2136. [CrossRef]
41. Bohren, C.F.; Huffman, D.R. *Absorption and Scattering of Light by Small Particles*; John Wiley & Sons: New York, USA, 2008; pp. 3–9.
42. Duke, C.; Guérif, M. Crop reflectance estimate errors from the SAIL model due to spatial and temporal variability of canopy and soil characteristics. *Remote Sens. Environ.* **1998**, *66*, 286–297. [CrossRef]
43. Shin, H.; Yu, J.; Jeong, Y.; Wang, L.; Yang, D.-Y. Case-Based Regression Models Defining the Relationships Between Moisture Content and Shortwave Infrared Reflectance of Beach Sands. *IEEE J. Sel. Top. Appl. Earth Obs. Remote Sens.* **2017**, *10*, 4512–4521. [CrossRef]
44. Wagner, W.; Ullrich, A.; Ducic, V.; Melzer, T.; Studnicka, N. Gaussian decomposition and calibration of a novel small-footprint full-waveform digitising airborne laser scanner. *ISPRS J. Photogramm. Remote Sens.* **2006**, *60*, 100–112. [CrossRef]
45. Höfle, B.; Pfeifer, N. Correction of laser scanning intensity data: Data and model-driven approaches. *ISPRS J. Photogramm. Remote Sens.* **2007**, *62*, 415–433. [CrossRef]
46. Fang, W.; Huang, X.; Zhang, F.; Li, D. Intensity correction of terrestrial laser scanning data by estimating laser transmission function. *IEEE Trans. Geosci. Remote Sens.* **2015**, *53*, 942–951. [CrossRef]
47. Tan, K.; Cheng, X. Intensity data correction based on incidence angle and distance for terrestrial laser scanner. *J. Appl. Remote Sens.* **2015**, *9*, 094094. [CrossRef]
48. Tan, K.; Cheng, X.; Ding, X.; Zhang, Q. Intensity data correction for the distance effect in terrestrial laser scanners. *IEEE J. Sel. Top. Appl. Earth Obs. Remote Sens.* **2016**, *9*, 304–312. [CrossRef]
49. Coren, F.; Sterzai, P. Radiometric correction in laser scanning. *Int. J. Remote Sens.* **2006**, *27*, 3097–3104. [CrossRef]
50. Errington, A.F.; Daku, B.L.; Prugger, A.F. A model based approach to intensity normalization for terrestrial laser scanners. In Proceedings of the International Symposium on Lidar and Radar Mapping 2011: Technologies and Applications [illegible]

system. *Remote Sens.* **2015**, *7*, 6336–6357. [CrossRef]
53. Kaasalainen, S.; Hyyppa, H.; Kukko, A.; Litkey, P.; Ahokas, E.; Hyyppa, J.; Lehner, H.; Jaakkola, A.; Suomalainen, J.; Akujarvi, A. Radiometric calibration of LIDAR intensity with commercially available reference targets. *IEEE Trans. Geosci. Remote Sens.* **2009**, *47*, 588–598. [CrossRef]
54. Ahokas, E.; Kaasalainen, S.; Hyyppä, J.; Suomalainen, J. Calibration of the Optech ALTM 3100 laser scanner intensity data using brightness targets. *Int. Arch. Photogramm. Remote Sens. Spat. Inf. Sci.* **2006**, *36*, 1Á6.
55. Bitenc, M. Evaluation of a laser Land-Based Mobile Mapping System for Measuring Sandy Coast Morphology. Master Thesis, Delft University of Technology, Delft, The Netherlands, 8 March 2010.

56. Ding, Q.; Chen, W.; King, B.; Liu, Y.; Liu, G. Combination of overlap-driven adjustment and Phong model for LiDAR intensity correction. *ISPRS J. Photogramm. Remote Sens.* **2013**, *75*, 40–47. [CrossRef]
57. Tan, K.; Cheng, X. Specular reflection effects elimination in terrestrial laser scanning intensity data using Phong model. *Remote Sens.* **2017**, *9*, 853. [CrossRef]
58. Xu, T.; Xu, L.; Yang, B.; Li, X.; Yao, J. Terrestrial laser scanning intensity correction by piecewise fitting and overlap-driven adjustment. *Remote Sens.* **2017**, *9*, 1090. [CrossRef]
59. Tan, K.; Cheng, X. Correction of incidence angle and distance effects on TLS intensity data based on reference targets. *Remote Sens.* **2016**, *8*, 251. [CrossRef]
60. Tan, K.; Chen, J.; Qian, W.; Zhang, W.; Shen, F.; Cheng, X. Intensity Data Correction for Long-Range Terrestrial Laser Scanners: A Case Study of Target Differentiation in an Intertidal Zone. *Remote Sens.* **2019**, *11*, 331. [CrossRef]
61. Kaasalainen, S.; Krooks, A.; Kukko, A.; Kaartinen, H. Radiometric calibration of terrestrial laser scanners with external reference targets. *Remote Sens.* **2009**, *1*, 144–158. [CrossRef]
62. Tan, K.; Zhang, W.; Shen, F.; Cheng, X. Investigation of tls intensity data and distance measurement errors from target specular reflections. *Remote Sens.* **2018**, *10*, 1077. [CrossRef]
63. Hofle, B. Radiometric correction of terrestrial LiDAR point cloud data for individual maize plant detection. *IEEE Geosci. Remote Sens. Lett.* **2014**, *11*, 94–98. [CrossRef]
64. Pesci, A.; Teza, G. Effects of surface irregularities on intensity data from laser scanning: An experimental approach. *Ann. Geophys.* **2008**, *51*, 839–848.
65. Kaasalainen, S.; Vain, A.; Krooks, A.; Kukko, A. Topographic and distance effects in laser scanner intensity correction. *Int. Arch. Photogramm. Remote Sens. Spat. Inf. Sci.* **2009**, *38*, 219–222.
66. Deronde, B.; Houthuys, R.; Henriet, J.P.; Lancker, V.V. Monitoring of the sediment dynamics along a sandy shoreline by means of airborne hyperspectral remote sensing and LIDAR: A case study in Belgium. *Earth Surf. Process. Landf. J. Br. Geomorphol. Res. Group.* **2008**, *33*, 280–294. [CrossRef]
67. Stal, C.; Incoul, A.; de Maeyer, P.; Deruyter, G.; Nuttens, T.; de Wulf, A. Mobile Mapping and the Use of Backscatter Data for the Modelling of Intertidal Zones of Beaches. In Proceedings of the 14th SGEM GeoConference on Informatics, Geoinformatics and Remote Sensing, Sofia, Bulgaria, 19–25 June 2014.
68. Qinsy Support Desk. Available online: https://confluence.qps.nl/qinsy/latest/en/how-to-calibrate-a-multibeam-echosounder-35587229.html (accessed on 15 August 2019).

Article

Multi-Temporal Cliff Erosion Analysis Using Airborne Laser Scanning Surveys

Dagmara Zelaya Wziątek [1], Paweł Terefenko [2,*] and Apoloniusz Kurylczyk [3]

1. Faculty of Geodesy and Cartography, Warsaw University of Technology, 00661 Warsaw, Poland; dagmara.zelaya@gmail.com
2. Institute of Marine and Environmental Sciences, University of Szczecin, 70383 Szczecin, Poland
3. Institute of Spatial Management and Socio-Economic Geography, University of Szczecin, 70383 Szczecin, Poland; apoloniusz.kurylczyk@usz.edu.pl

* Correspondence: pawel.terefenko@usz.edu.pl; Tel.: +48-91-444-23-54

Received: 18 October 2019; Accepted: 13 November 2019; Published: 14 November 2019 

Abstract: Rock cliffs are a significant component of world coastal zones. However, rocky coasts and factors contributing to their erosion have not received as much attention as soft cliffs. In this study, two rocky-cliff systems in the southern Baltic Sea were analyzed with Airborne Laser Scanners (ALS) to track changes in cliff morphology. The present contribution aimed to study the volumetric changes in cliff profiles, spatial distribution of erosion, and rate of cliff retreat corresponding to the cliff exposure and rock resistance of the Jasmund National Park chalk cliffs in Rugen, Germany. The study combined multi-temporal Light Detection and Ranging (LiDAR) data analyses, rock sampling, laboratory analyses of chemical and mechanical resistance, and along-shore wave power flux estimation. The spatial distribution of the active erosion areas appear to follow the cliff exposure variations; however, that trend is weaker for the sections of the coastline in which structural changes occurred. The rate of retreat for each cliff–beach profile, including the cliff crest, vertical cliff base, and cliff base with talus material, indicates that wave action is the dominant erosive force in areas in which the cliff was eroded quickly at equal rates along the cliff profile. However, the erosion proceeded with different rates in favor of cliff toe erosion. The effects of chemical and mechanical rock resistance are shown to be less prominent than the wave action owing to very small differences in the measured values, which proves the homogeneous structure of the cliff. The rock resistance did not follow the trends of cliff erosion revealed by volume changes during the period of analysis.

Keywords: cliff coastlines; cliff retreat; time-series analysis; airborne laser scanner

---

1. Introduction

According to current estimates, 80% of the world's coastlines is composed of cliffs [1]. Despite [illegible] contemporary research focuses on sedimentary coasts including mostly beaches, considering the high importance from sociological and economical perspectives [2–9]. It is popularly believed that rocky cliffs are characterized by slow rates of erosion and are only moderately vulnerable to global sea level changes [1,10,11]. This aspect explains the superficial treatment of the problem of cliff erosion.

Faster rates of beach coastline erosion are undeniable; however, the consequences of this phenomenon are more predictable and are less catastrophic than those in the case of cliff erosion caused by mass movement. Thus, researchers agree that the process of cliff erosion is important. Numerous models of chalk and limestone cliff retreat have been presented by authors from all over the world. Their works indicate that the factors influencing erosion vary among study areas. For example, studies conducted on chalk cliffs in East Sussex in the United Kingdom have identified geology as the main factor controlling the location and scale of cliff erosion, whereas studies on chalk cliffs in Pas de Calais, France, suggest that cliff stability is more relevant than other factors in cliff erosion [12]. Another element considered is marine action, which in some cases is the main reason for cliff erosion [13,14]; in other cases, its influence is restricted to debris removal [15]. In addition, the effect of rainwater is noted for its significant influence. Furthermore, sub-aerial processes have also been found to be relevant in the cliff retreat process [12].

Despite efforts made to the describe cliff erosion mechanisms, the relationships among factors such as precipitation, geology, cliff stability, and sub-aerial processes are highly complex and have been inadequately explained thus far. An essential question remains: Which of these factors is critical in initiating erosion processes? If a cliff is composed of hard, dolomitized, and compressed chalk, heavy rain will not be as erosive as high-power waves. Conversely, soft chalk can be easily saturated by rainwater and thus erodes at a high rate without the influence of other factors.

Identification of the factors influencing cliff erosion is a challenge owing to the high complexity of this mechanism. Even when applying the available techniques, this task remains very difficult and demands prolonged observation and correlation of many factors.

The use of high-accuracy three-dimensional (3D) spatial data is necessary for such sophisticated analysis. Tools and methods such as light detection and ranging (LiDAR), structure-from-motion (SfM) photogrammetry, or video imaging provide the quickest and most accurate and detailed data available for topographic analysis [16–20]. In the present study, a cliff system in Jasmund National Park in Rugen, Germany, located in the southern Baltic Sea, was monitored by using multi-temporal LiDAR data comparison. This method enabled us to track the annual cliff surface changes in these well-known, spectacular white chalk cliffs. On the basis of the gathered data, precise calculation of the rates of erosion and volumetric changes was performed, and ongoing processes were analyzed. The results, when incorporated with wave, hydrological, and geological data, provided an overview of interdependencies and influences of coastal erosion processes. However, LiDAR data are not free of errors. Thus, creation of a digital terrain model (DTM) with suitable accuracy is another challenge that demands data pre-processing. The vertical absolute accuracy of LiDAR surveys used for coastal analysis is generally about $\pm$ 0.15 m [21].

The analyzed part of this coast has experienced erosion since the Pleistocene era. However, a coastal monitoring program using an airborne LiDAR scanner has recently revealed intensification of these processes.

Therefore, the main objective of the present study was to identify the possible correspondence among cliff erosion rates, cliff exposure to wave action, and cliff rock resistance based on multi-temporal LiDAR data. These data, in addition to those of the volumetric changes along the cliff–beach profile, were used to produce final reproducible solutions for analyzing the relationship among the erosion rate on coastal cliffs and selected variables such as wave action, rock resistance, the hydrological regime, and geological structures.

2. Materials and Methods

2.1. Study Sites

Cliff erosion analysis was performed for a non-tidal basin of the Baltic Sea, a subdivision of the shallow Arkona Basin, bordered by Borholm, Falster, and Zealand islands. The analyzed cliff formations have long been the subject of widespread interest since they became part of Jasmund National Park in 1990. These formations are known worldwide for their distinctive, high, white chalk cliffs, with the highest occurring along the southern Baltic coastline [22]. In June 2011, the beach forest in Jasmund Park, which at only 30 km^2 is the smallest German National Park, became part of the United Nations Educational, Scientific and Cultural Organization (UNESCO) World Heritage list.

In this region, Cretaceous chalk rock has formed steep cliffs that often exceed 100 m in height [23]. The highest position in this area reaches 118 m above sea level and it is located in Königsstuhl near Sassnitz. The elevated chalk cliffs are separated by lower, gently dipping parts consisting mostly of Pleistocene deposits [23].

The morphology of the cliffs varies along the coastline and depends on the predominant building material. The upper part of the cliffs forms a slightly concave, smooth slope with an almost vertical profile. The majority of the chalk cliffs contain "apron fans of chalk rubble" at the base [22]; this talus was produced by erosion processes. In the other parts of the cliffs, where the sea surface and the waves interact with the cliff, wave-cut notches are a visible indicator of sea erosion. Beaches of Jasmund National Park are covered by flint pebbles of different sizes the have formed shingle beaches (Figure 1).

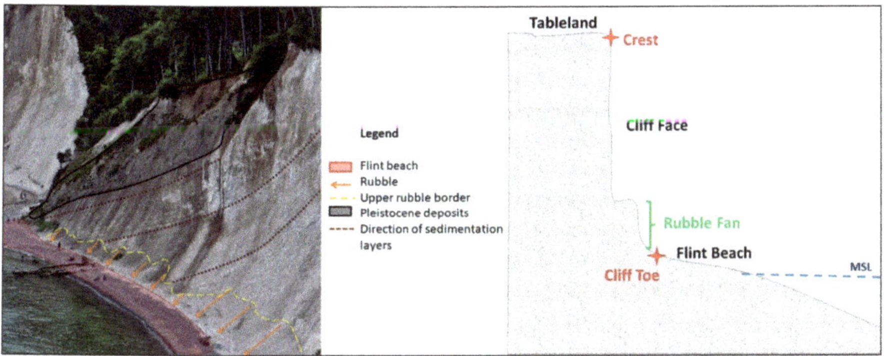

Figure 1. Major morphological components of the coast cliff system in Jasmund National Park.

For this research, two study areas were chosen for analysis. The first located in the central part of the national park coastline on the east-facing cliff, and the second is situated at the southern end of the park close to the city of Sassnitz, with cliffs facing southeast.

Selection of test sites with different coastline orientations was crucial for this study. The power of the waves reaching the cliffs changes depending on mean angle of the study area. Therefore, to analyze the influence of waves on cliff erosion, the power of waves reaching the cliffs in both study areas should be adjusted to reflect local conditions. Because wave power depends on the angle between the refracted wave and the shoreline, the two study areas with different orientations were assumed to receive different levels of wave power. The two selected study areas are located approximately 4 km apart.

The cliffs of both study sites dip steeply seaward. The cliffs are lower and the slope is moderate only on the sides of the cliffs, where bluffs occur. The top parts of these cliffs are covered by Pleistocene glacial sediments. The beach slopes in [illegible] large-scale cliff failure according to the kinematic rock slope evaluation of Grunther and others [23]. In particular, the northern part of the cliff qualifies as having very high/high susceptibility to chalk rock failure. The northern and southern parts of the chalk cliff in this area border layers of glacial sediments; the presence of this material corresponds with landslides occurring in the northern part of the study area after 2000 and in the part bordering the study area to the south before that year. In the middle part of the analyzed area, the chalk cliff experienced mass failure in 1994.

Study Area 2 is located on the Gakower coastline (*Gakower Ufer*). Although the analyzed cliff did not experience rock failure, landslides of the glacial sediments have occurred in neighboring areas to the north and south before 2000. Small-scale landslides occurring after that year have been recorded only in one part of the southern cliff side. The central part of study area has been characterized as having moderate and low susceptibility to rock failure [23].

2.2. Data

Recent technical developments have enabled the broad use of new remote sensing techniques such as Light Detection and Ranging (LiDAR) in topographic survey and coastal process monitoring [24–26].

In the case of cliff areas, airborne LiDAR surveys have one important advantage such that laser scanning can be performed from the seaside. This enables collection of data from positions that were previously inaccessible when using terrestrial LiDAR and traditional methods. This enables much more flexibility and allows for the collection of data with greatly expanded coverage. However, the data accuracy is usually lower owing to the very high laser beam incident angle; therefore, this method does not always enable sufficient data collection, particularly from slanted surfaces. This problem is also associated with the high-plain altitude: to access data with high vertical accuracy with all notch concavity penetrated, the airplane equipped with the LiDAR instrument must fly at very low altitudes and almost perpendicular to the cliff.

The data used in this study cover two airborne LiDAR campaigns. The first was performed in April 2007 as part of a Federal Institute for Geosciences and Natural Resources (BGR) project in which a 3D Optech ALTM3100 laser scanner was used to scan the area of Jasmund National Park. Data with a horizontal point distribution of 0.5 m were obtained during 10 fly routes with flight strip swath widths of about 9 km × 4.5 km. The resulting point cloud of the scanned area comprised 12.1 million points. The data were divided into sub-areas of 1 km × 1 km in extent and were saved as separate files. The average point density for the coastline area was equal 2.2 points per m^2, and the average point resolution was 0.67 m.

The second scanning campaign was performed in April 2012 by the National Board of Agriculture and Environment of Central Mecklenburg (STALUMM) Coastal Group Department. The main purpose of this LiDAR data collection was to create coastal and river flood simulation and coastal hazard maps. Many flight routes were performed that sometimes covered the same area two or three times. As a result, the data density varied between 1 and 25 points per m^2.

For the purpose of this work, only the data from the coastline were used. Average point density equals 24.5 points per m^2. Average point resolution equals 0.2 m. The property of signal reflection enabled categorizing, separating, and filtering out points representing terrain surface and overlying points representing buildings or vegetation. The primary returns, including the intermediate and first of many returns, represent vegetation, whereas the secondary returns, including the last returns, correspond to the ground surface. Single point returns were also categorized as primary returns; however, they were considered as a bare earth because their high return signal intensity resulted in small error. It should be noted that the signal intensity and time of return is only one of many available methods used for data classification and further filtration. Depending on concepts and research objectives, data filters can be based on the morphology, progressive densification, surface, or segmentation [27].

The wave data used in the present study were obtained from the Federal Maritime and Hydrographic Agency of Germany (Bundesamt für Seeschifffahrt und Hydrographie, BSH) as a part of the Western Baltic Sea Monitoring Program (MARNET). The records of the wave parameters were obtained from the station Arkona Becken located at a depth of 45 m at 54°53′ N and 13°52′ E. The station is equipped with water quality sensors to measure salinity, temperature, and radioactivity; water movement sensors to detect wave height, wave periods, wave direction, and current; and meteorological sensors to measure temperature, wind speed, and wind direction.

Data of significant wave height, wave period, and wave mean propagation direction were provided for the period corresponding to the period in which the LiDAR data were obtained: 1 January 2007–31 December 2012. The wave data were recorded hourly. The continuity of the data was disturbed only on a few occasions.

The study area inventory revealed the presence of objects that change cliff morphology and could influence cliff erosion. Two types of obstacles were identified: log pilings located close to the coastline on the sea side and fallen trees placed on the beach toe close to the coastline. These can be classified as major breakwater structures.

Two locations with fallen trees were identified in the first study test area. Trees most likely grew on the cliff tableland and then fell to the beach after cliff mass movement. Greater accumulation of the trees was observed in the southern part of the study area, where they were situated across the beach (Figure 2a). In the second area, a fallen tree was located on the beach toe parallel to the cliff (Figure 2b). This single tree acted as cliff protection by shielding talus and beach material from erosion. It is not clear whether the position of the tree is a result of natural processes or human intervention. The location of Study Area 1 in the national park should exclude any type of human intervention in the natural environment. However, wooden log pilings were found in front of the coastline (Figure 3). This small, definitely anthropogenic breakwater structure is commonly used for cliff protection. Any type of obstacle located in front of a cliff decreases the wave power, which in turn decreases the erosive influence of the wave. Thus, the presence of these structures is expected to significantly affect the results. Presumably, the location of the city of Sassnitz and the considerable tourist traffic on the beach area justify the presence of this structure. It is likely that, without protection, narrow beach sections 3–10 m in width would disappear completely, which would make the area less attractive for tourism.

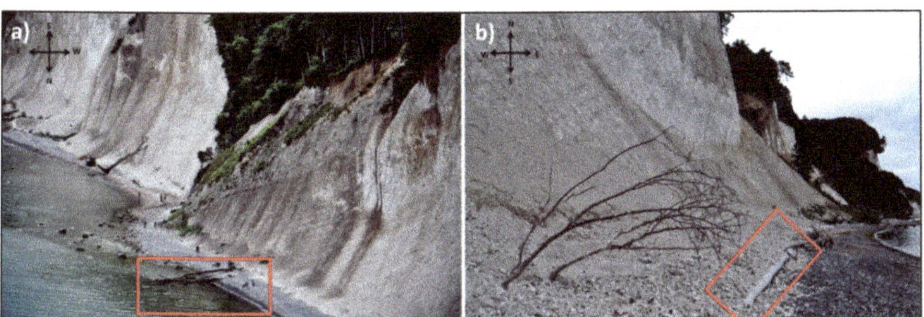

Figure 2. Inventory base of Study Area 1: (**a**) trees lying across the beach; and (**b**) tree located parallel to the cliff.

Figure 3. Inventory base of Study Area 2, showing log pilings.

2.3. Methodology

Due to representation of the surface and features by LiDAR raw data as a point cloud, "hydrographically coherent surfaces" such as river banks or shorelines are not accurately represented [28]. To provide accurate understanding of the data, a good digital elevation model (DEM) and analysis are essential for recognizing morphologically important features [29]. Depending on the study objective, four major line indicators are extracted from LiDAR-derived DEMs: shorelines, cliff bases with talus, vertical cliffs without talus, and cliff crests.

In both datasets, the shoreline was created on the basis of an isoline 0.3 m above sea level. Owing to the low slope and almost constant values below 0.3 m, this value is considered to be a practical border for the division of beach and sea level data. However, the 0.3 m isoline indicates the presence of debris and large stones on some parts of the coast, resulting in almost closed curves. For these areas, the isoline was manually smoothed. Supervised changes were made to hill shade maps in which two rasters were created: the first had an azimuth of 315° and an attitude of 45°, and the second had an azimuth and attitude both equal to 45°.

For sea level recognition, the aspect raster was used as the water surface because the presence of waves and ripples reveal significant aspect changes in small areas. Each wave or ripple consists of one crest and two depressions (wave troughs) that divide the wave surface aspect in at least two different directions. Therefore, it was easy to distinguish the water surface from the smother surface of the beach, which has smaller aspect changes, or even from single waves.

The process of identifying the cliff base line was more challenging than that used to distinguish the shoreline. Numerous studies on cliffs assume manual delineation of the cliff baseline by relying mainly on aerial photographs, topographic maps, and in situ surveys [30–33]. Some examples of automatic delineation were reported by Palaseanu-Lovejoy and others [34], who used generalized coastal shoreline vectors, and by Terefenko and others [33], who considered a simplified methodology of rapid changes in altitude. In our work, cliff base line evaluation was done based on slope and hill shade analysis. The slopes of cliffs are greater than those of beaches; therefore, areas adjacent to the coastline showing sudden and significant changes in the slope were classified as the base of the cliff. However, in some parts of a cliff, talus material concentrated in front on the cliff toe changes the morphology by smoothing the slope close to the beach. These areas are recognized in LiDAR data as characteristic cone embankments, sometimes with the flat top surfaces showing a step-like shape. On parts of a cliff containing obvious talus material, a second break line referred to as a cliff base with talus, with an azimuth of 315° and a horizon of 44°, was created on the basis of the slope and hill shade map. To ensure the presence of cone or step-like talus, additional profiles of the cliff were created for problematic areas. If the profiles indicated the presence of this form, a second break line was created. Moreover, evaluation of the talus material for the dataset from 2012 is supported by notes from field work and onsite photo documentation. The presence of this talus material at the base of a cliff is not guaranteed or continuous. Areas in which the beach is separated from the cliff's surface by a sudden and high slope were identified by creating a third break line known as a vertical cliff base.

Owing to the relatively high resistance of the rock-building chalk cliffs in Jasmund National Park, vertical cliffs forming steep surfaces are easily recognized. In the case of a cliff crest, the same technique as that used in creating the second and third break lines was employed, in which the cliff rest border was determined in areas of high steepness. The tableland close to the cliff is flat, which facilitates recognition even without comparison with a hillshade raster. Mapping the cliff crest and its migration over time is one of the most common methodologies used for investigating cliff recession with both manual hand-digitized procedures [35] and automatic extraction [34].

Finally, to explore the changes in the entire cliff system that occurred between LiDAR surveys, line indicator migration and volumetric changes were analyzed. The total volume of eroded material was calculated on the basis of the indicated differences between surveys.

The average values of cliff recession were calculated on the basis of differences in the x direction (in meters) between two mean center points, where one line represents one feature in both datasets.

The mean center point identifies the geographic center of the line based on average x and y values and thus is assumed to effectively represent the average line indicator location.

The differences in distance between two central points on each line indicator reveal the shift distance. The results indicate the total shift per analyzed time period of six years and as the average value of line shift in meters per year.

The wave capacity for eroding a cliff depends on the power of the wave action on the cliff surface. The energy and power of waves depend on the wave incident angle, in which a smaller angle is related to higher power in the wave reaching the cliff. This relation among wave erosion capacity, its power, and its incident angle is assumed to be an indicator of cliff exposure. The calculation of cliff exposure to wave action consists of four major phases: (a) calculation of the deep water wave parameters based on data from buoys; (b) calculation of the wave parameters for the near-shore location, at 5 m in depth; (c) calculation of the cliff exposure based on the wave refraction angle; and (d) calculation of the total wave power for a wave acting on 1 m of cliff, measured in watts.

The height and direction of refracted wave propagation are functions of the propagation angle, depth, height, and period of wave initiation for offshore wave. Following this relation, the wave refraction and height for near-shore conditions were calculated using offshore wave data calculated for waves affecting the shoreline. The calculation was performed using MATLAB software including equations for wave refraction and height modification derived from the Coastal Engineering Manual of the U.S. Army Corps of Engineers [36].

For evaluating the offshore wave parameters affecting the Jasmund National Park coastline, a wind rose for the wave period (Tp) and wave significant height (Hs) was created. The final wave parameters were calculated only for waves with significant Tp and Hs that affect the coastline.

After determining the offshore wave parameters, including Tp, Hs, and wave angle oriented to the coastline calculated for each coastline segment, it is possible to calculate the refraction of the wave at the 5 m depth by using Snell's law [36]. The height of the near shore wave was calculated for each part of the coast line based on Equation (1), where Ho is offshore wave height, Ks is shoaling coefficient, and Kr refraction coefficient.

$$Hnearshore = Ho\ Ks\ Kr \tag{1}$$

Wave refraction depends on offshore wave parameters as well as the coastline orientation. Therefore, the wave refraction angle will change along the coastline depending on its angle. The phenomenon of wave refraction angle change according to variation in the mean shoreline angle for both study areas is illustrated in Figure 4. Even with the same parameters or offshore wave, the wave angle after refraction at Study Area 1 was different from that at Study Area 2.

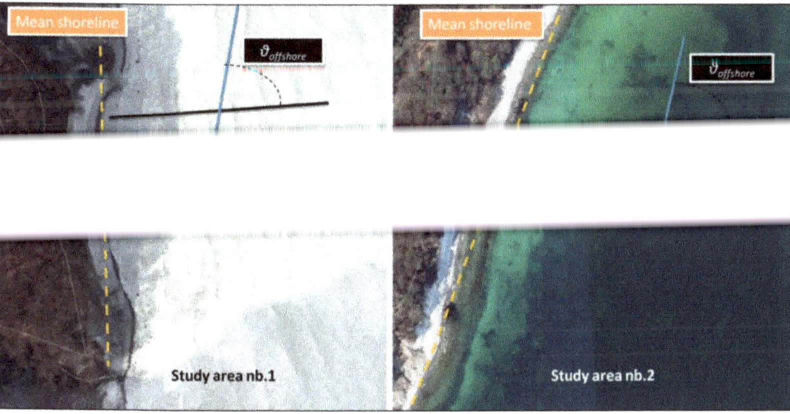

Figure 4. Wave refraction changes according to mean shoreline angle.

The wave power was then calculated based on the results of wave refraction and near-shore wave height. This parameter is related to the wave energy and wave celerity for the wave group and wave incident angle.

The cliff exposure affected by wave action was evaluated according to the wattage of the wave acting on 1 m of the cliff, where greater power is related to more cliff exposure. As a result, five classes of wave exposure were identified using equal intervals between maximum and minimum observed values, as presented in Table 1.

Table 1. Cliff Exposure Classification Based on Wave Power.

Class	Exposure	Wave Power Range (W/m)
1	Very exposed	2460.1–2780
2	Exposed	2140.1–2460
3	Moderately exposed	1820.1–2140
4	Moderately sheltered	1500.1–1820
5	Sheltered	1178–1500

Finally, to rate the chemical and mechanical resistance against the wave action creating mechanical and chemical erosion, rock samples were collected and analyzed. The content of the calcium carbonate ($CaCO_3$) in the sample was used to evaluate of the rock's chemical resistance: The rock's resistance decreases with an increase in $CaCO_3$. For this analysis, the commonly used Scheibler calcimeter was employed owing to its ease of use and rapid results [37].

The analysis revealed that the studied cliff sections include dolomitized chalk. The presence of dolomite is known to significantly increase the strength of chalk [38]. Dolomite is formed from the carbonate rocks during the dolomitization process under high temperature and high pressure. As a result of this metamorphosis, rocks lose volume but gain mechanical resistance [39]. Even though calcite and dolomite have similar chemical and structural properties, calcite undergoes plastic deformation, while dolomite is "brittle and very strong" [38]. Cleven [38] proved that, although dolomite effectively increases the "structural capacity of the rock to compress and store", the calcite-bearing network is still dominant.

Considering these factors, the dolomite content in the rock sample, calculated during the calcimeter test as part of the $CaCO_3$, was assumed to be an indicator of the rock's mechanical resistance.

3. Results and Discussion

3.1. Volumetric Changes

The calculations of total volume change between 2007 and 2012 revealed progressive erosion in both analyzed study areas. During this period, the total volume of the cliff measured in 2007 was decreased by 6% (form 322.566 to 303.470 m^3) and 2.6% (from 638.649 to 621.801 m^3) in Study Areas 1 and 2, respectively. Therefore, the rate of coastline erosion in Study Area 1, at 0.01 m^3/year, was faster than in Study Area 2, at only 0.004 m^3/year.

The areas of the highest cliff activity were evaluated based on the volume changes occurring between 2007 and 2012. In Study Area 1, the amount of material erosion in the upper part of the cliff averaged 2.2–9 m^3/0.25 m^2, and the maximum value was 36 m^3/m^2. In the middle and bottom parts, the average was 0.01–2.1 m^3/0.25 m^2, and the maximum value was 9.6 m^3/m^2. It is worth noting that, in addition to the differences in activity across the cliff profile, an increasing tendency for erosion was evident on the northern part of the cliff, whereas the southern part of the study area is characterized by smaller volume changes and thus lower cliff activity.

The cliff section in Study Area 2 revealed significant changes in the material volume at the bottom of the cliff, with average volume changes of 3.6–6.9 m^3/0.25 m^2 and a maximum erosion value of 27 m^3/m^2. In the upper part of the cliff, only small changes in volume were found, with averages of

0.01–2.8 $m^3/0.25\ m^2$ and a maximum value of 11.2 m^3/m^2. In addition to the areas of high volume change in the lower part of the cliff, two other areas were shown to be significantly eroded. The first is the southern part of the cliff near the crest, and the second area is in the middle of the cliff in the northern part. This indicates susceptibility to linear erosion, which occurs at the bottom to the upper part of the cliff.

The erosion–accumulation ratio of the vertical cliff surface in both study areas indicates significant erosion. Slightly more than 98% and 92.5% of the cliff surface measured in 2007 had been eroded by 2012 in Study Areas 1 and 2, respectively. However, the progressive cliff erosion calculated in both study areas does not always correspond with the beach–talus development (Table 2).

Table 2. Volumetric Changes and Dominant Morphological Processes of Vertical Cliff, Talus Material, and Beach. The 95% Confidence Interval was Used as a Critical Threshold for Volumetric Calculations.

			Volume (m^3)	Area m^2	Volume Change (m^2)	Percentage of Total Area	Ratio (Erosion/Accumulation)
AREA 1	Cliff	Erosion	19260.35	3503.50	5.50	98.10	51.52
		Accumulation	15.60	68.00	0.23	1.90	
	Talus	Erosion	238.73	289.50	0.82	38.96	0.64
		Accumulation	251.36	53.50	0.55	61.04	
	Beach	Erosion	33.64	443.50	0.08	27.07	0.37
		Accumulation	238.73	1194.75	0.20	72.93	
AREA 2	Cliff	Erosion	15167.41	12314.75	1.23	92.52	12.38
		Accumulation	315.34	995.00	0.32	7.48	
	Talus	Erosion	818.93	942.25	0.87	91.26	10.44
		Accumulation	6.08	90.25	0.07	8.74	
	Beach	Erosion	2768.97	3247.25	0.85	92.98	13.25
		Accumulation	42.89	245.00	0.18	7.02	

According to the erosion–accumulation rate, accumulation was observed in Study Area 1, particularly in the beach and talus areas. Material accumulation on the cliff toe is a direct effect of mass movement of the upper part of the cliff. After a cliff collapse, the beach area is enriched by fresh debris material that provides additional cliff protection against wave erosion. In the case of Study Area 2, erosion was predominant across the entire cliff–beach profile.

Analysis of the recession rate, calculated separately for each part of the cliff, revealed that the cliff retreated at varying speeds across the profile (Table 3). In Study Area 1, the difference between the recession rates of the cliff crest and base was insignificant, at 0.004 m/year, and an increasing tendency was revealed in the top part. In Study Area 2, the opposite occurred such that the cliff toe retreated more quickly than the crest. In addition, the difference in erosion between the cliff top and toe was significant, at 0.46 m/year. These results are within the average values presented in various scientific works [40–43].

Cliff crest	2.70	0.45	0.63	0.11
Cliff base (vertical)	2.6	0.43	1.15	0.19
Cliff base (with talus)	2.45	0.41	3.43	0.57
Average	2.58	0.43	1.74	0.29

3.2. Exposure to Erosion

The analysis of the amount of cliff exposed to wave action revealed relatively high exposure in both study areas. The average values of wave power received by the cliffs in Study Areas 1 and 2 were 2665.6 and 2535 W/m, respectively. With such a high value of wave power acting per cliff unit, the influence of waves in the cliff recession processes is expected to be significant.

Study Area 1 has greater exposure to wave action than that in Study Area 2. In addition, the wave power amplitude in Study Area 2 is roughly three times smaller than in Study Area 1, with values of 1601.5 and 521.1 W/m, respectively. This means that the coastline of Study Area 2 is more diverse, and the refracted waves approach it from different directions with varied power. Finally, the cliff parts of Study Area 1 were classified as exposed and highly exposed to wave action (Figure 5), whereas those in Study Area 2 included all classes, from sheltered to very exposed.

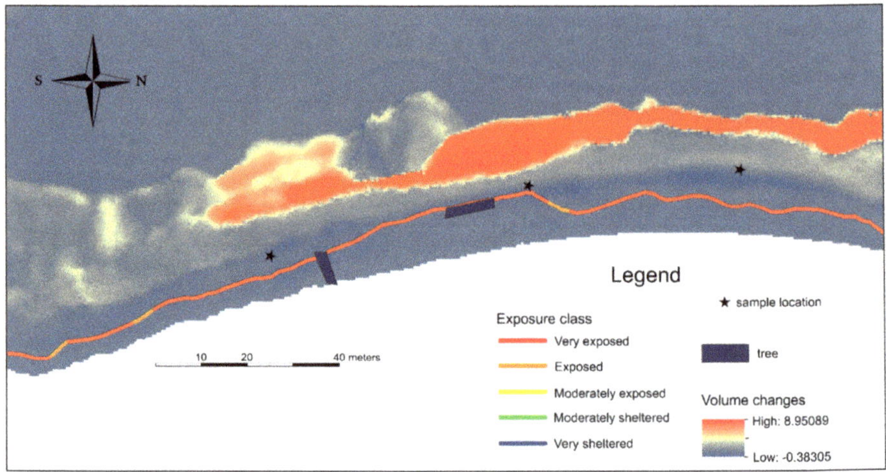

Figure 5. Cliff exposure to wave action corresponding to volumetric changes in Study Area 1.

3.3. Rock Resistance

The chemical resistance of the cliff in Study Area 1 decreases from south to north, although the largest variance between measured values was relatively small, at 8%. This result confirms a homogeneous chalk structure with variations in $CaCO_3$ content occurring along the cliff.

In Study Area 2, the resistance increases along the cliff from southwest to northeast. However, an extremely high $CaCO_3$ value in one of the samples, at 99%, significantly influenced the trend. Even though the fluctuations in $CaCO_3$ content in the cliff indicate geological differences, the difference between the maximum and minimum content was only 5%. Therefore, the cliff is assumed to be homogeneous.

The measurements of mechanical rock resistance revealed large differences between both study sides. The cliff in Study Area 1 is composed mostly of pure chalk with a high $CaCO_3$ content and no dolomitization and thus has low resistance to mechanical wave action. Only one cliff part, located in the northern area, revealed the presence of dolomite and is considered to be moderately resistant. The cliff in Study Area 2 is classified as being highly resistant to mechanical action, designated as Class 3, with an average value of [illegible] dolomite content and a northward trend for decreased resistance.

3.4. Correspondence to Morphological Changes

As described above, material accumulation in the beach area is explained by mass movement. However, in the analyzed study areas, slight accumulation was also noticed along the vertical cliff surface. The amount of accumulated material is insignificant and can be explained by the presence of vegetation, particularly near the cliff crest, and material movement along the structural breaks.

This phenomenon is clearly visible in Study Area 2. Large amounts of erosion on the upper part of the cliff were observed in correspondence with changes in the geomorphological structure identified in the same area by Niemeyer and others [44]. These changes structure clearly proceeded downslope during the analysis period, causing significant erosion as well as accumulation in specific parts. In addition, the linear orientation of the high cliff activity and its location over the middle part of the cliff profile in this study area most likely corresponds to a structural break. However, only erosion was observed in this case.

Owing to the small changes in wave power amplitude occurring along the coastline of Study Area 1, most of the cliff sides were classified as very exposed and exposed. The variability of the exposure is very small; thus, it can be assumed that the waves acting on the cliff were similar for all parts of Study Area 1 and had very high intensity. Analysis of the retreat rate also indicated that no significant changes occurred along the cliff profile.

Study Area 1 is an example of a cliff retreat process caused mainly by wave action: high coastal exposure to wave action accelerates cliff erosion. When the cliff toe and beach are eroded, the cliff stability decreases until reaching a critical value, at which point sudden mass movement occurs. Debris material from the collapsed cliff enriches the talus and beach area and provides additional protection against wave erosion. However, owing to the high power of the waves, this material is again eroded and washed offshore by currents, which leaves the cliff vulnerable to erosion.

The higher degree of erosion on cliff crest than that on the base can be easily explained. First, the presence of accumulated material at the cliff base indicates the occurrence of a cliff collapse between 1997 and 2012. According to Gunther and others [23], toppling is the most common type of mass movement in Rugen. During this type of erosion, the top of the cliff experiences the greatest amount of material loss. The volume of the collapsed material can be so high that the material loss after a single mass movement event is greater than that caused by constant erosion occurring at the cliff base. As a result, the average recession rate of the cliff crest will have a higher value than the rate of ongoing cliff base erosion, particularly because it is also enriched by material from the cliff collapse.

In Study Area 2, the area of cliff exposure to wave action corresponds with the areas of greatest volume change between the analyzed years. Cliff sections composed predominantly of exposed and moderately exposed sections, as well as the section classified as sheltered, have not been significantly eroded. On the contrary, small accumulation was observed. Relatively small wave power reaching the cliff has been further reduced by breakwater structures. Presumably, the presence of pilings facilitates material accumulation.

Sections containing predominantly very exposed cliffs appeared to be eroded and showed the highest rate of erosion of all analyzed cliff profiles, at 0.57 m/year. Particularly strong cliff toe erosion is visible owing to the very high degree of cliff exposure to wave action. Erosion of the material on the vertical cliff surface was caused by the presence of structural features.

In Study Area 1, only parts of the area revealed trends of strong cliff erosion that do not correspond with exposure. Owing to the predominance of mostly exposed cliffs, [illegible]

The cliff in Study Area 1 has been categorized as moderately resistant to chemical erosion but not resistant to mechanical erosion. In this case, the downwearing process caused mostly by freshwater action is expected to be slower than the backwearing process resulting in erosion of the cliff toe by wave action. However, the volume changes indicate the exact opposite trend: the erosion is stronger at the cliff top than at the cliff base. Only the northern part of the cliff has high mechanical resistance and low chemical resistance, and the upper part of the cliff has become eroded. This result indicates no connection of the mechanical and chemical resistance to location of the active erosion areas across the cliff profile.

The chemical resistance in the cliff decreases in the northern direction. This trend agrees with the trend of increasing cliff erosion northward. Therefore, it can be assumed that waves are acting on the cliff and causing erosion by both mechanical action depending on the wave power and chemical reaction through dissolution.

Low chemical resistance and high mechanical resistance characterize the cliff in Study Area 2. According to the factors mentioned above, the cliff is expected to be eroded more quickly by the downwearing process at the cliff top than by the mechanical wave action at the cliff base. However, volume changes indicate the exact opposite trend: The erosion is stronger at the cliff bottom. As discussed above, mechanical and chemical rock resistance are not good indicators of downwearing/backwearing processes. In addition, variations in the chemical resistance along the coastline location do not indicate changes in the volume of eroded material. Unfortunately, all samples collected from Study Area 2 correspond only to the second sector of cliff exposure. This fact prevents correlation of cliff resistance and wave influence.

3.5. Model of the Cliff Retreat

Considering the high wave activity and its correspondence with active, eroded areas identified across the cliff–beach profile, a wave-induced cliff recession model with four stages is proposed to explain the mechanisms of cliff erosion of the Jasmund National Park coastline (Figure 6). Stage 1 occurs when the vertical cliff is in equilibrium. No accumulation occurs on the beach and no talus material appears. In this stage, waves act on the cliff surface directly, although only during storm and high-water level events. With the progression of cliff erosion, the cliff base experiences material loss (Stage 2). As a result, the cliff is undercut by wave-induced mechanical and chemical erosion, and the slope stability decreases. In this stage, the cliff base retreats more quickly than the still-balanced cliff crest. After reaching a critical value, the cliff collapses (Stage 3). From a geomorphometry perspective, the upper part of the cliff incurs the greatest loss of material. As a result, the top part of the cliff shows a high retreat rate. Material lost from the upper part of the cliff accumulates at the cliff base as talus material. Because the waves now act directly on the accumulated talus material, it provides the vertical cliff with additional protection from further erosion. In time, talus material is eroded (Stage 4) until it is completely removed or transported offshore, and the cliff returns to a vertical profile with slope equilibrium (Stage 1).

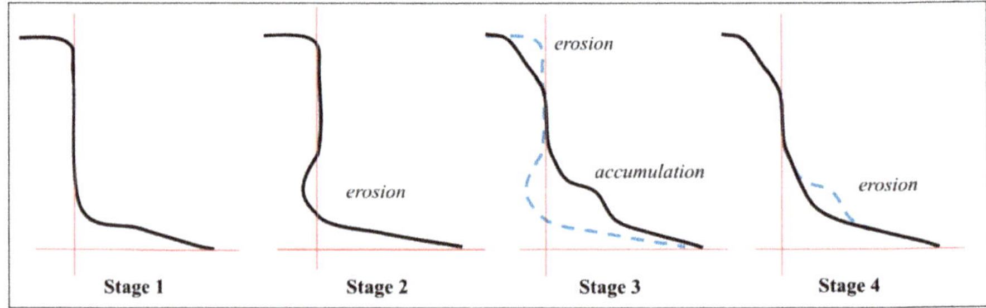

Figure 6. Model of cliff retreat induced by wave action.

According to the described wave-induced cliff retreat model, Study Area 1 is at Stage 4. The significant material accumulation on the cliff base and the large material loss at the cliff crest indicate the occurrence of a recent cliff collapse, most likely by toppling. According to the model, the cliff should be in Stage 3; however, the erosion–accumulation ratio calculated for the talus–beach area indicates accumulation as the dominant process, with small amounts of erosion present. Finally, the amount of material accumulated on the talus–beach area is not equal to the amount of eroded material from the upper part of the cliff. This indicates that progressive erosion of the accumulated beach and talus material has already begun, and the cliff in Study Area 1 is at Stage 4.

According to the observed progressive erosion of the cliff base and thus the much faster retreat rate of the cliff bottom, the cliff in Study Area 2 is at Stage 2, characterized by progressive erosion. The lack of material accumulation at the bottom of the cliff can be explained by the low rate of retreat, at 0.004 m/year. The erosion process is not as fast as that in the case of Study Area 1; thus, mass movement likely occurs less often in this area. Therefore, the beach is not supplied by the material from the cliff.

Assuming the factors described above and considering such large wave power acting on the cliff, the changes along the cliff are expected to be similar for the wave influenced model of cliff erosion. However, in Study Area 1, although the cliff is affected the waves with similar power, a trend of high material erosion on the north part of the cliff is evident. Therefore, in addition to the predominant influence of the wave action indicated by similar retreat rates along the cliff profile, fast erosion rates, and high wave power, other influencing factors that facilitate erosion cannot be neglected. After incorporating the resistance of the cliff rocks, it appeared that the faster erosion in the northern sector was induced by lower chemical resistance. As a result, it is assumed that the wave action is causing mechanical and chemical erosion through hydraulic and dissolution processes, respectively. However, dissolution at the cliff bottom can also be an effect of freshwater action through which precipitation is concentrated between the flint pebbles accumulated on the beach or on concave parts of the talus material. Nevertheless, this contribution to cliff erosion in Study Area 1 must be further evaluated and was thus excluded in the final cliff retreat model developed in this study.

The spatial distribution of the cliff erosion in Study Area 1 appears to agree with the map of cliff vulnerability to collapse evaluated by Gunther and others [23], based on the kinematic cliff slope stability. According to their evaluation, the northern part of the cliff in Study Area 1 is particularly prone to cliff collapse. The present study shows that the northern section indeed experiences mass movement. The cliffs in Study Area 2 were also evaluated by Gunther and others [23] and were determined to have moderate to low susceptibility to cliff collapse. In the present study, the same cliff revealed a low rate of retreat, indicating a low degree of cliff exposure to wave action. Thus, the low wave erosion identified in the present study is reflected in the low susceptibility for cliff collapse [23].

The model results of cliff retreat caused by waves presented in this study combined with those of erosion susceptibility based on cliff stability reported by Gunther and others [23] have proved the causes and effects of the cliff erosion mechanism. Although the results of both models indicate areas prone to erosion, the topic is evaluated from different perspectives. The results demonstrate that cliff slope stability is an effect of cliff erosion caused primarily by wave action.

4. Conclusions

The number of studies on remote coastal morphology have increased in recent years and include high temporal resolution in terrestrial laser scanning, structure-from-motion photogrammetry, and/or video imaging survey. Our study demonstrates a clear advantage of using airborne LiDAR scanning for studies of rocky cliffs characterized by slow rates of erosion. By using this technique, several factors such as wave action and [illegible]

variables of each segment along the profile. Wave action is classified as the primary force in cliff erosion. According to the differences in the rate of retreat based on features along a cliff–beach profile, the wave erosion-dominated cliff retreat model with the following four stages effectively represents the analyzed study areas in Jasmund National Park.

- Stage 1: Cliff in equilibrium. The vertical cliff has balanced stability, and no talus or material is accumulated on the beach. Waves acts directly on the cliff surface.

- Stage 2: Progressive erosion. Erosion of the cliff toe is caused by wave action. The cliff stability is decreased, and a high degree of erosion occurs along the cliff base.
- Stage 3: Cliff collapse, most likely by toppling. The beach is enriched by material removed from the upper part of the cliff. High erosion of the cliff crest and accumulation on the cliff base occur.
- Stage 4: Talus reduction: Waves act on talus and beach areas to remove accumulated material. Progressive erosion of the talus and beach material occurs until complete removal.

For both analyzed study areas, volume changes occurring in the cliff–beach profile between 2007 and 2012 show strong correlation to cliff exposure by wave action. The cliff of Study Area 1, which is highly exposed, retreats more quickly in comparison to the moderately exposed parts of the cliff located in Study Area 2. As a result, the rate of retreat of the cliff in Study Area 1 is 0.43 m/year, whereas that in Study Area 2 is only 0.29 m/year. Since 2007, 6% of the cliff in Study Area 1 and 2.6% of that in Study Area 2 have been eroded. All mentioned values of cliff retreat are within the retreat range typical for chalk cliffs, between 1 cm and 10 m per year.

The predominance of accumulation processes occurring on the beach–talus sector in Study Area 1 and the significant material loss from the cliff crest indicate the recent occurrence of a cliff collapse.

Although the cliff located in Study Area 2 is classified as moderately exposed, one analyzed part indicates a high degree of cliff erosion. This anomaly, which does not correspond to a high degree of cliff exposure, is attributed to changes in geomorphological structures that descend on the cliff, revealing the high importance of cliff structural breaks in cliff retreat–erosion processes.

The single trees present on the beach side were not found to be important obstacles for cliff erosion. These features are likely not large enough to cause a significantly reduction in the power of waves affecting the cliff. Only breakwater structures in the form of log pilings are likely to decrease the wave power and thus the degree of cliff erosion. However, owing to the ubiquitous presence of such obstacles along the shore in Study Area 2, its influence was difficult to evaluate.

Owing to the homogeneous geological structure of the analyzed cliffs and thus the very small differences in chemical and mechanical cliff resistance, this parameter did not closely follow the trends of cliff erosion revealed by volume changes during the period of study. Only one correlation was found in Study Area 1 between the chemical resistance of the cliff and the erosion trend; thus, the influence of chemical erosion caused by waves was proved.

The model results indicate that the cliffs in both study areas are in different stages of cliff erosion. Study Area 1 shows material accumulation on the cliff bottom and significant material loss in upper part of the cliff as well as the onset of talus material erosion; therefore, it in Stage 4: talus material reduction. However, the cliffs in Study Area 2 show progressive erosion on the cliff base; therefore, it is in Stage 2: progressive erosion.

It has been shown that wave erosion is the most active ongoing process triggering the retreat of the rocky coastline in Jasmund National Park. However, two additional factors cannot be neglected: both structural breaks and chemical erosion were found to play significant roles in the cliff retreat model.

Author Contributions: Conceptualization, D.Z.W.; Methodology, D.Z.W., P.T. and A.K.; Software, D.Z.W. and P.T.; Validation, D.Z.W. and P.T.; Formal analysis, D.Z.W.; Investigation, D.Z.W. and P.T.; Resources, D.Z.W.; Data curation, D.Z.W., P.T. and A.K.; Writing original draft Preparation, D.Z.W. and P.T.; Writing-Review & Editing, P.T.; Visualization, D.Z.W. and P.T.; Supervision, P.T.

Funding: This research received no external funding.

Conflicts of Interest: The authors declare no conflicts of interest.

References

1. Naylor, L.A.; Stephenson, W.J.; Trenhaile, A.S. Rock coast geomorphology: Recent advances and future reasearch directions. *Geomorphology* **2009**, *114*, 3–11. [CrossRef]

2. Furmańczyk, K.; Andrzejewski, P.; Benedyczak, R.; Bugajny, N.; Cieszyński, Ł.; Dudzińska-Nowak, J.; Giza, A.; Paprotny, D.; Terefenko, P.; Zawiślak, T. Recording of selected effects and hazards caused by current and expected storm events in the Baltic Sea coastal zone. *J. Coast. Res.* **2014**, *70*, 338–342. [CrossRef]
3. Paprotny, D.; Andrzejewski, P.; Terefenko, P.; Furmańczyk, K. Application of Empirical Wave Run-Up Formulas to the Polish Baltic Sea Coast. *PLoS ONE* **2014**, *9*, e105437. [CrossRef] [PubMed]
4. Paprotny, D.; Terefenko, P. New estimates of potential impacts of sea level rise and coastal floods in Poland. *Nat. Hazards* **2017**, *85*, 1249–1277. [CrossRef]
5. Aniśkiewicz, P.; Łonyszyn, P.; Furmańczyk, K.; Terefenko, P. Application of Statistical Methods to Predict Beach Inundation at the Polish Baltic Sea Coast. In *Interdisciplinary Approaches for Sustainable Development Goals: GeoPlanet: Earth and Plan*; Zielinski, T., Sagan, I., Surosz, W., Eds.; Springer: Berlin, Germany, 2017; pp. 73–92.
6. Deng, J.; Harff, J.; Zhang, W.; Schneider, R.; Dudzińska-Nowak, J.; Giza, A.; Terefenko, P.; Furmańczyk, K. The Dynamic Equilibrium Shore Model for the Reconstruction and Future Projection of Coastal Morphodynamics. In *Coastline Changes of the Baltic Sea from South to East: Past and Future Projection*; Harff, J., Furmańczyk, K., VonStorch, H., Eds.; Springer: Berlin, Germany, 2017; pp. 87–106.
7. Novikova, A.; Belova, N.; Baranskaya, A.; Aleksyutina, D.; Maslakov, A.; Zelenin, E.; Shabanova, N.; Ogorodov, S. Dynamics of Permafrost Coasts of Baydaratskaya Bay (Kara Sea) Based on Multi-Temporal Remote Sensing Data. *Remote Sens.* **2018**, *10*, 1481. [CrossRef]
8. Uścinowicz, G.; Szarafin, T. Short-term prognosis of development of barrier-type coasts (Southern Baltic Sea). *Ocean Coast. Manag.* **2018**, *165*, 258–267. [CrossRef]
9. De Sanjosé Blasco, J.J.; Gómez-Lende, M.; Sánchez-Fernández, M.; Serrano-Cañadas, E. Monitoring Retreat of Coastal Sandy Systems Using Geomatics Techniques: Somo Beach (Cantabrian Coast, Spain, 1875–2017). *Remote Sens.* **2018**, *10*, 1500. [CrossRef]
10. Andriani, G.F.; Walsh, N. Rocky coast geomorphology and erosional processes: A case study along the Murgia coastline South of Barri, Apulia—SE Italy. *Geomorphology* **2006**, *87*, 224–238. [CrossRef]
11. Terefenko, P.; Zelaya Wziątek, D.; Dalyot, S.; Boski, T.; Pinheiro Lima-Filho, F. A High-Precision LiDAR-Based Method for Surveying and Classifying Coastal Notches. *ISPRS Int. J. Geo-Inf.* **2018**, *7*, 295. [CrossRef]
12. Cherith, M.; Robinson, D. Chalk coast dynamics: Implications for understanding rock coast evolution. *Earth Sci. Rev.* **2011**, *109*, 63–73.
13. Wziątek, D.; Vousdoukas, M.V.; Terefenko, P. Wave-cut notches along the Algarve coast, S. Portugal: Characteristics and formation mechanisms. *J. Coast. Res.* **2011**, *64*, 855–859.
14. Terefenko, P.; Terefenko, O. Determining the role of exposure, wave force, and rock chemical resistance in marine notch development. *J. Coast. Res.* **2014**, *70*, 706–711. [CrossRef]
15. Lageat, Y.; Henaff, A.; Costa, S. The retreat of the chalk cliffs of the Pays-de-Caux (France): Erosion processes and patterns. *Z. Fur Geomorphol.* **2006**, *44*, 183–197.
16. Terefenko, P.; Giza, A.; Paprotny, D.; Kubicki, A.; Winowski, M. Cliff retreat induced by series of storms at Międzyzdroje (Poland). *J. Coast. Res.* **2018**, *85*, 181–185. [CrossRef]
17. Fonstad, M.A.; Dietrich, J.T.; Courville, B.C.; Jensen, J.L.; Carbonneau, P.E. Topographic structure from motion: A new development in photogrammetric measurement. *Earth Surf. Process Landf.* **2013**, *38*, 421–430. [CrossRef]
18. James, M.R.; Quinton, J.N. Ultra-rapid topographic surveying for complex environments: The hand-held mobile laser scanner [illegible]

[illegible]
example from the southern Baltic Sea coast. *Baltica* **2019**, *32*, 10–21.
21. Sallenger, A.H., Jr.; Krabill, W.B.; Swift, R.N.; Brock, J.; List, J.; Hansen, M.; Holman, R.A.; Manizade, S.; Sontag, J.; Meredith, A.; et al. Evaluation of Airborne Topographic Lidar for Quantifying Beach Changes. *J. Coast. Res.* **2003**, *19*, 125–133.
22. Schwarzen, K.; Horst, S. Book section: Germany. In *Encyclopedia of the World's Coastal Landforms*; Bird, E.C.F., Ed.; Springer: Berlin, Germany, 2010; Volume 1.
23. Gunther, A.; Thiel, C. Combined rock slope stability and shallow landslide susceptibility assessment of the Jasmund cliff area (Rugen Island, Germany). *Nat. Hazard Earth Syst.* **2009**, *9*, 687–698. [CrossRef]

24. Buckley, S.J.; Howell, J.A.; Enge, H.D.; Kurz, T.H. Terrestrial laser scanning in geology: Data acquisition, processing and accuracy. *J. Geol. Soc.* **2008**, *165*, 625–638. [CrossRef]
25. Vousdoukas, M.; Kirupakaramoorthy, T.; Oumeraci, H.; de la Torre, M.; Wübbold, F.; Wagner, B.; Schimmels, S. The role of combined laser scanning and video techniques in monitoring wave-by-wave swash zone processes. *Coast. Eng.* **2014**, *83*, 150–165. [CrossRef]
26. Almeida, L.P.; Masselink, G.; Russell, P.E.; Davidson, M.A. Observations of gravel beach dynamics during high energy wave conditions using a laser scanner. *Geomorphology* **2015**, *228*, 15–27. [CrossRef]
27. Korzeniowska, K.; Łącka, M. Generating DEM from LIDAR data-comparison of availabe software tools. Archives of Photogrammetry. *Cartogr. Remote Sens.* **2011**, *22*, 271–284.
28. Campbell, J.B.; Wynne, R.H. *Introduction to Remote Sensing*; The Guilford Press: New York, NY, USA, 2011.
29. Hengl, T.; Reuter, H.I. *Geomorphometry: Concepts, Software, Application*; Elsevier: Amsterdam, The Netherlands, 2009.
30. Rosser, N.J.; Brain, M.J.; Petley, D.N.; Lim, M.; Norman, E.C. Coastline retreat via progressive failure of rocky coastal cliffs. *Geology* **2013**, *41*, 939–942. [CrossRef]
31. Johnstone, E.; Raymond, J.; Olsen, J.M.; Driscoll, N. Morphological Expressions of Coastal Cliff Erosion Processes in San Diego County. *J. Coast. Res.* **2016**, *76*, 174–184. [CrossRef]
32. Warrick, J.A.; Ritchie, A.C.; Adelman, G.; Adelman, K.; Limber, P.W. New techniques to measure cliff change form historical oblique aerial photographs and structure-for-motion photogrammetry. *J. Coast. Res.* **2017**, *33*, 39–55. [CrossRef]
33. Terefenko, P.; Paprotny, D.; Giza, A.; Morales-Nápoles, O.; Kubicki, A.; Walczakiewicz, S. Monitoring Cliff Erosion with LiDAR Surveys and Bayesian Network-based Data Analysis. *Remote Sens.* **2019**, *11*, 843. [CrossRef]
34. Palaseanu-Lovejoy, M.; Danielson, J.; Thatcher, C.; Foxgrover, A.; Barnard, P.; Brock, J.; Young, A. Automatic Delineation of Seacliff Limits using Lidar-derived High-resolution DEMs in Southern California. *J. Coast. Res.* **2016**, *76*, 162–173. [CrossRef]
35. Kostrzewski, A.; Zwoliński, Z.; Winowski, M.; Tylkowski, J.; Samołyk, M. Cliff top recession rate and cliff hazards for the sea coast of Wolin Island (Southern Baltic). *Baltica* **2015**, *28*, 109–120. [CrossRef]
36. U.S. Army Corps of Engineers. *Coastal. Engineering Manual—Chapter 3: Estimation of Nearshore Waves*; Corps of Engineers U.S. Army: Washington, DC, USA, 2002.
37. Fonnesbeck, B.B.; Boettinger, J.L.; Lawley, J.R. Improving a Simple Pressure-Calcimeter Method for Inorganic Carbon Analysis. *Soil Sci. Soc. Am. J.* **2012**, *77*, 1553. [CrossRef]
38. Cleven, R.N. *The Role of Dolomite Content on the Mechanical Strength and Failure Mechanisms in Dolomite-Limestone Composites*; University of British Colombia: Vancouver, BC, Canada, 2008.
39. Czubla, P.; Mizerski, W.; Świerczewska-Gładysz, E. *Przewodnik Do Ćwiczeń Z Geologii*; Wydawnictwo Naukowe PWN: Warszawa, The Netherlands, 2005.
40. Masselink, G.; Hughes, M.G. *Introduction to Coast. Processes & Geomorphology*; Hodder Education: London, UK, 2003.
41. Dornbusch, U.; Robinson, D.A.; Moses, C.; Williams, R.; Costa, S. Retreat of Chalk cliffs in the eastern English Channel during the last century. *J. Maps* **2006**, *2*, 71–78. [CrossRef]
42. Letortu, P.; Costa, S.; Maquaire, O.; Delacourt, C.; Augereau, E.; Davidson, R.; Suanez, S.; Nabucet, J. Retreat rates, modalities and agents responsible for erosion along the coastal chalk cliffs of Upper Normandy: The contribution of terrestrial laser scanning. *Geomorphology* **2015**, *245*, 3–14. [CrossRef]
43. El Khattabi, J.; Carlier, E.; Louche, B. The Effect of Rock Collapse on Coastal Cliff Retreat along the Chalk Cliffs of Northern France. *J. Coast. Res.* **2018**, *341*, 136–150. [CrossRef]
44. Niemeyer, J.; Rottensteiner, F.; Kuhn, F.; Sorgel, U. Extraktion geologisch relevanter Strukturen auf Rügen in Laserscanner-Daten. *DGPF Tag. Wien* **2010**, *19*, 298–307.

Article

Application of Multiple Geomatic Techniques for Coastline Retreat Analysis: The Case of Gerra Beach (Cantabrian Coast, Spain)

José Juan de Sanjosé Blasco [1,*], Enrique Serrano-Cañadas [2], Manuel Sánchez-Fernández [1], Manuel Gómez-Lende [2] and Paula Redweik [3]

1 Department of Graphic Expression, INTERRA Research Institute for Sustainable Territorial Development, NEXUS Research Group Engineering, Territory and Heritage, University of Extremadura, Avenida de la Universidad s/n, 10003 Cáceres, Spain; msf@unex.es
2 Department of Geography, PANGEA Research Group Natural Heritage and Applied Geography, University of Valladolid, Plaza del Campus s/n, 47011 Valladolid, Spain; serranoe@fyl.uva.es (E.S.-C.); manuelglende@hotmail.com (M.G.-L.)
3 Department of Engenharia Geográfica, Geofísica e Energia and Instituto Dom Luiz, Faculdade de Ciências, University of Lisbon, 1649-004 Lisbon, Portugal; pmredweik@fc.ul.pt
* Correspondence: jjblasco@unex.es

Received: 12 October 2020; Accepted: 7 November 2020; Published: 9 November 2020 

Abstract: The beaches of the Cantabrian coast (northern Spain) are exposed to strong winter storms that cause the coastline to recede. In this article, the coastal retreat of the Gerra beach (Cantabria) is analyzed through a diachronic study using the following different geomatic techniques: orthophotography of the year 1956; photogrammetric flights from 2001, 2005, 2010, 2014, 2017; Light Detection and Ranging (LiDAR) survey from August 2012; Unmanned Aerial Vehicle (UAV) survey from November 2018; and terrestrial laser scanner (TLS) through two dates per year (spring and fall) from April 2012 to April 2020. With the 17 observations of TLS, differences in volume of the beach and the sea cliff are determined during the winter (November–April) and summer (May–October) periods, searching their relationship with the storms in this eight-year period (2012–2020). From the results of this investigation it can be concluded that the retreat of the base of the cliff is insignificant, but this is not the case for the top of the cliff and for the existing beaches in the Cantabrian Sea where the retreat is evident. The retreat of the cliff top line in Gerra beach, between 1956 and 2020 has shown values greater than 40 m. The retreat in other beaches of the Cantabrian Sea, in the same period, has been more than 200 m. With our measurements, investigations carried out on the retreat of the cliffs on the Atlantic coast have been reinforced, where the diversity of the cliff lithology and the aggressive action of the sea (storms) have been responsible for the active erosion on the face cliff. In addition, this research applied geomatic techniques that have appeared commercially during the period (1956–2020), such as aerial photogrammetry, TLS, LiDAR, and UAV and analyzed the results to determine the precision that could be obtained with each method for its application to similar geomorphological structures.

1. Introduction

Coastal changes, at a geological scale, are very slow and the morphology of beaches and cliffs retain features of earlier sea levels, but there are changes, at a human scale, that can be easily detected along the coastline. Two main findings from the study of sea cliff evolution highlight the temporal change in cliff top line recession mode, and the effect of beach sediment at the cliff toe on cliff erosion. [1,2] pointed out that our ability to quantify sea cliff retreat rates and their variability through time was the first step to understanding the sea cliff erosion processes and also the responses to environmental and climate changes. Information about the factors responsible for triggering gravitational landslides (rockfalls, slides, and debris falls) has been provided by high resolution and high frequency monitoring.

Measures on sea cliff recession show distinct behaviors at different geographical locations and geological structures, and the spatial variations of the cliff retreat rates have been explained by changes in the geological structure, cliff collapses, or anthropogenic obstacles [3,4]. Twelve years (1998–2010) of coastal cliff erosion and retreat, measured by airborne light detection and ranging (LiDAR), of the California coast have shown that less than 50% of the coast was active and the mean recent retreat rates were 52–83% lower than mean historical retreat rates. Locations with elevated historical retreat had low levels of recent retreat and locations with elevated recent retreat were preceded by low rates of historical retreat [5]. In addition, in the Mediterranean, recent short-term (i.e., annual to decadal) cliff top retreat rates exceeded longer term rates of "background" (i.e., centennial to millennial) retreat by one to two orders of magnitude [1], however, the authors pointed out that an inherent sampling bias in rate estimates inferred from shorter observation intervals could also have explained such a pattern. In the Atlantic coast, the retreat rates obtained from historical maps, aerial photographs, recent TLS and photogrammetry monitoring were −10 to −50 cm/yr [3]. The behavior of cliffs characterized by landslides on beaches are related to several factors, depending on climate, structural geology, lithology, and sea exposure. On European Atlantic coasts, they are also under the dominant influence of precipitation and the evolution of the groundwater level [3].

Coastline and sea cliff retreats have been checked for the Cantabrian Coast (northern Spain) over the past 40 years [6–12], where major sand changes and cliff failure events have occurred at several points along the beaches and sea cliff systems. Since the mid-twentieth century, storms have been of major importance and are an important public concern, reaching inhabited areas and causing damage to private property on sand systems and cliff tops. Strong wind and storm wave events imply bigger and faster spatial and temporal changes in the cliffs. Retreat and changes occur due to sudden impulses related to storms where interactions between rock strength, structure, lithology and wave action, rainfall, temperature variations, and runoff generate cliff processes such as rock falls, topples, and slides, as well as sand movements [12–15]. These facts are of significant relevance to public authorities because the public safety of residential areas along the cliffs is threatened. Therefore, the processes and causes of sea cliff instability and quick changes must be understood at a detailed scale in order to provide authorities with data-driven models of cliff erosion in time and space [16–18]. The concern about shoreline response to climate change has also increased in the social and scientific environment [19–21].

The applied techniques for costal studies are as diverse as the objectives of the studies. Synthetic aperture radar (SAR) images have been applied for an automatic extraction of the shoreline [22]. For the determination of wave run-up on a beach, video footage from two cameras has been processed [23]. The relevance of the width of the observation time window for monitoring coastal region dynamics has led researchers to resort to very different data originating from older records and newer acquisition techniques. Traditionally, the most common methodologies to evaluate cliff changes and retreat have been the analysis of historical aerial photographs, topographic maps, and survey plans covering several decades. Retreat rates have usually been quantified on the cliff top, yielding incomplete information because the data accuracy of available aerial images has been metric or decimetric and the point of view has only been vertical looking downwards [16,17]. Terrestrial laser scanner (TLS) has been applied in recent years for the same purpose, due to the rapid progress regarding maximum scanning range, spatial resolution, and accuracy and the fact that it allows a higher frequency of geomatic surveys [24]. This technique can monitor the evolution cliffs and beaches, and it can be used to investigate processes on the cliff top, cliff faces, and the cliff toe, as well as beaches located at the base of the cliffs [12,16,17,25–33].

The aim of the present work is to detect changes in the cliff top, face, and base, and to map the deformation of the cliff by using four different geomatic techniques (aerial photogrammetry, light detection and ranging (LiDAR), unmanned aerial vehicle (UAV), and terrestrial laser scanning

(TLS). Available data on the 20th century Gerra beach consist of photogrammetric flights which had low precision (high flight height). Fortunately, the same did not happen with the photogrammetric flights during the 21st century, which were used for the stereophotogrammetric plotting. Geomatic techniques that have been developed since the beginning of the 21st century, such as LiDAR, TLS, and UAV, have also been used. Specifically, TLS has been used, for eight years, leading to a continuous series of 17 surveys for estimating erosive changes at a detailed scale and for understanding the processes involved in the retreat of the coastline. The TLS surveys took place before and after the storm period (December–March) and in this way the behavior of the beach could be analyzed against the occurrence of storms. The data collected for this project were used for different types of analyses, some of them in a time span of 64 years, which was remarkably long as compared with most projects described in literature. The longer the time span, the sounder the conclusions about cliff erosion.

This research is part of a long-term monitoring program to better understand and quantify geomorphic behavior and local retreat rates of the Cantabrian coast. The ultimate goal is to provide useful information to the local planning authorities, as the cliff tops close to the Gerra beach suffer similar slope processes and cliff erosion. Until now, there has been no data on cliffs' instability in this study area or on the Cantabrian cliffs.

2. The Study Area

The Cantabria coast is located in northern Spain, facing the Cantabrian Sea. It is a mesotidal environment with a tidal range of 4 m. Cantabria has a maritime west coast climate with average winter temperatures around 14–15 °C. Rainfall is distributed over the year (ca. 1000–1200 mm/year). The main coast landforms are steep cliffs, which occupy 77% of the Cantabria coast, from 20 m to 70 m height on average, alternated by embayed beaches and small estuaries, both with aeolian dune fields. The main beaches are located out of estuaries, as the San Vicente-Merón beaches, where the Gerra beach is located (Figure 1) [34]. The cliff top was shaped by the sea processes, as marine deposits and ancient abrasion platforms, locally named "rasas", raised between 4 and 285 m height. Together with the wide platforms, some erosional features on cliffs and slopes are related to marine erosion and sedimentation [35–38].

beach is located at the north facing sea cliff in the eastern part of the San Vicente-Merón beaches system (formed by El Rosal Beach, Merón Beach, Bederna Beach, and Gerra Beach), protected by the West Point (or Merón Point) and located 700 m NW towards Gerra village. The study site is only a part of the San Vicente-Merón beaches system (Figure 1), which is a tourist beach which extends for 3 km. It is an important focus for surfers and bathers, who have access to the beach by a dirt road, but human influence is restricted to a minimum and the cliffs are unprotected.

Gerra beach (43°24′03′′N, 4°21′18′′W) is a composite beach with sea cliff. Therefore, the beach consists of an upper backshore composed of cobbles and boulders, and a foreshore of sand modeled by the waves and nearshore and offshore currents. The onshore cliffs rise to a maximum height of 40 m above the Cantabrian Sea. In the areas of Gerra and San Vicente-Merón beaches, it is possible to differentiate the following three old sea levels: two "rasas" at 40–60 m and 65–75 m height and one small level located at 5–6 m height [34,37,39]. On the beach, depositional sequences have been described consisting of basal gravels, clays, and eolian sands related to sea level changes and dated as glacial (Figure 1) [34] and preglacial age (71,570 ± 13,400 years BP) [40]. Along the San Vicente-Merón beaches, naturally driven mass movements are a frequent phenomenon [37,40] of various magnitudes, from 0.05 to 1 ha. Processes located on the cliff are minor rock falls and slope landslides, ranging in magnitude with volumes of several thousand of m^3. Such slope instabilities have recently affected the cliff top settlements of the El Puntal neighborhood (Figure 2), a new housing development and road. The sea cliffs of Gerra beach are composed of two areas with different characteristics. The east side is formed by two turbiditic geological units (Figure 2).

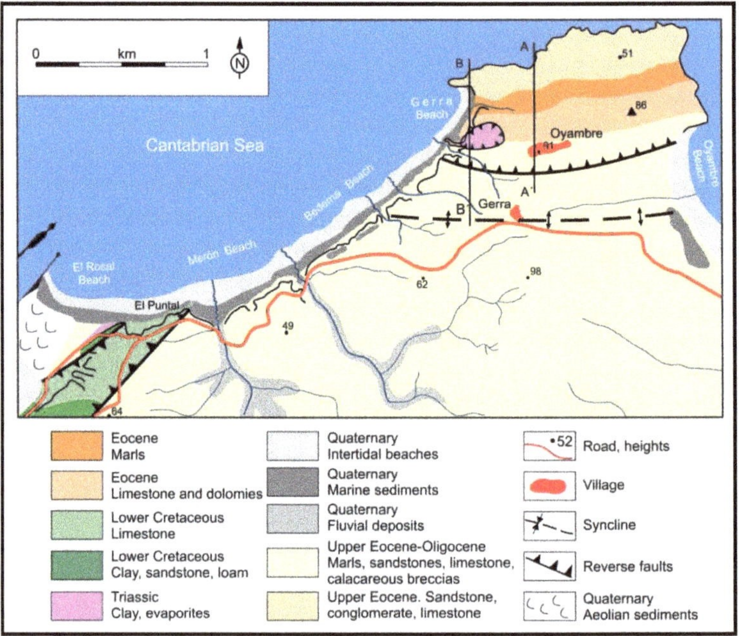

Figure 2. Geological schematic of the SanVicente-Merón coastal system [41,42].

Historic data and field evidences have shown that cliff erosion and landslide mode have been controlled by geology, as [43] noted in other coastal areas of northern Spain. The steeply inclined turbidic cliff sections have mainly collapsed due to sudden falls and landslides along structural discontinuities, such as joints or stone beds. This has led to small rock falls, and clay flows with volumes reaching between 500 and 2500 m^3.

3. Data Acquisition and Methodology

In order to measure the retreat at Gerra beach, we began by digitizing orthophotos from 1956 obtained from the American Flight of the Iberian Peninsula, available on the web of Instituto Geográfico Nacional (IGN, National Geographic Institute). Using the photogrammetric flights from 2001, 2005, 2010, 2014, and 2017, maps, at a scale of 1:5000, were made through stereophotogrammetry. A LiDAR (light detection and ranging) flight from IGN, dating from the summer of 2012, was considered, as well as an unmanned aerial vehicle (UAV) flight completed in November 2018. A significant quantity of data was obtained with a terrestrial laser scanner (TLS) since surveys were made on the beach twice a year. These TLS observations began in 2012 and continued uninterrupted to the present (last one in March 2020).

3.1. Aerial Acquired Data (1956–2018)

The existent photogrammetric flights over Gerra beach date from 1956 (B-series of the American Flight of the Iberian Peninsula), 2001, 2005, 2010, 2014, and 2017. The 1956 flight has no calibration certificate available, but the corresponding orthophoto generated by the IGN does exist. In this orthophoto, the dividing line between the upper part of the sandy slope and the stable zone of the surrounding plots is evident (Figure 5a). Therefore, this upper line of the sandy slope has been represented by digitizing it in the orthophotography. Obviously, the 1956 flight does not have the same precision as the subsequent photogrammetric flights, but its digitization allowed us to have baseline beach information from 1956 [44].

From the other flights, all originating from the Plan Nacional de Ortofotografia Aérea (PNOA, National Plan of Aerial Orthophotography), maps at a scale of 1:5000 were produced. For these flights, camera calibration certificates are available (focal length, fiducial marks coordinates, principal point coordinates). There is also information about each flight. The ground sampling distance (GSD) of the scanned photographs is 0.22 m and for the orthophotos it is 0.25 m, the planimetric precision is ≤0.5 m for the orthophotos and the altimetric precision for the digital elevation model (DEM) is ≤1 m [45].

A set of 10 ground control points was determined in the official geodetic reference system of Spain (ETRS89) using Leica 1200 GNSS (Global Navigation Satellite System) receivers in real-time kinematics (RTK) modus with postprocessing achieving a positioning precision of ±2 cm. From the 2001, 2005, 2014 and 2017 PNOA flights, maps of the Gerra beach were made. In order to determine the scale that could be used to produce the maps, the coordinates of four check points were measured in the stereo pairs (Table 1). The maximum differences at check points were ±0.99 m in planimetry and ±0.63 m in altimetry. Therefore, these values conditioned the scale and the equidistance of contour lines in the map to be produced [46].

Table 1. Check points measured in the photogrammetric models of the different flights. Measured coordinates and maximum differences to ground measured points. Maximum difference of coordinates (photogrammetric restitution) are X (# 3) 0.99 m, Y (# 2) 0.97 m, and Z (# 2) 0.63 m.

Point	Coordenate	Year 2001 (m)	Year 2005 (m)	Year 2010 (m)	Year 2014 (m)	Year 2017 (m)	Difference in X (m)	Difference in Y (m)	Difference in Z (m)
1	X	390,237.57	390,237.21	390,237.02	390,237.10	390,237.06	0.55 m		
	Y	4,806,117.60	4,806,117.49	4,806,117.90	4,806,117.18	4,806,117.65		0.72 m	
	Z	4.91	4.73	4.39	5.00	4.97			0.61 m
2	X	390,044.87	390,044.19	390,044.75	390,045.10	390,044.55	0.91 m		
	Y	4,805,844.06	4,805,845.03	[illegible]	[illegible]	[illegible]		[illegible]	
	Z	45.31	45.27	45.90	45.53	45.45			0.63 m
3	X	389,796.87	389,796.10	389,796.40	389,797.09	389,796.61	0.99 m		
	Y	[illegible]	[illegible]	[illegible]	[illegible]	[illegible]			

In the summer of 2012, the IGN performed a LiDAR survey of the Cantabrian coast (Gerra beach included) for the PNOA. These flights had a point density of 0.5 points/m^2 and the point clouds had a planimetric/altimetric mean square error of RMSE x, y, and z ≤ 0.2 m.

To complete the map series of the project, a UAV photogrammetric flight was done in November 2018, using an eBee Classic from SenseFly. The eBee is a fixed wing UAV equipped with a calibrated Sony VX RGB camera with a 6.16 × 4.63 mm sensor, 18.2 Mp resolution, and 4.57 mm focal length.

The flight was planned with forward and side overlap of 80% between photographs, at a flying height of 100 m adapting to the terrain through the geoid model STM. The software eMotion Version 3 from SenseFly was used for flight planning. According to the camera parameters and the flight plan, the expected GSD was 2.75 cm. A total of 173 images were obtained. Previously, 20 ground control and check points were measured using a GNSS Leica 1200 in RTK modus, with corrections from the real-time positioning service of the IGN from the network solution and a RMSE x, y, and z ≤ 20 mm was achieved.

The images were photogrammetrically processed with the Pix4D Mapper pro version 3.1.18 software. The predefined configuration in the software was used to process the photogrammetric surveys carried out with UAVs. The relative orientation of the model started from the position of the cameras given by the metric GNSS system loaded in the UAV and an adjustment of the position of the cameras was performed by running the calculation algorithm of the software itself. The absolute orientation was done by measuring the ground control points of the model in the images. From this orientation, an optimization of the parameters of the relative orientation was carried out depending on the control points. Twenty ground control and check points were used and a model with RMSEx ≤ 4.2 cm, RMSEy ≤ 2.9 cm, and RMSEz ≤ 3.6 cm was obtained. The distribution of the control and check points was mostly in the beach area (Figure 3). Once the photogrammetric model was defined, the dense point cloud and the orthoimage were obtained as cartographic products necessary for our research.

Figure 3. Distribution of the control points for the unmanned aerial vehicle (UAV).

3.2. Terrestrial Laser Scanning Data (2012–2020)

The application of terrestrial laser scanning (TLS) for monitoring and analyzing geomorphological dynamics on sandy coasts with millimetric precision is relatively recent and very efficient even at detecting annual changes [47], as well as complementing traditional analytical methods. TLS has functional sensitivity under adverse meteorological conditions [48] and a versatility that permits monitoring at any time. Its reliability has been proven on the coasts of the Cantabrian Sea in the study area, on confined beaches nearby, and in the ongoing observation of sand banks [11,49]. The scanner measured the three-dimensional (3D) position of data points in the survey area, the x, y, and z coordinates, and also collected the reflection intensity of each point [17]. The spatial position accuracy of the points measured with the Leica Scan Station C10 was given as ±6 mm at a distance of 100 m, although this value could increase, influenced by the type of element in the cliff that reflected the laser pulse (for instance, moving vegetation during measurement). This uncertainty added to the positioning error of the targets which was determined by GNSS (±20 mm) and yielded an estimated position uncertainty ≤ 3 cm. Then, the spatial point information of the resulting point cloud could be used to derive accurate digital elevation models (DEMs) [50,51].

Since May, 2012, until the present, TLS surveys have been made twice a year, in spring (April–May) (Figure 4a) and in the fall (October–November) (Figure 4b), in order to be able to obtain 3D data of Gerra beach before and after the winter storms of the Cantabrian Coast.

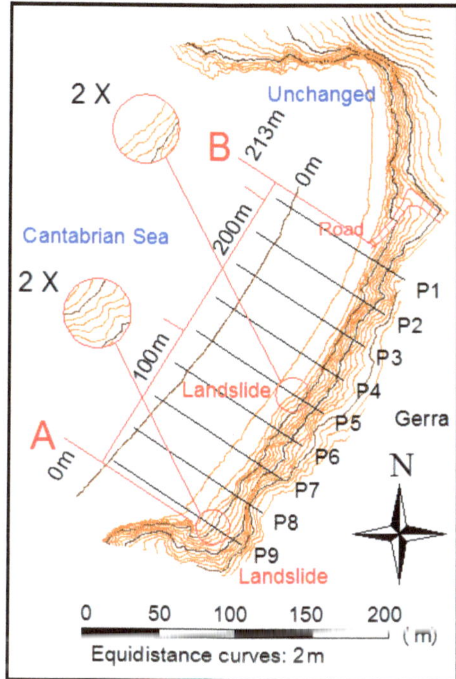

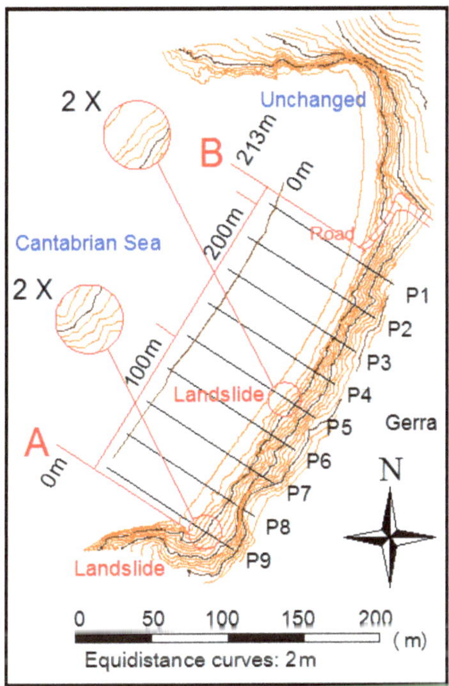

[illegible] (a) Digital elevation model (DEM) of October 2018; (b) DEM of April 2019.

In each campaign, six or seven scans were performed using at least 4 targets to register neighboring scans. The whole point cloud contained more than 250 million points. With the point clouds, DEMs were generated. Difference models for specific periods were later calculated by comparing successive point clouds and surface models. In addition, nine profiles (P1, ..., P9) were also determined. This procedure provided visualization of areas subject to temporal surface changes, and therefore permited monitoring and quantifying changes.

3.3. Analysis Evolution of the Beach and Coastline on the Top and Toe of the Cliff

To compare the upper and lower coastlines of the sandy slope, a 1/5000 scale map was made from the aerial photographs of the PNOA flights by stereophotogrammetry. With this scale, a planimetric error tolerance of 1 m (0.2 mm times the scale denominator) and an equidistance between contours of 2 m (1/3 of the contours equidistance allowing maximal errors of 0.67 m) was obtained. This means that, in the produced maps, the probability of obtaining planimetric accuracy better than 1 m and altimetric accuracy better than 0.67 m was very low. These errors were bigger than those estimated for LiDAR (maximum error ±20 cm), for TLS (maximum errors of ±3 cm), and for UAV surveys (maximum errors of ±4 cm), indicating that the map representations defined the method with less accuracy for the present purpose (Table 2).

Table 2. General information of initial data and obtained results for Gerra beach.

Date (Year)	Information	Maximum Error (m)	Result
1956	Ortophotography	2 m	Digitization
200–2005–2010–2014–2017	Aerial photogrammetry	1 m	Cartography by restitution photogrammetric
August 2012	LiDAR	0.20 m	DEM
November 2018	UAV	0.04 m	DEM
2012–2020 (Semiannual measurements, spring and fall)	TLS	0.03 m	DEM

With the other of the techniques used in this project (LiDAR, UAV, TLS), the upper and lower coastline of the sandy slope were not extracted. From the generated DEMs (LiDAR, UAV, TLS), different DEMs and profiles were made in order to study the evolution of the cliff. Currently, there is only one LiDAR flight in the north of the Iberian Peninsula (summer of 2012), belonging to the IGN public body. As for UAV, there was only the information of one flight (November 2018). With TLS, two measurements per year were performed (spring and fall) for the period 2012–2020, and the respective DEMs were generated, nine parallel profiles 25 m apart and perpendicular to the coastline were extracted, and their evolution studied (Table 2).

4. Results

4.1. Applying Aerial Photogrammetry (1956–2017)

Through analyzing the digitized top and toe lines of the 1956 and the five maps done by photogrammetric methods for the (2001–2017) period, no significant height changes were detected (Figure 5). The planimetric elements that were extracted from the map were the road for the access to the beach and the cliff top line, as well as the contour of the landslides (Figure 6a). The height contours were also represented with an equidistance (normal curves) of 2 m and master curves every 10 m.

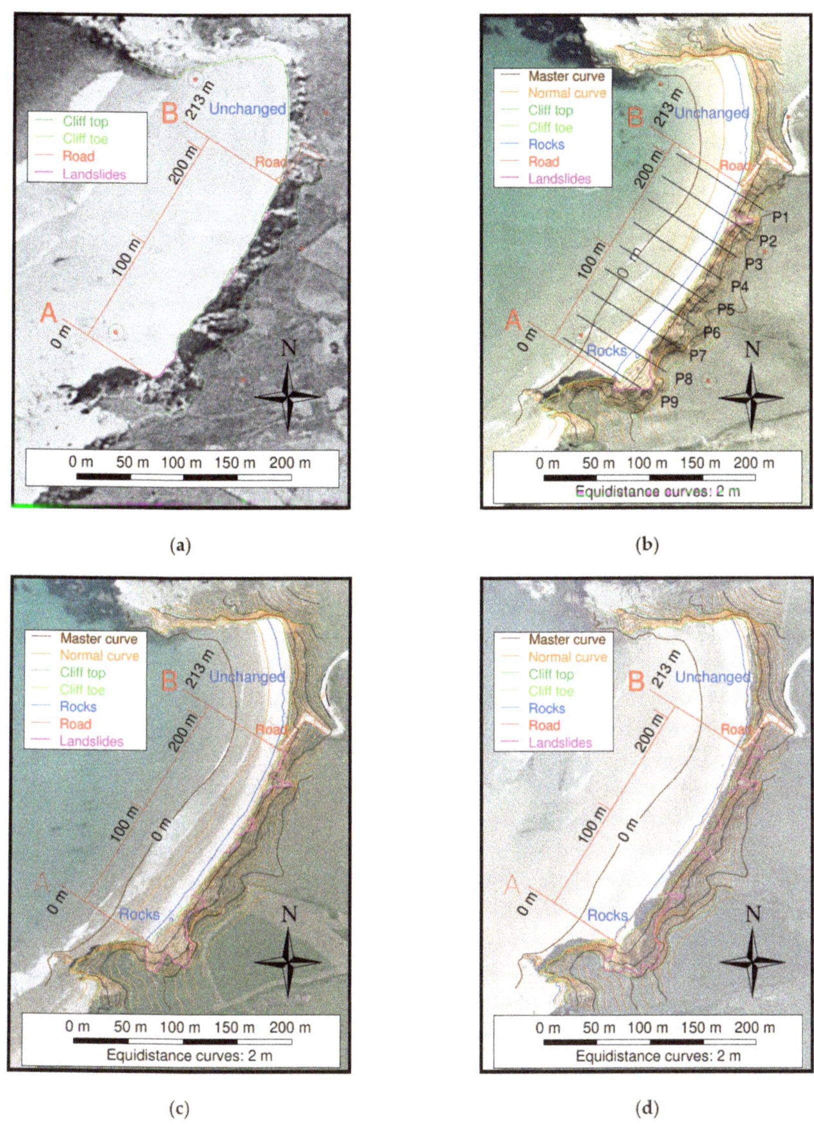
Cliff top
Cliff toe
Road
Landslides
A
B
0 m
100 m
200 m
213 m
Unchanged
N
0 m 50 m 100 m 150 m 200 m
(a)
Master curve
Normal curve
Cliff top
Cliff toe
Rocks
Road
Landslides
P1
P2
P3
P4
P5
P6
P7
P8
P9
Equidistance curves: 2 m
(b)
(c)
(d)

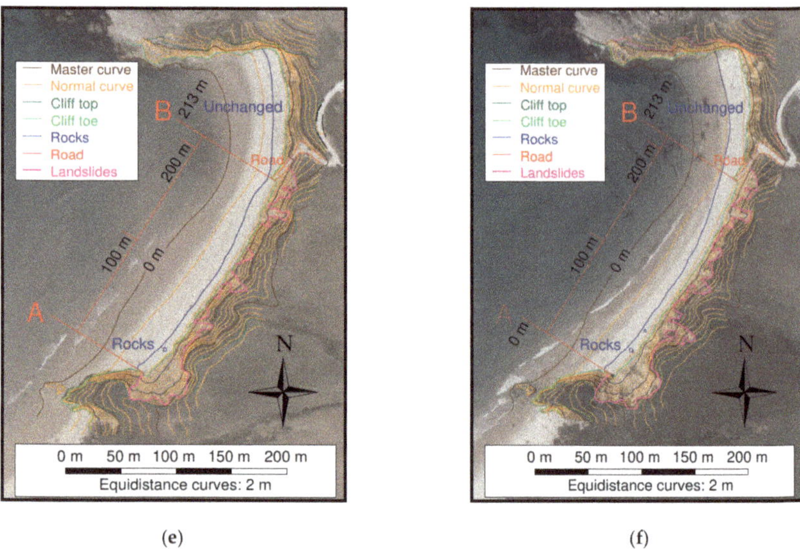

(e) (f)

Figure 5. (**a**) Delimitation of the coastline in the orthophoto of the American flight (1956). DEMs (height contours every 2 m) for the years (**b**) 2001; (**c**) 2005; (**d**) 2010; (**e**) 2014; (**f**) 2017.

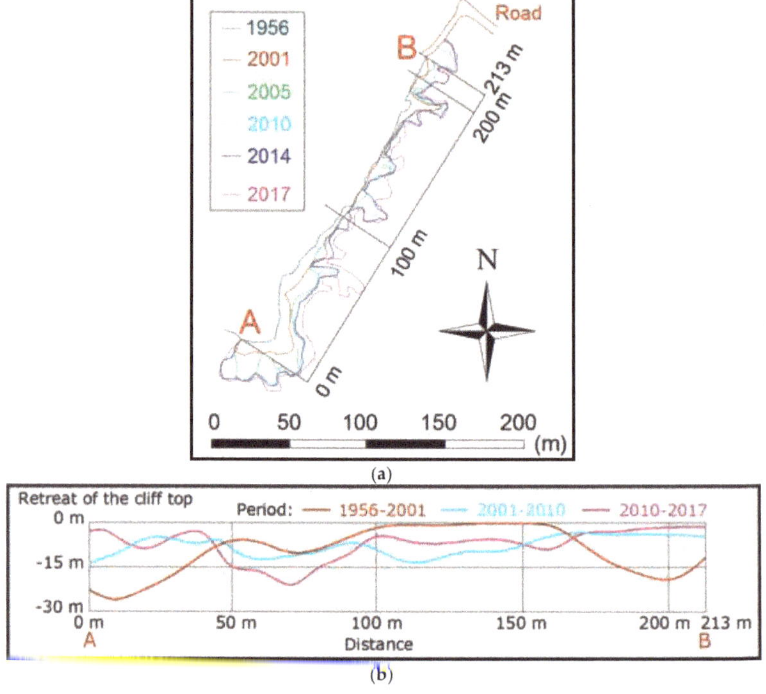

Figure 6. (**a**) Evolution of the cliff top between A and B. Lines of the years 1956, 2001, 2005, 2010, 2014 and 2017; (**b**) Comparison of the retreat of the cliff top in the periods 1956–2001, 2001–2010, and 2010–2017.

The retreat of the cliff baseline was not very meaningful for the occurring process, since there were no big changes during the analyzed period. The same cannot be said regarding the cliff top line, where landslides took place and significant retreat values could be determined. Therefore, the extracted cliff top lines (line A–B of 213 m) from the 1956 orthophoto and from the photogrammetric plotting were compared and the evolution graphics were produced for the following periods: 1956–2001, 2001–2010, and 2010–2017 (Figure 6b). For the three periods analyzed (1956–2001, 2001–2010, and 2010–2017) the upper horizontal line (retreat of the cliff top, 0 m) indicates the initial position of the cliff top line for the year of origin of each period (1956, 2001, and 2010). The curved lines (red, blue, and purple) represent the differences between the situation of the cliff top line between the oldest and the most recent date in each period. In all cases, the cliff top line is pushed inland.

4.2. Applying Light Detection and Ranging (LiDAR) (August 2012) and Unmanned Aerial Vehicle (UAV) (November 2018)

Collecting the available information from the 2012 LiDAR flight, the cartography was made with an equidistance of 4 meters (normal curve). The 0 m height contour from the TLS survey of April 2012 was incorporated. The location of the beach access road was also represented (Figure 7a).

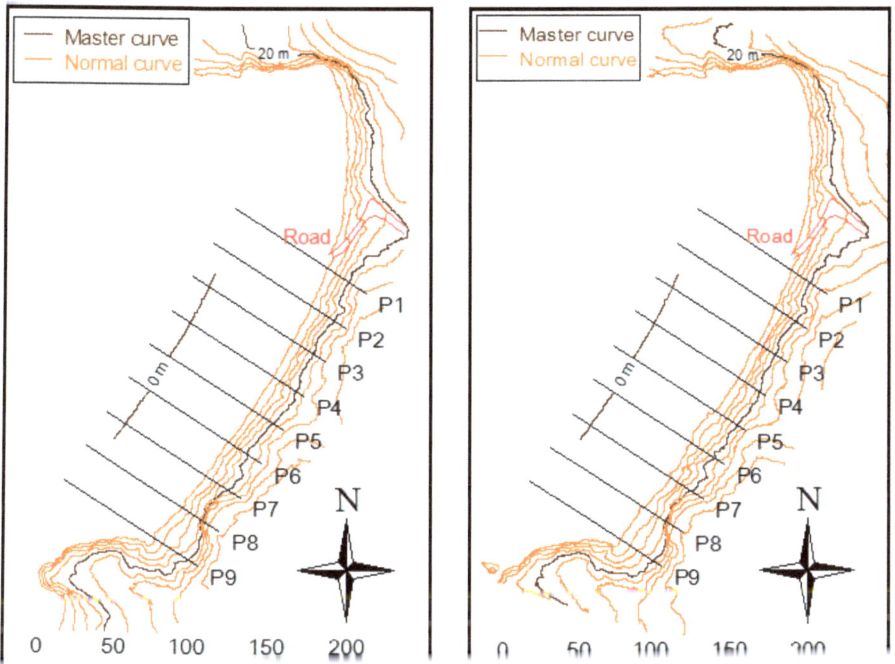

Figure 7. *Cont.*

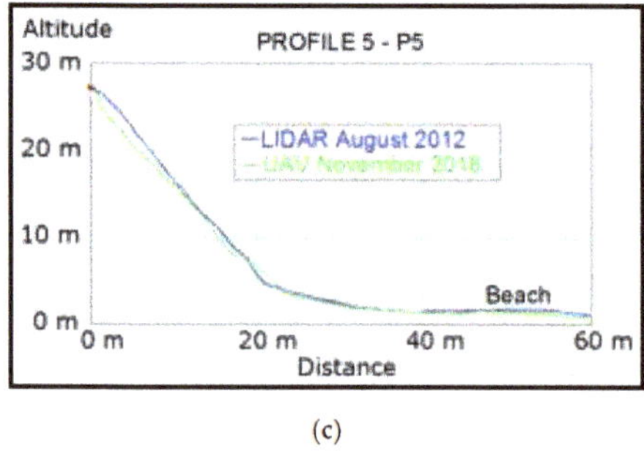

(c)

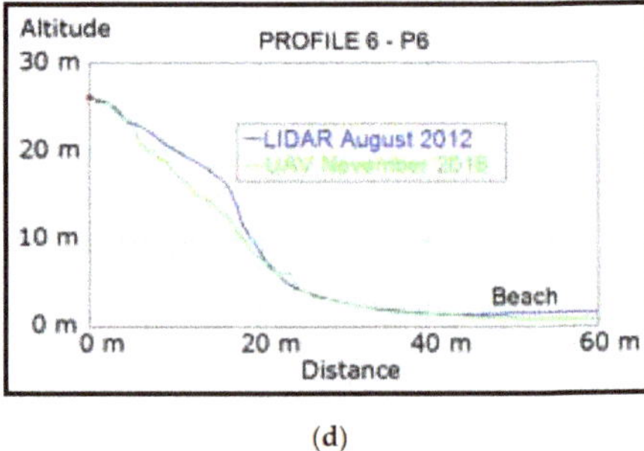

(d)

Figure 7. (**a**) Cartography light, detection, and ranging (LiDAR) (National Geographic Institute (IGN)) (August 2012); (**b**) Height contours from DEM of UAV (November 2018); (**c**) Comparison of Profile 5 from LiDAR and from UAV; (**d**) Comparison of Profile 6 from LiDAR and UAV.

From the results of the UAV flight, the height contours were interpolated from the DEM and the most significant elements of planimetry were obtained by digitization, representing the base and top line of the cliff, landslides, and access road. Later, an orthophoto mosaic with a GSD (ground sampling distance) of 4 cm was also done (Figure 7b).

To check the variations in the cliff, Profile 5 (P5) and Profile 6 (P6) (Figure 7c,d) were selected from those indicated in Figure 7a,b. While in P5 there was a small landslide during this period, in P6 there was a large landslide, specifically between autumn 2017 and spring 2018 (Figure 9). In the beach area, there were hardly any variations in elevation for these two dates, although the measurements carried out every six months verified a variation in the elevation (Figure 8).

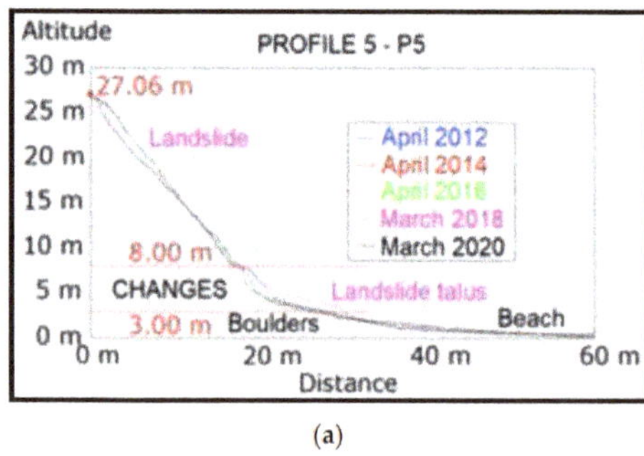

(a)

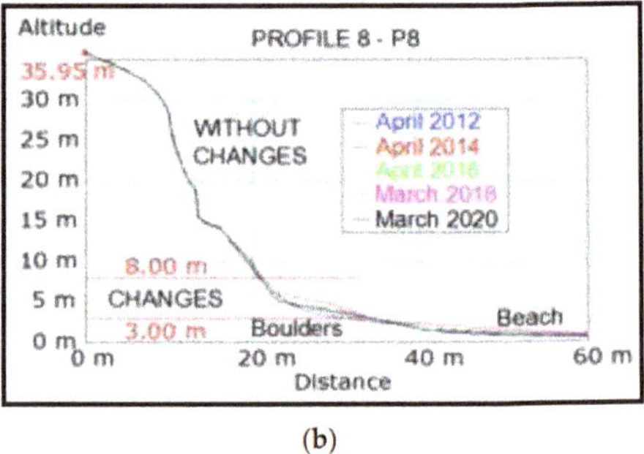

(b)

Figure 8. Cliff profile evolution between the more significant campaigns. (**a**) Profile 5 shows a landslide that occurred in March 2018; (**b**) Profile 8 contains a rocky substrate.

4.3. Applying Terrestrial Laser Scanning (2012–2020)

In Figure 8, two profiles are drawn (P5 and P8), P5 is a typical profile of the cliff, but in March 2018 there was a landslide there (Figure 8a). Profile P8 has a rocky substrate, and therefore is very stable (Figure 8b). There was a set of 17 TLS surveys (2012–2020), therefore, for each of the profiles (P1 to P9) the respective heights were extracted from the 17 DEMs produced from the surveys. In Figure 9, however, only the initial (2012) and the final (2020) situations are shown together with the corresponding years when there were significant differences between successive profiles (2014, 2016, and 2018). The profiles corresponding to non-significant changes between surveys are not represented. These changes mainly occurred at an altitude above the mean sea level between 3 and 8 m, since this was the area hit by the waves.

During the measurement period with TLS (2012–2020), the stability of the cliff was verified, despite the occurrence of storms on the Cantabrian coast. In fact, in the winter of 2013 there were four major storms [38], and therefore it was expected that large changes could be detected between the measurement of November 2013 and April 2014. However, although there were changes, the changes were not of the magnitude of what happened on other Cantabrian beaches [12,38]. The erosion of the cliff was more significant after the winter period, detected in the TLS observations of March or April (spring), and especially in 2014, 2016, and 2018 (Figure 8).

In addition to the profiles, the DEMs produced from the TLS survey data were also analyzed. Regarding the interpretation of the volumetric evolution of the DEMs, shown in Figure 9, we deduced the following:

- In general, the semiannual altimetric variations were within the interval of 0 to ±0.5 m. In rare situations there were differences between ±0.5 and ±1 m, and differences higher than ±1 m are very rare.
- Small and large landslides were detected in DEMs. Large landslides, in some cases were caused by storms, for example, between the fall of 2013 and the spring of 2014; in other cases, the landslides were caused by the instability of the cliff.
- The total accumulated volume in the study area (beach and cliff), from the spring of 2012 to the present (April 2020), indicated a material gain of 399.66 m^3. It was a very small, almost negligible, gain value for the eight-year period. But, depending on the campaigns, there could be greater differences (gains or losses) in the beach area. Thus, for example, on the one hand, between the spring of 2016 and the fall of 2016, there was a sand gain of 3467.39 m^3, and on the other hand, between the fall of 2015 and the spring of 2016, there was a sand loss of −3040.04 m^3.
- The volumetric values that occurred above the base line of the cliff (3 meters above sea level) were also analyzed. At 3 meters altitude, small landslides occurred (for example, between the fall of 2013 and the spring of 2014). There were also large landslides across the cliff (for example, between the fall of 2017 and the spring of 2018). Throughout the study period, the total loss of material that occurred in the cliff area was −3633.32 m^3.

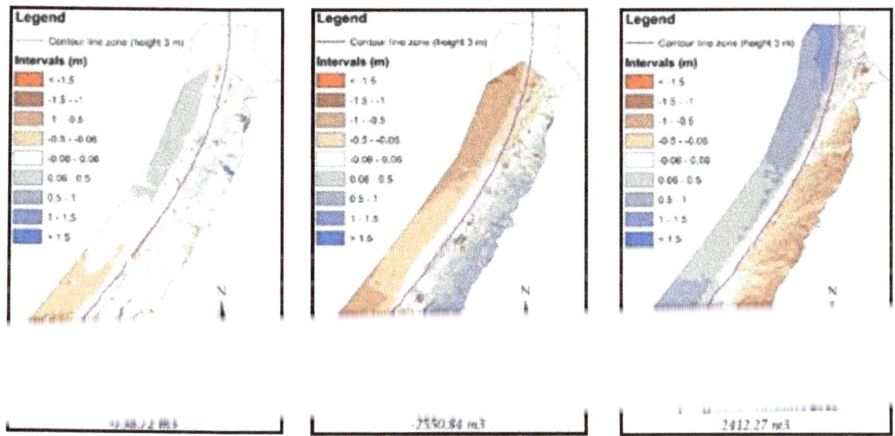

Figure 9. *Cont.*

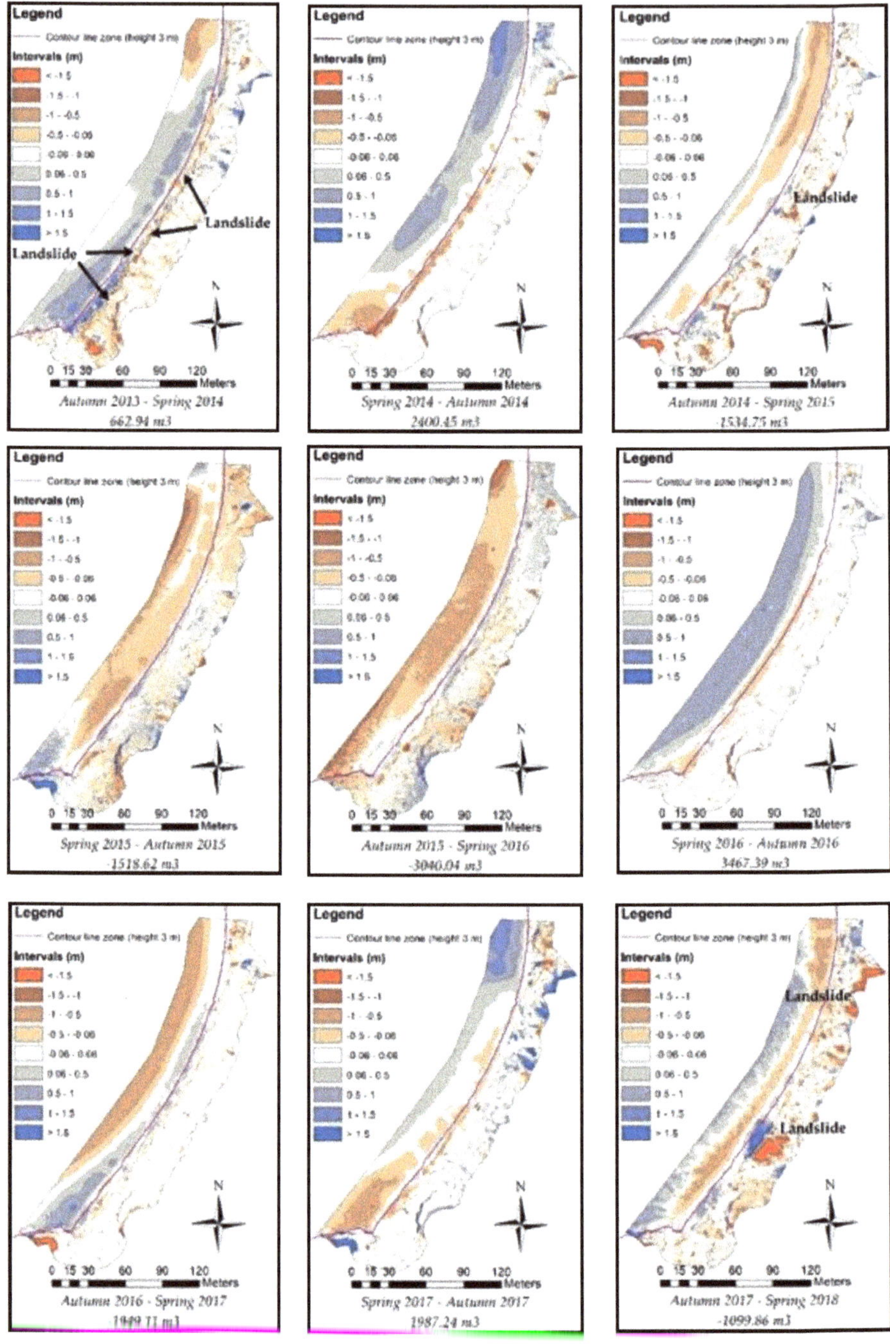

Figure 9. *Cont.*

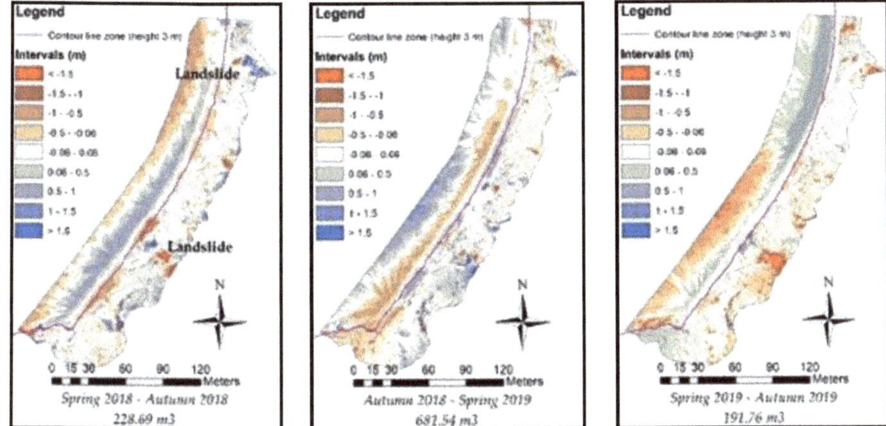

Figure 9. Volumetric evolution between the spring of 2012 and the fall of 2019.

5. Discussion

Factors related to cliff behavior are the same on the Atlantic and Pacific coasts. Landslides on the cliff face, the geological structure and lithology, and the location of alternating claystone, sandstone, and limestone imply important differential erosion, and finally, sea exposure [3,5]. Furthermore, on European coasts, the dominant influence of precipitation, in a humid climate with more than 1000 mm/year, is a determinant factor in cliff top evolution and groundwater level [3]. Contact between claystone and turbiditic facies implies water drains quickly from the top of the cliff and the claystone becomes saturated, causing landslide and rock fall.

Changes in the three sectors of the sea cliff, top, base, and face, can differ significantly over short time periods. On analogous cliffs formed by claystone and sandstone, it is possible that wave action can cause cliff base retreat, but no cliff top change, and also the cliff face erosion can occur without changes to the cliff top or base [4]. In Gerra beach, between 1956 and 2001 (45 years) the retreat occurred mainly in the area of the beach access road, possibly caused by its construction, since, in 1956, it was a very narrow dirt road, and now it is a larger asphalt road. The road building generated artificial slopes and unbalance causing new landslides. Cliff collapses and landslides can be attributed to anthropogenic factors, as seen on the Atlantic coast [3]. During the period 1956–2001 there has also been significant cliff retreat south of the beach (near point A) (Figure 6a), where several landslides are located. In this case is the lithology, a big outcrop of claystone, and the geological structure, the limit of a small diapiric structure, the cause of successive landslides, sometimes occupying the cliff base. The cliff areas with less retreat between 1956 and 2001 were those that have been more intensely affected during the period 2001–2017 (16 years), with average values around 20 m of retreat (Figure 6b).

During the period 1956–2017, the maximum retreat of the cliff top was 42 m, with the average value for th [illegible]

[illegible] data collection. The elements are both common on the top and face of the cliff when the stratigraphy and the changing lithologies support differential erosion, as has been pointed out in California or Bretagne cliffs [3,5]. Sometimes, advances of the cliff top can happen when landslides are generated in the upper part before they break and fall. In Gerra beach, there was a balance between the offshore and inland displacement. The cliff base appeared to have accreted, although this process occurred very rarely, only when slope deposits were not removed between surveys.

As the results show, the retreat of the cliff baseline is not meaningful, with small changes during the studied period. On the Atlantic coast, retreat rates of −10 to −50 cm/year have been observed [3,14], but these have increased whenever human action was present [6–10,14,15]. The retreat rates of Gerra beach are low, in accordance with low human pressure, but this inactivity could change if human activity increases [7,9,12]. As mentioned in the Introduction, on the California coast, recent retreat rates have been lower than historical retreat rates in some regions and dynamic studies have also shown that regions with high levels of recent retreat had low historical retreat [5]. This is the opposite of what has happened on Mediterranean coasts, but the authors have pointed out that an inherent sampling bias could have occurred due to shorter observation intervals [1]. On the Gerra beach, the activity at the base of the cliff does not imply a retreat, and therefore it is considered to be inactive. This fact does not mean that all the coast is inactive, since very active Cantabrian coast portions have been observed [6,7,12]. As has been checked on the California coast, there is an alternation between regions of active and inactive coast, and more than 50% of the coast is inactive [5].

Spatial variations and changes between top, face, and base cliff retreat rates can be explained by common factors of sea cliff dynamics, such as geological structure, cliff collapses on top and base cliff, anthropogenic intervention [1–3], and storm succession, commonly affecting beaches on the Cantabrian coast [10–12,40]. Storms are an important factor, as it is well known that cliff erosion with subsequent deposition of large boulder debris is driven by large storms [52,53], although rainfall, groundwater pore pressure, and long-term vibrational disturbance from fair-weather waves also have an influence [25].

The turbiditic and claystone lithostratigraphic units are affected by wave energy during storm events overlapping with high tide, but ordinary events occur throughout the year and they rework and reshape sediments and beach landforms. During wave periods without a storm the coarse-grained beach and deposit of the storm berm protect the base of the cliff and there is no erosion.

Regarding the influence of storms on the evolution of the coastline, one must indicate that in the period 2000–2014 there were continuous storms [11,12]. There were four major storms in the winter of 2013 (from December 2013 to March 2014) [54]. However, during this period no significant differences in the evolution of the lower baseline of the cliff were detected. Measurements after the storms of March 2014 and subsequent years (2015–2019) showed an increase in landslides. The lower base line of the cliff remained stable, while the upper line of the cliff top was affected by landslides (Figure 6a). Therefore, there is a relationship between the influence of the waves on the lower part of the cliff that makes the cliff recede with an accumulation of terrestrial rainfall that causes the landslides. Therefore, landslides and rockfalls accumulate at the base of the cliff, while the boulder beach dissipates the energy and protects the base of the cliff. Between surveys there is no apparent change in the base line, while pushing the upper shoreline inland.

Geometric characterization has been carried out by different methods and data sources. On the one hand, series of aerial photographs have been used for evaluating coastal evolution over longer periods of time [12,55–58]. On the other hand, techniques based on obtaining massive data, such as LiDAR, UAV, and TLS, have demonstrated their effectiveness for studying changes in the coastal system [29,59–63]. High resolution orthophotos (4 cm pixel) have been obtained using UAV photogrammetric techniques, and from these a detailed analysis of the current limits of the top and base lines of the cliff has been possible. In addition, from UAV coverage, a point cloud was obtained of spatial resolution between 10 and 15 cm that defined the topography of both the beach and the cliff. TLS was used two times per year aimed at observing the topographic evolution of the beach and the cliff. This technique provided point clouds of spatial resolution <10 cm. This allowed us to evaluate, from the comparison of DEMs, the areas where landslides occurred on the cliff or the evolution of the level of sand on the beach.

To check the degree of precision of the different techniques used (aerial photogrammetry, LiDAR and UAV photogrammetry), a comparison of these techniques was made, with respect to the measurements with TLS (Figures 10–12). As previously stated, the maximum estimated error of the TLS measurements was ±3 cm. Profiles 1 and 9 (Figure 4a) were used for this analysis:

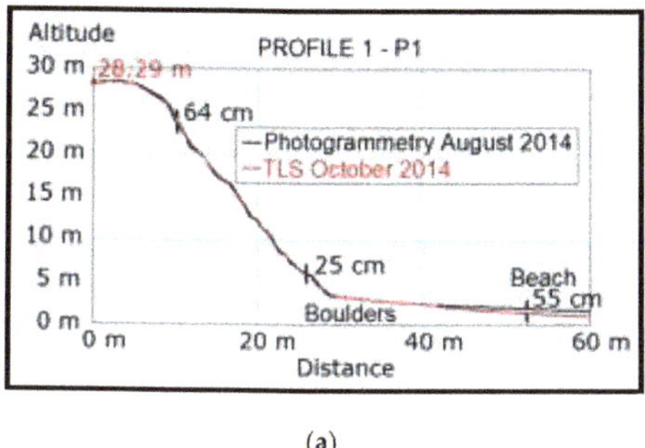

(a)

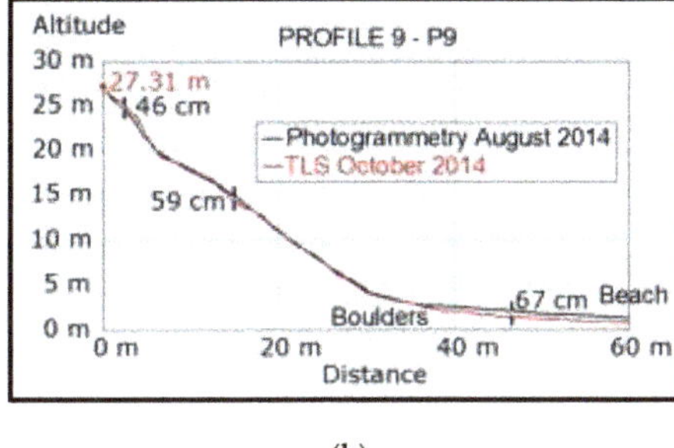

(b)

Figure 10. Comparison between the profiles obtained by photogrammetry and by TLS (2014). (**a**) Profile 1 (north area of the cliff); (**b**) Profile 9 (south area of the cliff).

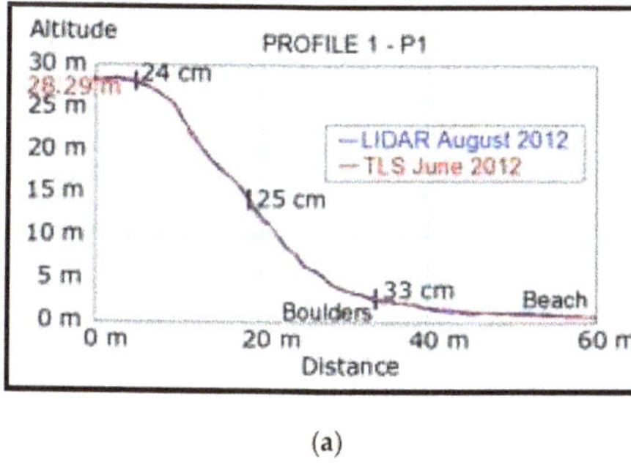

(a)

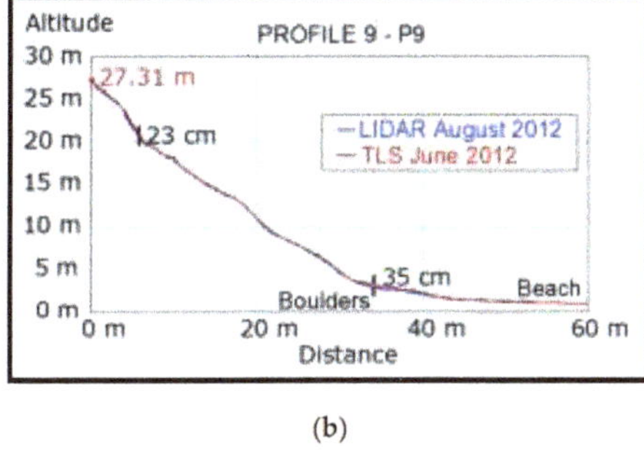

(b)

Figure 11. Comparison between the profiles obtained by LiDAR and by TLS (2012). (**a**) Profile 1 (north area of the cliff); (**b**) Profile 9 (south area of the cliff).

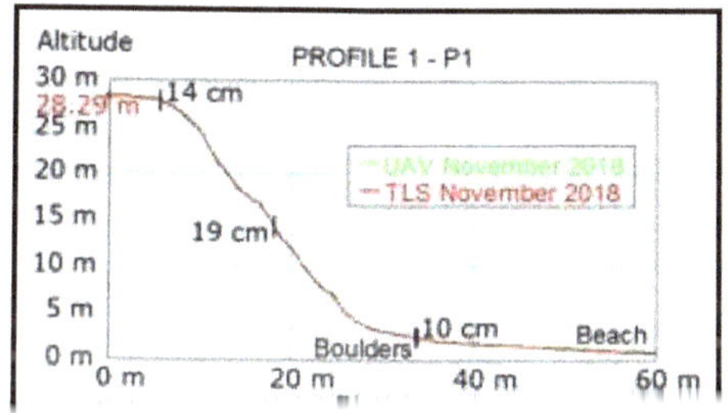

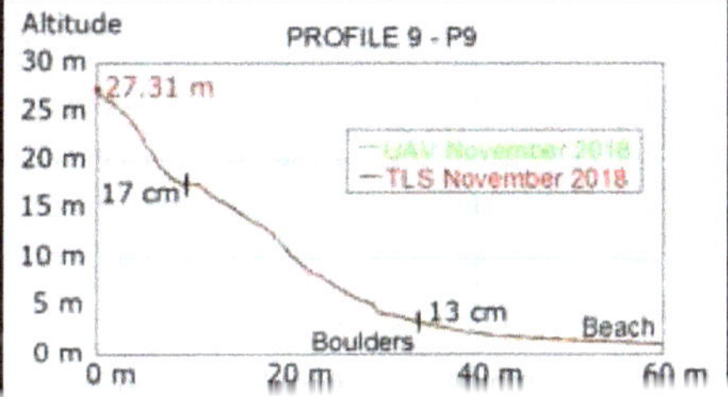

Figure 12. Comparison between the profiles obtained by UAV and by TLS (2018). (**a**) Profile 1 (north area of the cliff); (**b**) Profile 9 (south area of the cliff).

Comparison aerial photogrammetry and TLS There are many geomorphological investigations that compare the photogrammetric techniques supported by the structure from motion (SfM) technique with the TLS method [64–67], but this investigation makes a comparative analysis of the precision of the classical photogrammetric stereo restitution with TLS, as has been done by other authors [68–70]. This type of comparison is interesting due to the extensive mapping that was done using analytical or digital photogrammetry before the advent of automated SfM methods. In the case at hand, from the available photogrammetric flights, the one of August 2014 was chosen, since it approaches in time the TLS survey of October 2014. Therefore, the conditions of the Gerra beach in both DEMs are assumed to be similar (a difference of two months in the summer period without storms). In the case of the profiles shown in Figure 10, it can be seen that the maximum differences that the photogrammetry shows with respect to the TLS are ±0.67 m in the zone of the beach. The maximum differences are distributed both in the area of the beach and on the cliff, and therefore not generalized in a specific area of them.

The photogrammetric mapping is not automatic, since it has been performed by a human being (operator with experience in photogrammetric restitution), where the altimetry interpretation is difficult to determine due to the lack of details and little slope in the beach area. There are significant differences in the areas with the steepest slope, where the errors are consistent with the equidistance between contour lines (2 meters) and the tolerance is 1/3 of the equidistance, that is ±0.67 m. The photogrammetric flight of August 2017 was also compared with the TLS survey of November 2017, and they showed very similar results to those previously described (2014 flight and 2014 TLS).

LiDAR and TLS comparison The LiDAR flight of August 2012 was compared with the TLS survey of May 2012. It was previously indicated in the section on data collection and methodology with LiDAR that the maximum expected error for the LiDAR was ±20 cm. When comparing the DEM obtained from LiDAR and the DEM from the TLS survey, maximum errors of ±35 cm are found in the area of the base line of the cliff where there are a large number of boulders (Figure 11). The difference in this area is foreseeable, since the LiDAR density is 0.5 points/m^2 and the TLS obtains hundreds or thousands of points/m^2, therefore, when the DEMs are generated large differences occur between them. In the area of the beach sand the differences are less than ±20 cm.

UAV and TLS comparison The UAV flight of November 2018 was compared with the TLS measurement of the same day. The maximum differences shown by the DEM from the UAV flight with respect to the DEM from the TLS is ±19 cm in the cliff area, ±13 cm in the pebble area, while in the beach area there are no major differences (Figure 12). The maximum values expected with the UAV flight were ±4 cm (a very optimistic but not real value), since the topographic support determined by GPS positioning has an uncertainty of ±3 cm. Possibly, the differences obtained in the beach area, where they are ±5 cm, are more real.

UAV and TLS can generate errors around two very different types of terrain features, which must be taken into consideration:

1. The TLS can generate occlusions when the laser pulses hit a surface at an oblique angle from the scanner point of view (beach). The image processing algorithms used with the UAV photographs can generate occlusions on surfaces that have little inherent contrast or texture of color (beach area). This results in a dependency of the precision of each method on the landscape in question [71].
2. The TLS and UAV methods are capable of collecting data at very different spatial scales. This has a substantial influence on the application of these methods and needs to be considered when making a comparison between them. In this case, the scale of both methods is very similar, since the measurement errors are ± 4 cm.

Interpreting the maps obtained from the photogrammetric flights in Figure 5, it is stressed that there is always a retreat in the top line, while the base of the cliff does not go backwards. Between 1956 and 2001 (45 years) the setback occurred mainly in the area of the beach access road, possibly caused by its construction, since in 1956 it was a very narrow dirt road and now it is a larger asphalt road, which has generated landslides due to the existence of greater slopes. During this period, there has also been a great setback in the south of the beach (near point A) (Figure 6a), and it is in this area where there has been a set of various landslides. The areas of the coastline that had the least setback between 1956 and 2001 are those that have been affected the most in the period 2001–2017 (16 years), with the average values of 20 meters of retreat during this period (Figure 6b). In total, during the period 1956–2017, the maximum retreat of the coastline at the top of the cliff was 42 meters, and the average value for the entire cliff was 25 m.

Regarding the results obtained with TLS during the period 2012–2020, there were no major changes between the winter and summer periods. Therefore, the measurement made after the summer period (October–November) would not be necessary, and one annual measurement would be sufficient. However, two measurements per year allowed us to know in more detail the evolution of the beach level and of the base and top lines of the cliff. It also allowed us to make a volumetric control for the winter and summer periods (Figure 9) and to undertake detailed studies of the annual dynamics of the beach. The profiles generated from the TLS DEMs show no changes in the cliff baseline toe, excep after extreme storms that push this lower line landward with minor landslides (fall of 2013 and spring of 2014). Instead, the same DEMs show changes in the top line of the cliff, showing large landslides (for example, between the fall of 2017 and the spring of 2018) (Figure 9). The storms that occurred during the winter of 2013 have been the most aggressive on the Cantabrian coast so far in the 21st century, but the footprint left on Gerra beach has been much lower than that of other beaches in the Cantabrian Sea (Spain).

6. Conclusions

The application of different geomatic techniques (aerial photogrammetry, LiDAR, UAV photogrammetry, and TLS) to acquire data on the beach and cliff of Gerra has allowed us to obtain two main pieces of information. On the one hand, the precisions of the different techniques for this purpose, and on the other hand, the receding rhythms of the cliff coast. This information can be applied to other sections of the Cantabrian coast for empiric and numerical models of cliff retreat, considering factors such as waves, currents, beach and cliff location, onshore morphology (sand beach, pebble beach, shore platform), intensity and frequency of storms, wind directions, rainfall, geological structure, sedimentary deposits, lithology, geomorphic inheritance and present day processes (karstic, tectonic), and human use on the Cantabrian or Atlantic coast, both located in a temperate oceanic climate.

The combination of techniques and their adaptation to the geomorphological particularity and the orographic situation of the beach, serve as a reference for its application in the control of the dynamics of beaches and the retreat of cliffs. We can conclude that there is no single ideal geomatic technique for each situation, and different techniques must be complemented to solve the problem posed. In the case of the Gerra beach cliff, as long as there are no areas of occlusions caused by the orography of the cliff or vegetation, the most accurate remote sensor technique is the TLS followed by the UAV photogrammetry. However, if these occlusions occur, both techniques are complementary, and therefore occlusions measured from the beach with TLS can be avoided by adding 3D points obtained from the UAV flight. Aerial LiDAR is the third technique for precision, although its density was too low for this objective. Lastly, photogrammetric flights provide less precision, but they are the most useful source to obtain data on past evolution and over a longer term. Therefore, to obtain data for modeling future evolution of the coast, faced with theoretical assumptions such as increased number and strength of storms, sea level rise, changes in rainfall or human pressure on the coast, old photogrammetric flights are the best resource. This classification shows a direct link between the precision of the technique and the cartographic scale obtainable from the collected data, that is, the better the precision of the technique, the larger the scale

the retreat in the 21st century has been observed in relation to the 20th century. Throughout the period, high retreat rates have been found, as befits the European Atlantic cliff coasts. However, these variations of the Gerra beach coastline are insignificant with respect to the setbacks of other beaches located on the Cantabrian coast.

Large storms have been responsible for erosion on the lower cliff base line, where they have caused small landslides and rockfall. This erosion affects the setback of the cliff top line. Differential retreat rhythms have been found between the base line and the top line of the cliff, with retreat rhythms in the upper portion that double the rhythms of the lower one. This fact leads to a degradation of the cliff and a more irregular coastline, which together with the increase in the retreat rates implies greater instability that can affect human infrastructure and buildings. There are no infrastructures on the Gerra beach, but they are being affected on the nearby beach of San Vicente-Merón and other Cantabrian beaches. Coastal authorities and territorial managers must implement control plans based on accurate data on coastal dynamics, and geomatic techniques have been proven to be suitable to monitor the dynamic of cliffs.

Author Contributions: Conceptualization, J.J.d.S.B. and E.S.-C.; Methodology, J.J.d.S.B., E.S.-C., M.S.-F. and M.G.-L.; Software, J.J.d.S.B., M.S.-F. and P.R.; Validation, J.J.d.S.B., E.S.-C. and P.R.; Formal Analysis, J.J.d.S.B., E.S.-C., M.S.-F. and MGL; Investigation, J.J.d.S.B., E.S.-C., M.S.-F., M.G.-L. and P.R.; Resources, J.J.d.S.B. and E.S.-C.; Data Curation, J.J.d.S.B., E.S.-C., M.S.-F., M.G.-L. and P.R.; Writing—Original Draft Preparation, J.J.d.S.B., E.S.-C., M.S.-F., M.G.-L. and P.R.; Writing—Review & Editing, J.J.d.S.B., E.S.-C., M.S.-F., M.G.-L. and P.R. All authors have read and agreed to the published version of the manuscript.

Funding: This research was funded by Junta de Extremadura and European Regional Development Fund (ERDF) grant number GR18053 to the Research Group NEXUS (University of Extremadura) and also by FCT-project UIDB/50019/2020-Instituto Dom Luiz.

Conflicts of Interest: The authors declare no conflict of interest.

References

1. Mushkin, A.; Katz, O.; Porat, N. Overestimation of short-term coastal cliff retreat rates in the eastern Mediterranean resolved with a sediment budget approach. *Earth Surf. Process. Landf.* **2019**, *44*, 179–190. [CrossRef]
2. Sunamura, T. Rocky coast processes: With special reference to the recession of soft rock cliffs. *Proc. Japan Acad. Ser. B Phys. Biol. Sci.* **2015**, *91*, 481–500. [CrossRef] [PubMed]
3. Costa, S.; Maquaire, O.; Letortu, P.; Thirard, G.; Compain, V.; Roulland, T.; Medjkane, M.; Davidson, R.; Graff, K.; Lissak, C. Sedimentary Coastal cliffs of Normandy: Modalities and quantification of retreat. *J. Coast. Res.* **2019**, *88*, 46–60. [CrossRef]
4. Young, A.P.; Carilli, J.E. Global distribution of coastal cliffs. *Earth Surf. Process. Landf.* **2019**, *44*, 1309–1316. [CrossRef]
5. Young, A.P. Decadal-scale coastal cliff retreat in southern and central California. *Geomorphology* **2018**, *300*, 164–175. [CrossRef]
6. Losada, M.A.; Medina, R.; Vidal, C.; Roldan, A. Historical evolution and morphological analysis of "El Puntal" spit, Santander (Spain). *J. Coast. Res.* **1991**, *7*, 711–722.
7. Garrote, J.; Garzón, G.; Page, J. Condicionamientos antrópicos en la erosión de la playa de Oyambre (Cantabria). In Proceedings of the Actas V Reunión de Cuaternario Ibérico, Lisbon, Coambra, Portugal, July 2001; Volume 1, pp. 67–70.
8. Lorenzo, F.; Alonso, A.; Pagés, J.L. Erosion and accretion of beach and spit systems in Northwest Spain: A response to human activity. *J. Coast. Res.* **2007**, *2007*, 834–845. [CrossRef]
9. Flor-Blanco, G.; Pando, L.; Morales, J.A.; Flor, G. Evolution of beach–dune fields systems following the construction of jetties in estuarine mouths (Cantabrian coast, NW Spain). *Environ. Earth Sci.* **2015**, *73*, 1317–1330. [CrossRef]
10. Garrote, J.; Heydt, G.; Alcantara-Carrio, J. Influencia de Los temporales sobre el transporte de sedimentos en la Playa de Oyambre (Cantabria, N de España). In Proceedings of the Actas V Reunión Nacional de Geomorfología, Valladolid, Spain, 26 June 2002; pp. 361–371.
11. Sanjosé, J.D.; Serrano, E.; Berenguer, F.; González-Trueba, J.J.; Gómez-Lende, M.; González-García, M.; Guerrero-Castro, M. Evolución histórica y actual de la línea de costa en la playa de Somo (Cantabria), mediante el empleo de la fotogrametría aérea y escáner láser terrestre. *Cuaternario Geomorfol.* **2016**, *30*, 119–130. [CrossRef]
12. De Sanjosé Blasco, J.J.; Gómez-Lende, M.; Sánchez-Fernández, M.; Serrano-Cañadas, E. Monitoring retreat of coastal sandy systems using geomatics techniques: Somo Beach (Cantabrian Coast, Spain, 1875–2017). *Remote Sens.* **2018**, *10*, 25. [CrossRef]

13. Arteaga, C.; Juan de Sanjose, J.; Serrano, E. Terrestrial photogrammetric techniques applied to the control of a parabolic dune in the Liencres dune system, Cantabria (Spain). *Earth Surf. Process. Landf.* **2008**, *33*, 2201–2210. [CrossRef]
14. Letortu, P.; Costa, S.; Maquaire, O.; Delacourt, C.; Augereau, E.; Davidson, R.; Suanez, S.; Nabucet, J. Retreat rates, modalities and agents responsible for erosion along the coastal chalk cliffs of Upper Normandy: The contribution of terrestrial laser scanning. *Geomorphology* **2015**, *245*, 3–14. [CrossRef]
15. Letortu, P.; Costa, S.; Bensaid, A.; Cador, J.-M.; Quénol, H. Vitesses et modalités de recul des falaises crayeuses de Haute-Normandie (France): Méthodologie et variabilité du recul. *Géomorphologie Reli. Process. Environ.* **2014**, *20*, 133–144. [CrossRef]
16. Letortu, P.; Costa, S.; Cador, J.; Coinaud, C.; Cantat, O. Statistical and empirical analyses of the triggers of coastal chalk cliff failure. *Earth Surf. Process. Landf.* **2015**, *40*, 1371–1386. [CrossRef]
17. Kuhn, D.; Prüfer, S. Coastal cliff monitoring and analysis of mass wasting processes with the application of terrestrial laser scanning: A case study of Rügen, Germany. *Geomorphology* **2014**, *213*, 153–165. [CrossRef]
18. Young, A.P.; Ashford, S.A. Instability investigation of cantilevered seacliffs. *Earth Surf. Process. Landf. J. Br. Geomorphol. Res. Group* **2008**, *33*, 1661–1677. [CrossRef]
19. Toimil, A.; Losada, I.J.; Camus, P.; Díaz-Simal, P. Managing coastal erosion under climate change at the regional scale. *Coast. Eng.* **2017**, *128*, 106–122. [CrossRef]
20. Masselink, G.; Russell, P.; Rennie, A.; Brooks, S.; Spencer, T. Impacts of climate change on coastal geomorphology and coastal erosion relevant to the coastal and marine environment around the UK. *MCCIP Sci. Rev.* **2020**, *2020*, 158–189.
21. Martínez, C.; Contreras-López, M.; Winckler, P.; Hidalgo, H.; Godoy, E.; Agredano, R. Coastal erosion in central Chile: A new hazard? *Ocean Coast. Manag.* **2018**, *156*, 141–155. [CrossRef]
22. Bruno, M.; Molfetta, M.; Pratola, L.; Mossa, M.; Nutricato, R.; Morea, A.; Nitti, D.; Chiaradia, M. A Combined Approach of Field Data and Earth Observation for Coastal Risk Assessment. *Sensors* **2019**, *19*, 1399. [CrossRef]
23. Valentini, N.; Saponieri, A.; Danisi, A.; Pratola, L.; Damiani, L. Exploiting remote imagery in an embayed sandy beach for the validation of a runup model framework. *Estuar. Coast. Shelf Sci.* **2019**, *225*, 106244. [CrossRef]
24. Terefenko, P.; Paprotny, D.; Giza, A.; Morales-Nápoles, O.; Kubicki, A.; Walczakiewicz, S. Monitoring cliff erosion with LiDAR surveys and bayesian network-based data analysis. *Remote Sens.* **2019**, *11*, 843. [CrossRef]
25. Young, A.P.; Guza, R.T.; O'Reilly, W.C.; Burvingt, O.; Flick, R.E. Observations of coastal cliff base waves, sand levels, and cliff top shaking. *Earth Surf. Process. Landf.* **2016**, *41*, 1564–1573. [CrossRef]
26. Katz, O.; Mushkin, A. Characteristics of sea-cliff erosion induced by a strong winter storm in the eastern Mediterranean. *Quat. Res.* **2013**, *80*, 20–32. [CrossRef]
27. Dornbusch, U.; Robinson, D.A.; Moses, C.A.; Williams, R.B.G. Temporal and spatial variations of chalk cliff retreat in East Sussex, 1873 to 2001. *Mar. Geol.* **2008**, *249*, 271–282. [CrossRef]
28. Benumof, B.T.; Griggs, G.B. The dependence of seacliff erosion rates on cliff material properties and physical processes: San Diego County, California. *Shore Beach* **1999**, *67*, 29–41.
29. Rosser, N.J.; Petley, D.N.; Lim, M.; Dunning, S.A.; Allison, R.J. Terrestrial laser scanning for monitoring the process of hard rock coastal cliff erosion. *Q. J. Eng. Geol. Hydrogeol.* **2005**, *38*, 363–375. [CrossRef]
30. Young, A.P.; Ashford, S.A. Application of airborne LIDAR for seacliff volumetric change and beach-sediment budget contributions. *J. Coast. Res.* **2006**, *22*, 307–318. [CrossRef]
31. Marques, F. Rates, patterns, timing and magnitude [illegible]

[illegible] M.J.; Johnstone, E.; Driscoll, N.; Ashford, S.A.; Kuester, F. Terrestrial laser scanning of extended cliff sections in dynamic environments: Parameter analysis. *J. Surv. Eng.* **2009**, *135*, 161–169. [CrossRef]
34. Hernández-Pacheco, F.; Amor, I.A. Fisiografía y sedimentología de la playa y ría de San Vicente de la Barquera (Santander). *Estud. Geológicos* **1966**, *22*, 1–23.
35. Mary, G. Évolution de la Bordure Côtière Asturienne (Espagne) du Néogène á l´Actuel. Ph.D. Thesis, Université de Caen, Caen, France, 1979.
36. Mary, G. Evolución del margen costero de la Cordillera Cantábrica en Asturias desde el Mioceno. *Trab. Geol.* **1983**, *13*, 3–37.

37. González-Amucháscategui, M.J.; Serrano, E.; Edeso, J.M.; Meaza, G. Cambios del nivel del mar durante el Cuaternario y morfología litoral en la costa oriental cantábrica. (País Vasco y Cantabria). In *Proceedings of the Geomorfologia Litoral i Quaternari*; Sanjaume, E., Mateu, J., Eds.; Universitat de Valencia: Valencia, Spain, 2005; pp. 167–180.
38. Flor, G.; Flor-Blanco, G. Raised beaches in the Cantabrian coast. In *Landscapes and Landforms of Spain*; Springer: Berlin/Heidelberg, Germany, 2014; pp. 239–248.
39. Monino, M.; Diaz de Teran, J.R.; Cendrero, A. Variaciones del nivel del mar en la costa de Cantabria durante el Cuaternario. In Proceedings of the Reunión sobre el Cuaternario 7, Santander, Spain, 21–26 September 1987; pp. 233–236.
40. Garzón, G.; Alonso, A.; Torres, T.; Llamas, J. Edad de las playas colgadas y de las turberas de Oyambre y Merón (Cantabria). *Geogaceta* **1996**, *20*, 498–501.
41. IGME. *Mapa Geológico de España E. 1/50.000. Comillas, N° 33.*; Ministerio de Industria (España): Madrid, Spain, 2009.
42. IGME. *Mapa Geológico de España E. 1/50.000. Comillas, N° 33*; Ministerio de Industria (España): Madrid, Spain, 1990.
43. Gómez-Pazo, A.; Pérez-Alberti, A. Vulnerability of the Galician coast to marine storms in the context of global change. *Sémata Cienc. Sociais Humanid.* **2017**, *29*, 117–142.
44. Pérez, J.A.; Bascon, F.M.; Charro, M.C. Photogrammetric usage of 1956-57 usaf aerial photography of Spain. *Photogramm. Rec.* **2014**, *29*, 108–124. [CrossRef]
45. Soteres, C.; Rodríguez, A.F.; Martínez, J.; Ojeda, J.C.; Romero, E.; Abad, P.; Sánchez, A.; González, C.; Juanatey, M.; Ruiz, C.; et al. Publicación de datos LiDAR mediante servicios web estándar. In Proceedings of the II Jornadas Ibéricas de Infraestructuras de Datos Espaciales, Barcelona, Spain, 9–10 November 2011; Volume 2, p. 16.
46. Sanjosé, J.J.; Martínez, E.; López, M.; Atkinson, A.D.J. *Topografía Para Estudios de Grado*; Universidad de Extremadura. Servicio de Publicaciones: Bellisco, Spain, 2013.
47. Hoffmeister, D.; Tilly, N.; Curdt, C.; Aasen, H.; Ntageretzis, K.; Hadler, H.; Willershäuser, T.; Vött, A.; Bareth, G. Terrestrial laser scanning for coastal geomorphologic research in western Greece. *Int. Arch. Photogramm. Remote Sens. Spat. Inf. Sci.* **2012**, *39*, 511–516. [CrossRef]
48. Lindenbergh, R.C.; Soudarissanane, S.S.; De Vries, S.; Gorte, B.G.H.; De Schipper, M.A. Aeolian beach sand transport monitored by terrestrial laser scanning. *Photogramm. Rec.* **2011**, *26*, 384–399. [CrossRef]
49. González Amuchastegui, M.J.; Ibisate González de Matauco, A.; Rico Lozano, I.; Sánchez Fernández, M.; Sanjosé, J.J. Cambios geomorfológicos y evolución de una barra de arena en la desembocadura del río Lea, Lekeitio-Mendexa (Bizkaia). *Cuaternario Geomorfol.* **2016**, *30*, 75–85. [CrossRef]
50. Bremer, M.; Sass, O. Combining airborne and terrestrial laser scanning for quantifying erosion and deposition by a debris flow event. *Geomorphology* **2012**, *138*, 49–60. [CrossRef]
51. Jaboyedoff, M.; Oppikofer, T.; Abellán, A.; Derron, M.-H.; Loye, A.; Metzger, R.; Pedrazzini, A. Use of LIDAR in landslide investigations: A review. *Nat. Hazards* **2012**, *61*, 5–28. [CrossRef]
52. Hansom, J.D. Coastal sensitivity to environmental change: A view from the beach. *Catena* **2001**, *42*, 291–305. [CrossRef]
53. Hall, A.M.; Hansom, J.D.; Jarvis, J. Patterns and rates of erosion produced by high energy wave processes on hard rock headlands: The Grind of the Navir, Shetland, Scotland. *Mar. Geol.* **2008**, *248*, 28–46. [CrossRef]
54. Flor, G.; Flor-Blanco, G.; Flores-Soriano, C.; Alcántara-Carrió, J.; Montoya-Mpontes, I. Efectos de los temporales de invierno de 2014 sobre la costa asturiana. *VIII Jorn. Geomorfol. Litoral Geo-Temas* **2015**, *15*, 17–20.
55. Catalão, J.; Catita, C.; Miranda, J.; Dias, J. Photogrammetric analysis of the coastal erosion in the Algarve (Portugal). *Géomorphologie Reli. Process. Environ.* **2002**, *8*, 119–126. [CrossRef]
56. Esposito, G.; Salvini, R.; Matano, F.; Sacchi, M.; Troise, C. Evaluation of geomorphic changes and retreat rates of a coastal pyroclastic cliff in the Campi Flegrei volcanic district, southern Italy. *J. Coast. Conserv.* **2018**, *22*, 957–972. [CrossRef]
57. Marques, F. Regional scale sea cliff hazard assessment at sintra and cascais counties, western coast of Portugal. *Geoscience* **2018**, *8*, 80. [CrossRef]
58. Gómez-Pazo, A.; Pérez-Alberti, A.; Pérez, X.L.O. Recent evolution (1956–2017) of rodas beach on the Cíes Islands, Galicia, NW Spain. *J. Mar. Sci. Eng.* **2019**, *7*, 125. [CrossRef]

59. Earlie, C.S.; Masselink, G.; Russell, P.E.; Shail, R.K. Application of airborne LiDAR to investigate rates of recession in rocky coast environments. *J. Coast. Conserv.* **2015**, *19*, 831–845. [CrossRef]
60. Gonçalves, J.A.; Henriques, R. UAV photogrammetry for topographic monitoring of coastal areas. *ISPRS J. Photogramm. Remote Sens.* **2015**, *104*, 101–111. [CrossRef]
61. Long, N.; Millescamps, B.; Guillot, B.; Pouget, F.; Bertin, X. Monitoring the topography of a dynamic tidal inlet using UAV imagery. *Remote Sens.* **2016**, *8*, 387. [CrossRef]
62. Mancini, F.; Castagnetti, C.; Rossi, P.; Dubbini, M.; Fazio, N.L.; Perrotti, M.; Lollino, P. An integrated procedure to assess the stability of coastal rocky cliffs: From UAV close-range photogrammetry to geomechanical finite element modeling. *Remote Sens.* **2017**, *9*, 1235. [CrossRef]
63. Westoby, M.J.; Lim, M.; Hogg, M.; Pound, M.J.; Dunlop, L.; Woodward, J. Cost-effective erosion monitoring of coastal cliffs. *Coast. Eng.* **2018**, *138*, 152–164. [CrossRef]
64. Gómez-Gutiérrez, Á.; De Sanjosé-Blasco, J.J.; Lozano-Parra, J.; Berenguer-Sempere, F.; De Matías-Bejarano, J. Does HDR pre-processing improve the accuracy of 3D models obtained by means of two conventional SfM-MVS software packages? The case of the corral del veleta rock glacier. *Remote Sens.* **2015**, *7*, 10269–10294. [CrossRef]
65. Crawford, A.J.; Mueller, D.; Joyal, G. Surveying drifting icebergs and ice islands: Deterioration detection and mass estimation with aerial photogrammetry and laser scanning. *Remote Sens.* **2018**, *10*, 575. [CrossRef]
66. Jaud, M.; Kervot, M.; Delacourt, C.; Bertin, S. Potential of smartphone SfM photogrammetry to measure coastal morphodynamics. *Remote Sens.* **2019**, *11*, 2242. [CrossRef]
67. Hayakawa, Y.S.; Obanawa, H. Volumetric change detection in bedrock coastal cliffs using terrestrial laser scanning and uas-based SFM. *Sensors* **2020**, *20*, 3403. [CrossRef]
68. Lichti, D.D.; Gordon, S.; Stewart, M.; Franke, J.; Tsakiri, M. Comparison of digital photogrammetry and laser scanning. In Proceedings of the CIPA W6 International Workshop, Corfu, Greece, 1–2 September 2002; pp. 39–47.
69. Martin, C.D.; Tannant, D.D.; Lan, H. Comparison of terrestrial-based, high resolution, LiDAR and digital photogrammetry surveys of a rock slope. In Proceedings of the Proceedings 1st Canada-US Rock Mechanics Symp, British, DC, Canada, 27–31 May 2007; pp. 37–44.
70. Sturzenegger, M.; Stead, D. Close-range terrestrial digital photogrammetry and terrestrial laser scanning for discontinuity characterization on rock cuts. *Eng. Geol.* **2009**, *106*, 163–182. [CrossRef]
71. Seymour, A.C.; Ridge, J.T.; Rodriguez, A.B.; Newton, E.; Dale, J.; Johnston, D.W. Deploying Fixed Wing Unoccupied Aerial Systems (UAS) for Coastal Morphology Assessment and Management. *J. Coast. Res.* **2018**, *34*, 704–717. [CrossRef]

Publisher's Note: MDPI stays neutral with regard to jurisdictional claims in published maps and institutional affiliations.

Article

A Reduction Method for Bathymetric Datasets that Preserves True Coastal Water Geodata

Marta Wlodarczyk-Sielicka [1,*], Andrzej Stateczny [2] and Jacek Lubczonek [1]

1 Institute of Geoinformatics, Department of Navigation, Maritime University of Szczecin, 70-500 Szczecin, Poland
2 Department of Geodesy, Gdansk University of Technology, 80-233 Gdansk, Poland
* Correspondence: m.wlodarczyk@am.szczecin.pl; Tel.: +48-513-846-391

Received: 29 May 2019; Accepted: 3 July 2019; Published: 6 July 2019

Abstract: Water areas occupy over 70 percent of the Earth's surface and are constantly subject to research and analysis. Often, hydrographic remote sensors are used for such research, which allow for the collection of information on the shape of the water area bottom and the objects located on it. Information about the quality and reliability of the depth data is important, especially during coastal modelling. In-shore areas are liable to continuous transformations and they must be monitored and analyzed. Presently, bathymetric geodata are usually collected via modern hydrographic systems and comprise very large data point sequences that must then be connected using long and laborious processing sequences including reduction. As existing bathymetric data reduction methods utilize interpolated values, there is a clear requirement to search for new solutions. Considering the accuracy of bathymetric maps, a new method is presented here that allows real geodata to be maintained, specifically position and depth. This study presents a description of a developed method for reducing geodata while maintaining true survey values.

Keywords: big data applications; data processing; data visualization; neural networks; reduction; coastal waters

---

1. Introduction

Currently, bathymetric data are one of basic data types used in systems modeling of phenomena occurring in coastal zones. In general, the quality of these data is much more important than the physical models of the phenomena, as they have a major influence on the outcome of the simulation [1]. An example is the use of bathymetry for numerical analyses related to water quality prediction in coastal waters [2] or for coastal hydrodynamics modelling [3]. Undoubtedly, the use of bathymetric data in electronic navigational charts is also an important factor. The accuracy of these data in this case determines the safety of maritime transport, especially in coastal areas where shallow water is common. In this [illegible] model, a set of points with known parameters (i.e., location and depth) are required. Bathymetric data are, therefore, elaborated using appropriate processing methods and are then presented in final hydrographic products, which can be digital terrain model or charts. These products are then used during monitoring and analysis of the transformations of coastal water bottoms. A hydrographer is responsible for the proper collection, preparation, and presentation of bathymetric data, traditionally associated with labor-intensive processing that is generally time consuming [4–6].

Hydrographic remote sensors like echosounders are normally used to collect information about the depth of a given area of water where costal, especially, shallow waters are mostly selected. These devices utilize a hydro acoustic wave to measure the distance between the transducer and the sea bottom, or the location of objects on the latter. The use of multibeam echosounders allows the collection of large volumes of information in a relatively short period of time [7–12]. Even gathering data with a single-beam echosounder [13] collects a huge amount of data, which can be called spatial big data. The processing of LiDAR (Light Detection and Ranging) data is also a big data reduction problem [14–17]. Modern bathymetric data acquisition technologies enable the collection of huge volumes of points while ensuring full coverage across a tested water area [18]. Bathymetric big data are a collection of datasets so large that it is difficult to work with using traditional data processing algorithms. Indeed, this approach significantly exceeds the minimum accuracy requirements specified by the International Hydrographic Organization in either S-44 [19] or S-57 standards [20].

Existing reduction techniques for bathymetric data use interpolated values in the form of a regular rectangle GRID with an assumed mesh size [21]. The development and application of new measurement techniques and methods of data processing mean that it is now possible to present much more accurate positional values for a given point, including its depth. This is an important step forward in terms of determining real bottom characteristics in coastal areas instead of a commonly used interpolated seabed model. The efficient application of such approaches can be assessed on the basis of research results related to sea bottom shape reconstructions [22,23]. One interesting solution in this area might also include the use of artificial neural networks for modeling sea bottom shape, as these also continually implement a surface approximation process [24–26]. It is important to be able to present depth, as this maintains surface mapping accuracy as well as the position and value of depths. In Reference [27], the authors used bathymetry data and field data as inputs for GIS (Geographical Information System), GAM (Generalized Additive Models), and kriging methods to generate a series of maps that described bottom characteristics. An interesting approach using an artificial neural network for LiDAR data was presented in Reference [28], as well as in Reference [29] where a three-layer back propagation neural network is proposed to estimate bathymetry.

We emphasize the generalization of point object sets in this study, as these comprise bathymetric geodata. In earlier work, Li [30] divided data point operations into two basic groups, those comprising transformations of individual point objects and those made on groups. Bathymetric geodata can be included within a group of points whose transformations can be divided via the following processes:

- aggregation—combining more points into one;
- regionalization—drawing a border around a group of point objects and creating a new surface;
- selective omission—the selection of objects that are more important while omitting those of lesser importance;
- simplification—the removal of objects in order to correctly present the remainder, and;
- typography—the preservation of point object dominant source symbolization while removing points [30].

The problem addressed in this study can therefore be considered to be one of selective omission because of the choice of points with the smallest depths. This process is also a simplification because of the importance of point visualization on a bathymetric chart. As available methods utilize interpolated values, it is necessary to search for new solutions. The main assumption of the proposed method of reduction is the preservation of real data, without interpolation.

2. Materials and Methods

It is important to discuss the main assumptions that underlie the development of our method. The first assumption applied here was that bathymetric geodata should be subject to reduction, obtained using a multibeam echosounder. These datasets tend to be large and contain distributed characteristics. The second basic assumption utilized here was the position of geodata collected using a GPS RTK

(Global Positioning System Real Time Kinematic) system; these data were incorporated following correct preliminary processing such that their position and depth comprised true surveying values.

2.1. Proposed Solution: Initial Assumptions

The concept of a reduction method was developed in this study as the initial research stage. We assumed that this method comprised three basic stages: the pre-division of geodata into smaller subsets (pre-processing); data clustering using artificial neural networks; and the selection of a point or points at minimum depth for a given subset.

We designated a geodata border domain such that the range depended on the coordinates of the input area. In this approach, one of the following three conditions must be met: there is only one survey point within a given domain or there are none at all; the size of the side of a divided area must be smaller than the threshold value assumed by the operator; and the actual depth difference within a tested area must be smaller than that indicated by the operator. A domain will, therefore, either not be divided into smaller areas, or it will be cut into four equal squares. This process was then repeated until the assumptions discussed above were met; sets of data comprised the outputs from these iterations, their number varied depending on the characteristics of input area, and minimum and maximum values of location and depth range were associated with bottom shape. It has to be noted that the first stage of the proposed reduction method is intended for the initial division of bathymetric data and it prepares data for clustering; it is the pre-processing of big data. Big data problems were also taken from References [31–34].

The next step was data clustering. While working on this method, except for the artificial neural network (ANN), additional methods of clustering were taken into account: K-Nearest Neighbor (KNN) and traditional hierarchical algorithms. This stage is described in Reference [21]. From the analysis, it can be concluded that the best results were obtained by the KNN and ANN methods (Kohonen). This use of a Kohonen network for bathymetric data clustering has not yet been presented in the literature and is, therefore, innovative in this context. In addition, the choice of method also considered the fact that ANN is more sensitive to the depth value than the KNN method. Bathymetric data were then clustered so that all survey points were subdivided using a Kohonen network [35–37]. In the case of the Kohonen network's learning, there was no relationship between the input data and the output of the network. The competition between neurons provides the basis for updating values assigned to the weights. It can be adopted that x is the input vector, p is the number of data samples, as follows:

$$x = \left(x_1,\ x_2,\ x_3,\ \ldots,\ x_p\right)' \tag{1}$$

Variable w is the weight vector and connected with the node l, as follows:

$$w_l = \left(w_{l1},\ w_{l2},\ w_{l3},\ \ldots,\ w_{lp}\right)' \tag{2}$$

Various samples of the training dataset were presented to the network in random order. While neurons compete with each other, the one nearest to the input sources is a winner for the input dataset.

[illegible] the node l with the nearest distance is the winner. Then, the weights of the winner are updated using the learning rule, as follows [38]:

$$w_q^{s+1} = w_q^s\left(1-\alpha^s\right) + X_i\alpha^s = w_q^s + \alpha^s\left(X_i - w_q^s\right) \tag{3}$$

where α^s is the learning rate for the sth step of training. The weight vector for the lth node in the sth step of training can be characterized as w_l^s and the input vector for the ith training sample can be described as X_i. After several epochs, a training sample is selected, and the index q of the winning neuron is defined:

$$q = \underset{l}{arg\,min} \| w_l^s - X_i \| \tag{4}$$

As our focus was on the selection of network parameters for bathymetric data clustering, we selected a series of self-organizing settings for further research including a hexagonal topology and an initial neighborhood size of ten. We then applied a Euclidean distance analysis over 200 iterations via the Winner Take Most rule. All previous studies related to the use of ANN in the developed method have been published [21,39–42]. The stage related to the clustering is shown in Figure 1.

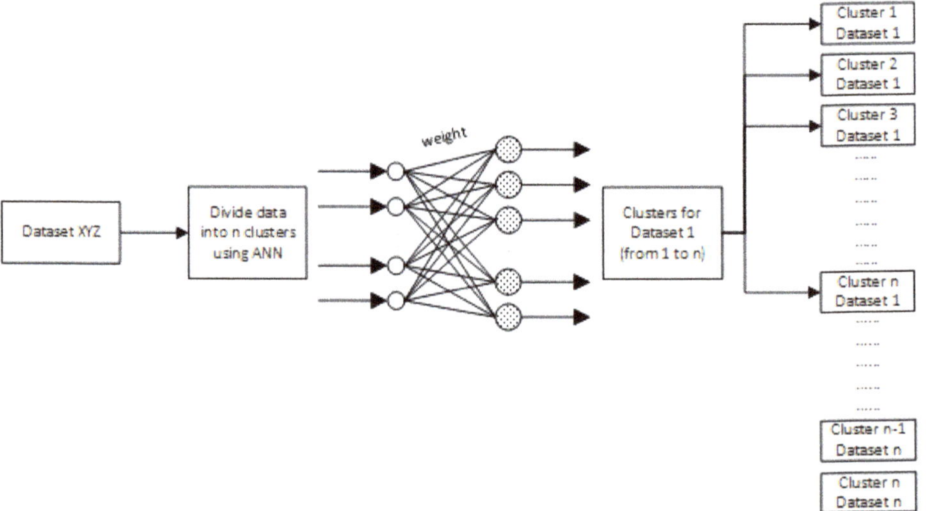

Figure 1. Clustering using an artificial neural network (ANN).

We then finally selected a series of survey points across previously designated clusters. In this context, the smallest depth (the shallowest) points will be the most important and inputs comprise collections of bathymetric geodata from individual clusters obtained in the previous phase of our method. A circle was then created around each XYZ point at a given size depending on characteristics. The radius of each circle was calculated in accordance with the authors' formula, which is presented, as follows:

$$R_i = C \times \frac{1}{Z} \tag{5}$$

In this expression, R_i denotes the radius at a given point, while C is a constant ($C > 0$), and Z is depth.

It is clear that points contained within a circle are of greater importance for analysis, as these indicate smaller depths (points with smaller values of Z) and will, therefore, be subject to reduction. This overall process was repeated until only districts containing the most significant objects remained and circles did not intersect. These reduced bathymetric geodata retained their real characteristics. In the next step, the initial assumptions of the proposed method were optimized on test data.

2.2. *The Experiment*

2.2.1. Test surfaces

Four distinct surface shape types were then determined in order to test our method that corresponded to various likely bottom surface forms. The test surfaces were built using the following mathematical functions:

- Test surface number one:

$$Z = \left(\left(\tfrac{3}{4}e^{-\left(\frac{(9X-2)^2+(9Y-2)^2}{4}\right)} + \tfrac{3}{4}e^{-\left(\frac{(9X+1)^2}{49}+\frac{(9Y+1)^2}{10}\right)} + \tfrac{1}{2}e^{-\left(\frac{(9X-7)^2+(9Y-7)^2}{4}\right)} - \tfrac{1}{5}e^{-\left((9X-4)^2+(9Y-7)^2\right)}\right) + 0.5\right) * 0.1 \tag{6}$$

- Test surface number two:

$$Z = \left(X * e^{\left(-X^2-Y^2\right)}\right) + 1 \tag{7}$$

- Test surface number three:

$$Z = \left(4 - 2.1 * X^2 + \frac{X^4}{3}\right)X^2 + X * Y + \left(-4 + 4 * Y^2\right)Y^2 \tag{8}$$

- Test surface number four:

$$Z = \tfrac{1}{2.427}\Big(\log\Big(1 + \left(\overline{X} + \overline{Y} + 1\right)^2 * \left(19 - 14\overline{X} + 3\overline{X}^2 - 14\overline{Y} + 6\overline{XY} + 3Y^2\right)\Big) * \Big(30 + \left(2\overline{X} - 3\overline{Y}\right)^2 * \left(18 - 32\overline{X} + 12\overline{X}^2 + 48\overline{Y} - 36\overline{XY} + 27Y^2\right)\Big)\Big) - 8.693\Big) \tag{9}$$

where: $\overline{X} = 4X - 2$ and $\overline{Y} = 4Y - 2$.

These surfaces illustrate the possibilities of modeling real sea bottom shapes and encapsulate irregularities (Figure 2).

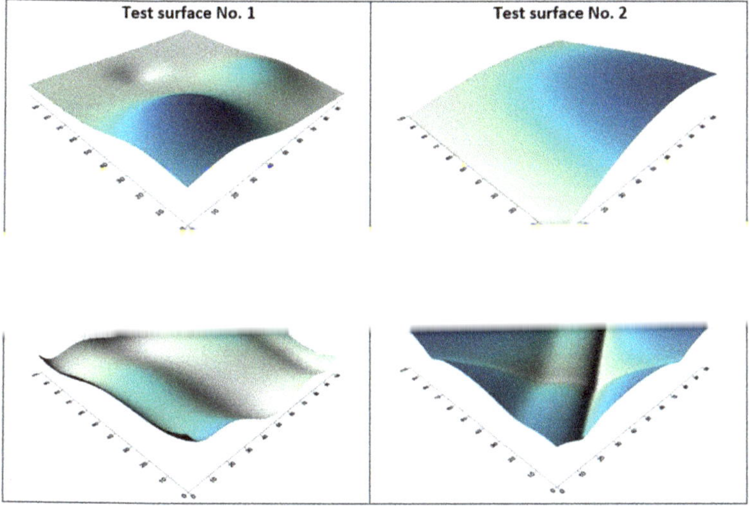

Figure 2. Alternative test surfaces considered in this analysis.

Test surface number one was characterized by irregular curvatures as it contained bottom faults, local elevations, and depressions, while test surface number two was more regular and is often seen in open water areas as it contains a gentle drop in depth associated with natural surface shapes. Test surface number three was characterized by larger changes in shape across its whole domain, while test surface number four had a shape that resembles waterway sea bottoms.

During the spatial distribution of test points on the surfaces, a simulation of scattered data was implemented, and all calculations were in the domain with sides from zero to one. Then for X and Y in this interval, Z was calculated according to the appropriate test function. In the next step, X, Y, and Z were scaled, i.e., X × 100, Y × 100, and, respectively, Z. The scaling was carried out in order to obtain sets of test points in the local metric system referring to the true measurements. Detailed characteristics of all test sets are presented in Table 1. The second and third columns of the table represent ranges of positions X and Y. In contrast, changes in the shape of the bottom were contained in appropriate intervals, which are specified in the fourth column. The last column also contains a histogram of the distribution of the depth value in a given set, where the *x*-axis shows the depth value and the *y*-axis the number of test points.

Table 1. Characteristics of test collections.

No.	X	Y	Z
1	X ϵ (0, 99.99)	Y ϵ (0, 99.99)	Zϵ (1, 13.18) 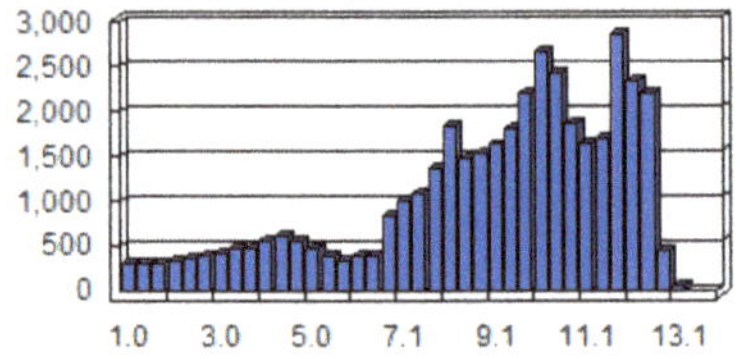
2	X ϵ (0, 99.99)	Y ϵ (0, 99.99)	Z ϵ (1, 5.28) 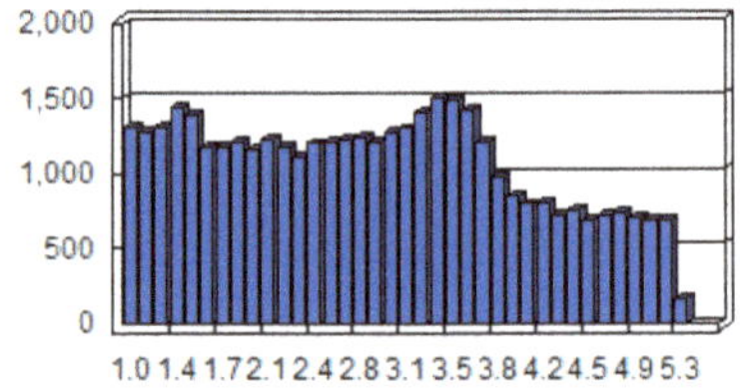
3	X ϵ (0, 100)	Y ϵ (0, 100)	Z ϵ (1, 7.71) 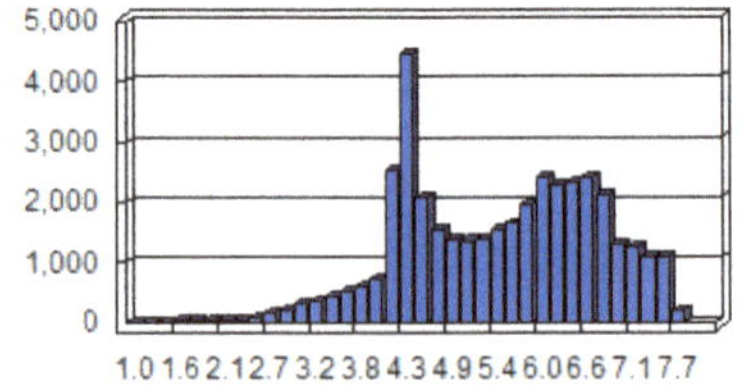
4	X ϵ (0, 99.99)	Y ϵ (0, 99.99)	Z ϵ (1, 6.24) 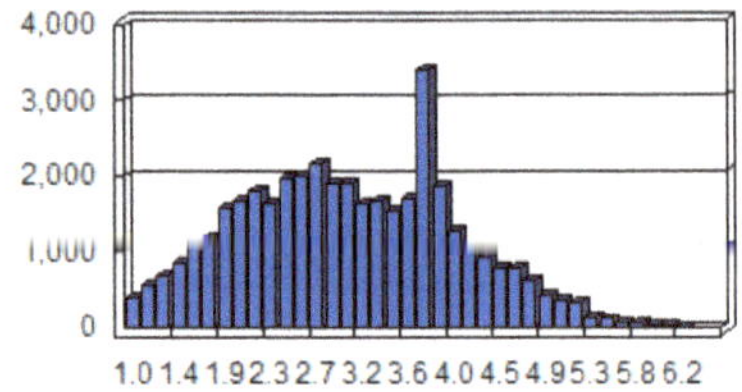

All tests were carried out using authoring scripts written using MatLab software and they can be found in the Supplementary Materials S1, S2, S3, and S4 Scripts. Thus, in order to fulfill the assumptions of our method, geodata must be collected using a multibeam echosounder, must exhibit distributed characteristics, and comprise large datasets. A total of 40,000 test points were, therefore, generated across each test surface and the test datasets can be found in the Supplementary Materials S1, S2, S3, and S4 Files.

2.2.2. Method Optimization

We tested the new approach developed here using four previously presented collections and evaluated each across three bathymetric chart scales, 1:500, 1:1000, and 1:2000. The choice of large scales was closely related to research in coastal areas where the accuracy of the obtained results is very important. The scope of this experiment is summarized in Figure 3; our approach enabled us to distinguish three basic research stages: geodata reduction optimization, the evaluation of applied criteria, and an evaluation and comparative analysis of data reduction versus existing methods.

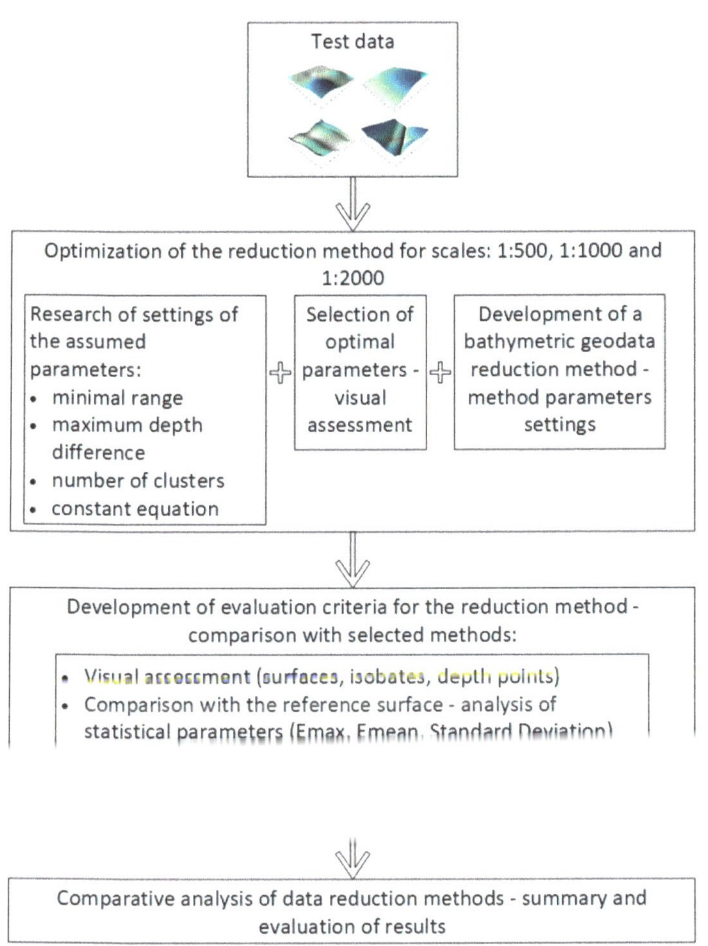

Figure 3. The research process used in this analysis.

The stage related to the optimization of the reduction method was the longest stage of the research. The four parameter settings used in our approach were optimized throughout this analysis such that:

- the threshold value in each case was related to the size of the area side under consideration—parameter value was entered on several levels from 1 to 100;
- the range between maximum and minimum depth in the study area—the minimum value was defined as 0.5 m and the maximum was equal to the maximum depth value in the tested set, changes were made every 0.5 m;
- the number of clusters—parameter value was changed from 4 to 289;
- the constant, C—as the minimum value was assumed 1, the maximum depending on the maximum depth value, changes were introduced every 0.5.

The list of the parameters examined for each of the sets is included in Table 2. The configuration of the four parameters was associated with carrying out different numbers of trials for each of the test datasets. In the case where in the specified intervals of one of the parameters the resulting number of points and their location did not change, other parameters of the method were introduced.

Table 2. List of the tested method's parameters.

Test Datasets	The Threshold Value	The Depth Range	The Number of Clusters	The Constant, C
1	6, 12, 18, 25, 50, 100 which gives 6 tested values	from 0.5 to 13.50 which gives 27 tested values	from 4 to 289 which gives 16 tested values	from 1 to 27 which gives 52 tested values
2	6, 12, 18, 25, 50, 100 which gives 6 tested values	from 0.5 to 5.50 which gives 11 tested values	from 4 to 289 which gives 16 tested values	from 1 to 13 which gives 24 tested values
3	6, 12, 18, 25, 50, 100 which gives 6 tested values	from 0.5 to 8.00 which gives 16 tested values	from 4 to 289 which gives 16 tested values	from 1 to 16 which gives 30 tested values
4	6, 12, 18, 25, 50, 100 which gives 6 tested values	from 0.5 to 6.50 which gives 13 tested values	from 4 to 289 which gives 16 tested values	from 1 to 14 which gives 26 tested values

Depending on the individual settings, a different number of output points were obtained—an example is shown in Table 3.

Table 3. An example of parameter method settings during tests—dataset number one.

Threshold Value	Depth Range ($minR$)	Number of Clusters (K)	Constant, C	Number of Points after Reduction
6	1	4	1	34,163
6	1	4	2	25,900
6	1	4	3	18,404
6	1	4	4	12,478
6	1	4	5	8341
6	1	4	5.5	6893
6	1	4	6	5756
6	1	4	6.5	4910
6	1	4	7	4225
6	1	4	7.5	3724
6	1	4	8	3364
6	1	4	8.5	3072
6	1	4	9	2852

Table 3. *Cont.*

Threshold Value	Depth Range (*minR*)	Number of Clusters (*K*)	Constant, *C*	Number of Points after Reduction
6	1	4	9.5	2678
6	1	4	10	2574
6	1	4	10.5	2495
6	1	4	11	2414
6	1	4	11.5	2367
6	1	4	12	2325
6	1	4	12.5	2296
6	1	4	13	2279

The table contains a small part of the research: a constant threshold equal to 6, *minR* equal to 1, *K* equal to 4, and *C* taking values from 1 to 13. The last column determines the number of XYZ points obtained after reduction. A total of nearly 800 tests were carried out for test set no. 1.

The number of points and the visual assessment in a given dataset were accepted as preliminary criteria. The point data that meet the criteria for visual assessments should be evenly distributed. Figure 4 shows examples of results that were taken into account in the visual assessment.

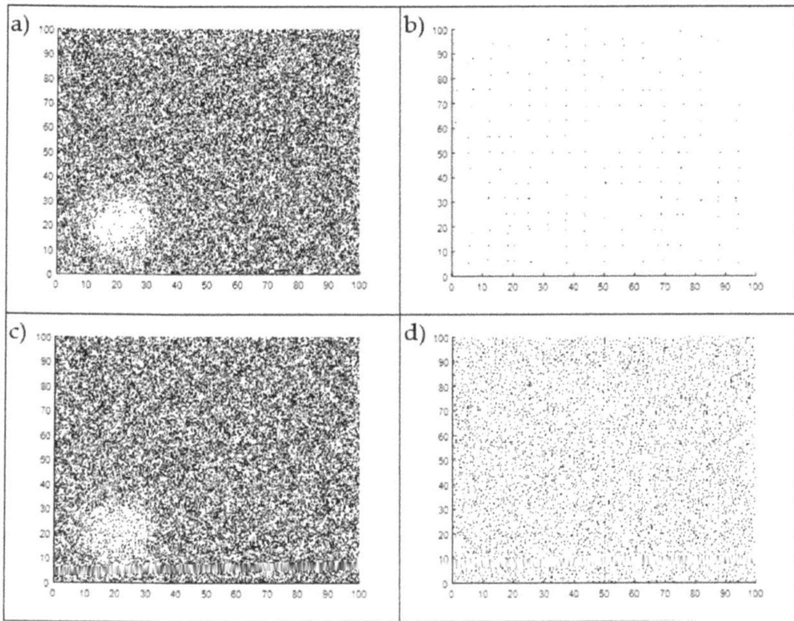

range between max and min depth: 0.5; number of clusters: 100 and constant C: 1; (**d**) threshold: 18; range between max and min depth: 0.5; number of clusters: 100 and constant C: 25.

For the parameter settings in Figure 4a, 33,942 points were obtained. It can be seen that the constant C was too small. It should be noted that in the case of test set no. 1, the depth ranged from 1 m to 13.18 m. In Figure 4b, when the example was extended to 25, only 267 points were obtained. In Figure 4c, with a minimum C value and a number of clusters equal to 100, 34,123 points were obtained;

for Figure 4a, similar results were obtained. During the Figure 4d trial, the collection was reduced to 6409 points—this collection meets the criteria of visual assessment.

Then, after rejecting the wrong results, the optimal parameter settings were selected, and the reduction method was developed in detail, which will apply to the bathymetric geodata. This enabled us to determine the optimal settings of our built-in reduction method following the detailed analysis of selected parameters including the simultaneous visualization of reduced geodata on bathymetric reporting site plans at different scales. The steps included in our reduction method with determined parameters are summarized in Figure 5.

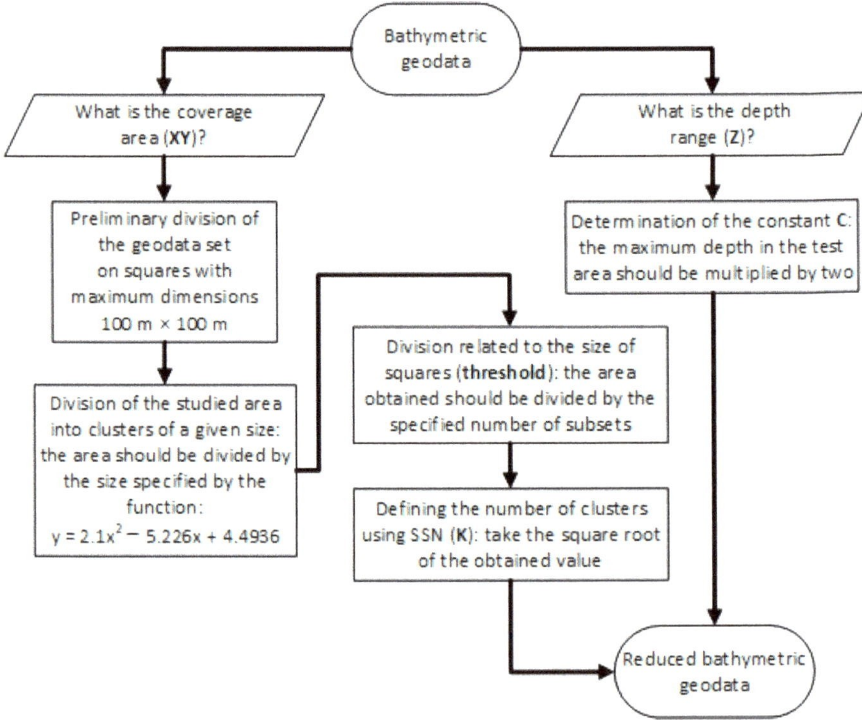

Figure 5. The bathymetric geodata reduction method used in this study.

We initially divided each coverage area into squares with the maximum dimensions of 100 m × 100 m because of the limitations inherent in processing large volumes of data. We then measured the extent of each area along x- and y-axes and assumed the smaller value to be the side of each square. Thus, if this value was greater than 100, we then divided it into smaller areas such that each received square would be subject to reduction. This means that the area covered by our data was divided depending on the desired scale of a given bathymetric chart, itself dependent on the cluster size obtained via a function with empirically selected coefficients, as follows:

$$f(x) = 2.1x^2 - 5.226x + 4.4936 \quad (10)$$

for which the graph is shown in Figure 6.

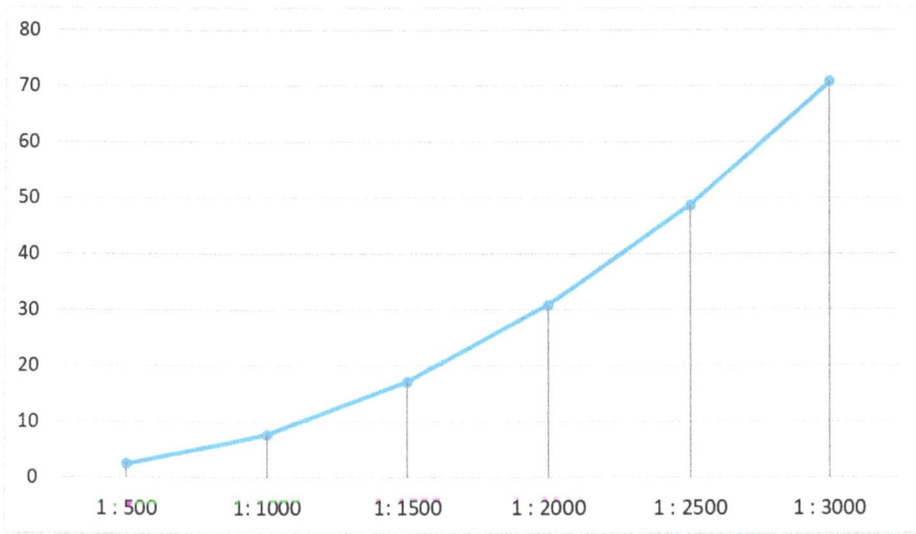

Figure 6. Graph of the cluster size function.

The horizontal axis represents the scale, the vertical axis shows the cluster size (m^2) through which the area covered by data will be divided.

This approach enabled us to select the best research results that met our visual assessment criteria and determined the functional relationship between map scale and cluster size. To this end, the value obtained using Equation (10) was then divided by a certain number of squares (Supplementary Materials Table S1); the number of subsets depended on the scope of input data and the scale of each result chart, obtained in pre-processing on the basis of each threshold value parameter for geodata areas. We then extracted the square root in each case from previously obtained values, including a value for parameter K that corresponds to the number of clusters obtained using an ANN. The last parameter in each case was the constant C, the optimal value of which corresponded to twice the maximum depth in each study area. We assumed that this constant (C) was an integer that could be obtained after rounding, while all applied parameter values in our bathymetric geodata reduction method depended on data scope and depth. The method was implemented in the Matlab environment.

2.3. Evaluation Criteria

We then established a series of criteria for assessing our reduction method and compared its operation with a selection of existing approaches. Using the reduction method, each test dataset was reduced for the three scales adopted in the tests, i.e., 1:500, 1:1000, and 1:2000—in total 18 sets of XYZ were created. In order to compare the approach developed in this study with alternatives that

[illegible] It is noteworthy that the reduction process implemented in method number one is similar to "rolling a ball" on a given surface at intervals defined by a vertex. Thus, once an input area was defined, it was then necessary to determine the scale of the resultant desired bathymetric plan; in this context, vertices corresponded to circle centers that each had a radius that equaled 1/100th of a predetermined scale. A surface was then obtained following buffering which was then re-buffered, this time in the reverse direction. This means that the final surface yielded via this process would be generalized and could, therefore, be used to build bathymetric charts at a given scale [43]. In contrast, method two reduced the points that overlapped on a resultant bathymetric plan at each assumed scale. The essence of this reduction is based on the removal of deeper (or shallower, depending on indications from an operator) values when their labels overlap; this method emphasizes just visualization, and therefore, does not consider sea bottom shape [43]. We reduced each of our test datasets at different scales for three resultant bathymetric charts. All received files were saved in the *.txt format and had three attributes: X position, Y position, and Z depth. With the help of Surfer 9 software, we modeled surfaces from all received output datasets. We compared all received surfaces with reference surfaces (test surfaces).

We applied a series of evaluation criteria when assessing the operation of our bathymetric geodata reduction method, as follows:

1. Visual assessment of the distribution of obtained points.
2. Visual assessment of obtained surfaces.
3. Visual assessment of isobaths obtained from evaluated surfaces.

Visual evaluation is subjective, but with such a number of examples, you can certainly classify them. Criteria related to the visual assessment of the distribution of reduced points was made for all test collections.

4. Comparison of obtained surfaces with our model via the analysis of statistical parameters, including:

    a. the maximum difference in Z-values among surfaces,
    b. mean difference in Z-value among surfaces, and
    c. standard deviation.

We used Surfer 9 software for the calculations. We calculated the differences in the Z-values among the test sets before the reduction and the values obtained for the created surfaces. In the next step, we estimated their absolute values and basic statistics, which would be analyzed.

5. Calculation of percentage data loss after reduction.

We calculated the percentage of the number of points lost after the reduction.

6. Percentage of the amount of data the XYZ coordinates preserved.

For this purpose, we used ArcGIS software with basic spatial analysis, which is selection by location.

3. Results

3.1. Test Datasets

We evaluated each of our test datasets, but due to the large amount of data, we present only a description of test surface No. 1 for the purposes of this paper. The surface from which data were generated in this case comprises an irregular shape containing shallow and deep areas (test surface No. 1 in Figure 1). Initial bathymetric plans of seawater at selected scales showing depth points were developed on the basis of each received set; our visual assessment of depth points in each case was focused primarily on their spatial distribution such that they were ideally regularly distributed and their descriptions did not overlap. Subsequent to analysis of all visualizations for the test surface No. 1, it was clear that the points obtained by our reduction did meet the visual assessment criteria as they only overlapped in a few places. Therefore, we concluded that in this case, a hydrographer's interference cannot be avoided during data development using our new approach. It was also the case that points remained clearly visible when method number one was applied, as these were interpolated values that had a regular distribution. The last case we examined also met our visual assessment criteria, although there was a tendency for values to cluster on isolines. One representative example of the results obtained at a scale of 1:1000 is presented in Figure 7.

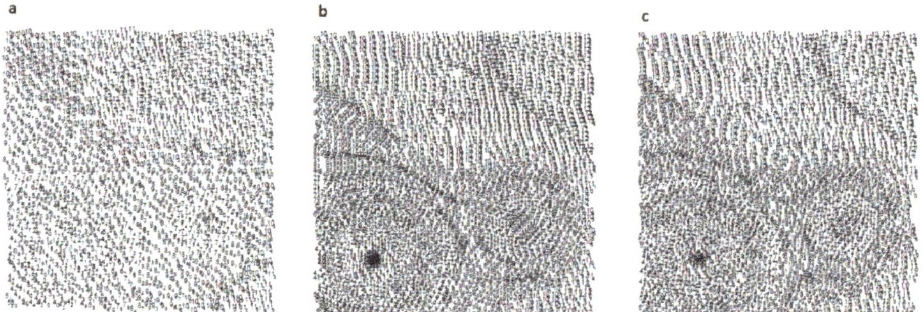

Figure 7. Depth points generated from test surface No. 1 at a scale of 1:1000. (**a**) Resolution based on our new method; (**b**) method number one; (**c**) method number two.

We also performed a visual assessment of the surface shapes generated from reduced point sets. Results showed that the surface we obtained at a scale of 1:500 from the collection reduced by our new method had almost no differences from that of the model in this case, with the only differences present at borders area. A perfect surface shape was obtained by maintaining real positions and depth values, as well as a sufficient number of points in the case of test surface No. 1. Surface roughness phenomena were seen when other methods were applied; these were associated with interpolated point values in method number one, while in method number two, these were due to an emphasis on the visualization of depth points and not on bottom shape mapping. We also illustrate the surfaces obtained at a 1:500 scale in Figure 8.

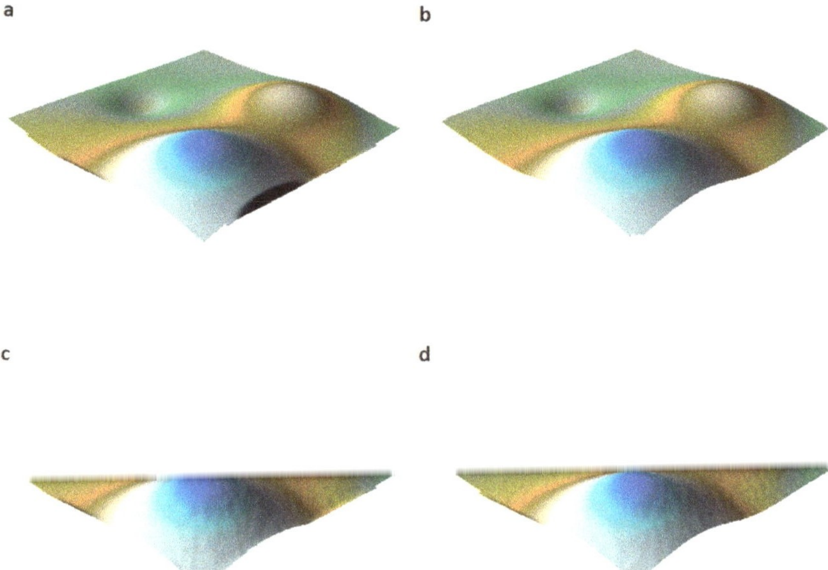

Figure 8. Surfaces obtained at a 1:500 scale using test surface No. 1. (**a**) Reference; (**b**) our method; (**c**) method number one; (**d**) method number two.

The results revealed roughness on all the surfaces obtained at scales of 1:1000 and 1:2000, likely closely related to the number of points in each set. Indeed, the surface obtained from points following reduction via our method at a scale of 1:1000 did not differ significantly from the model surface—the least of the methods studied. All tested methods generated very similar visualization results at a 1:2000 scale; however, overall, the surfaces obtained using points from our method very clearly illustrated surface shapes of shallower water but were less efficient at depth. The desirability of such an approach results from a need to enable correct bathymetry mapping in areas characterized by navigational danger.

The next step in this process comprised a visual assessment of the isobaths obtained from previously modeled surfaces. The results showed that those obtained via points generated from our reduction method almost completely coincided with surface shapes; indeed, there also seemed to be no clear difference between scales, although minimal variations were noticeable following a more detailed analysis and were indicative of very good test area mapping. The data also showed that the other methods we considered were unable to achieve such good results; the least accurately generated isobaths were obtained by our use of reduced points with method number two at a 1:500 scale. As in the case of the surfaces, this result was related to model specificity; isobaths were smoother (as is normal) at other scales; however, although their shapes were not entirely consistent with the standard. We also present a series of resultant isobaths at a 1:500 scale in Figure 9.

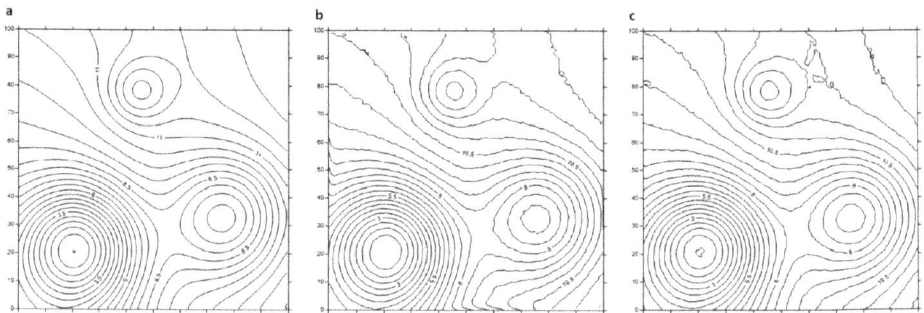

Figure 9. Isobaths at a 1:500 scale using test dataset number one. (**a**) Our method; (**b**) method number one; (**c**) method number two.

We also evaluated different reduction methods on the basis of their maximum and mean error as well as standard deviations from all tested variants (Supplementary Materials Table S2). These surfaces were then compared with the reference and all errors were calculated; in this context, maximum error can be helpful in capturing large deformations in received surface shapes. Data show that highest values of maximum error were seen in data obtained using method number one; values ranged between 67.90 cm and 72.09 cm. In contrast, the smallest errors were generated using method two and ranged between 7.03 cm and 18.42 cm in this case. Maximum errors from this approach were not much bigger, ranging between 8.38 cm and 28.32 cm; the values from each method increased in concert with scale, a natural phenomenon as the number of points decreased (Figure 10).

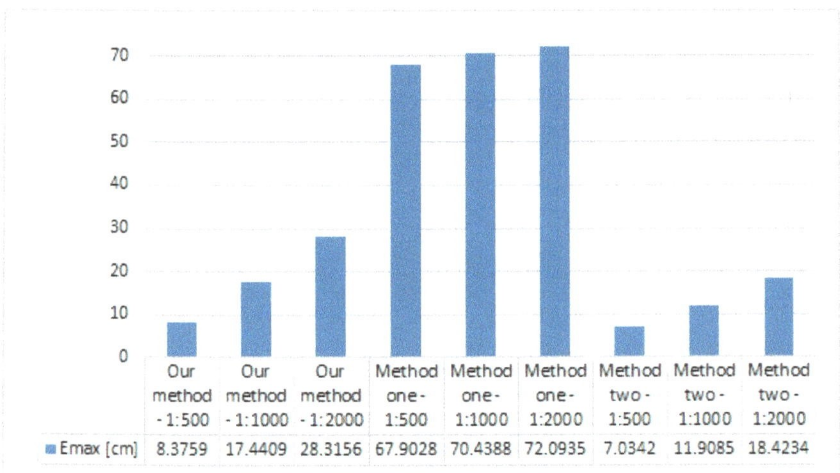

Figure 10. Maximum errors obtained for test dataset number one.

Indeed, the largest mean error at the level of 8.47 cm was obtained for points reduced using method one at a 1:2000 scale, while similar mean errors ranging between 1.32 cm and 1.82 cm were also generated using method number two. Smallest error values were obtained using our novel method; however, for 1:500 and 1:1000 scales, these errors were less than 0.50 cm and were just 1.24 cm at a 1:2000 scale (Figure 11).

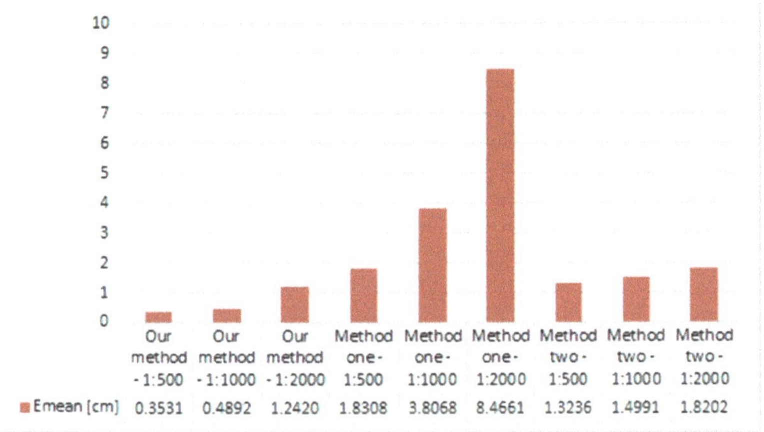

...ation results derived from test surface number one were very similar in their values and distribution to received average errors; we were, therefore, able to conclude that the best results overall can be obtained by applying our method for bathymetric data reduction, while the worst were derived from method number one. Errors in data turn out to be related to the interpolation of depth to form a square grid, as illustrated by similar research on other test datasets. The processing time during data transformation were similar and fluctuated from 15 s to 20 s for test areas (processor 2.3 GHz and RAM 16 GB).

The final evaluation criteria we applied referred to the percentage level of reduction of measurement points in each case as well as the percentage presentation of true values (Table 4).

Table 4. Comparison of data reduction levels and the number of points in each actual position in reduced datasets.

	Data Reduction Level	Number of Points in Real Positions	Data Reduction Level	Number of Points in Real Positions	Data Reduction Level	Number of Points in Real Positions	Data Reduction Level	Number of Points in Real Positions
	Test surface number one		Test surface number two		Test surface number three		Test surface number four	
Our method: 1:500	90%	100%	90%	100%	90%	100%	90%	100%
Our method: 1:1000	97%	100%	97%	100%	97%	100%	97%	100%
Our method: 1:2000	99%	100%	99%	100%	99%	100%	99%	100%
Method one: 1:500	79%	0%	70%	0%	75%	0%	73%	0%
Method one: 1:1000	95%	0%	92%	0%	94%	0%	93%	0%
Method one: 1:2000	99%	0%	98%	0%	98%	0%	98%	0%
Method two: 1:500	85%	81%	80%	81%	83%	80%	82%	82%
Method two: 1:1000	95%	81%	93%	81%	94%	80%	94%	81%
Method two: 1:2000	99%	80%	98%	81%	98%	78%	98%	81%

The analytical results of this study (Table 4) showed that our new method reduced the largest number of depth points. The data show that at a scale of 1:500, a 90% reduction can always be achieved regardless of the tested surface, while 97% and 99% were obtained at other scales.We also showed that more points were received following a reduction in the case of method one and method two; true measuring points (position and depth) were held at 100% taking into account the position and value of the depth of the reduced points when our method was applied. In contrast, the accuracy of method number one decreased following simultaneous interpolation, and therefore, depth points did not coincide with actual results. In method number two, however, just between 78% and 82% of the data overlapped despite a true position assumption in the software. It was also the case that the remainder were moved to enable a better presentation on our water area bathymetric plan; the source points for test surface one as well as those for individual scales are shown in Figure 12.

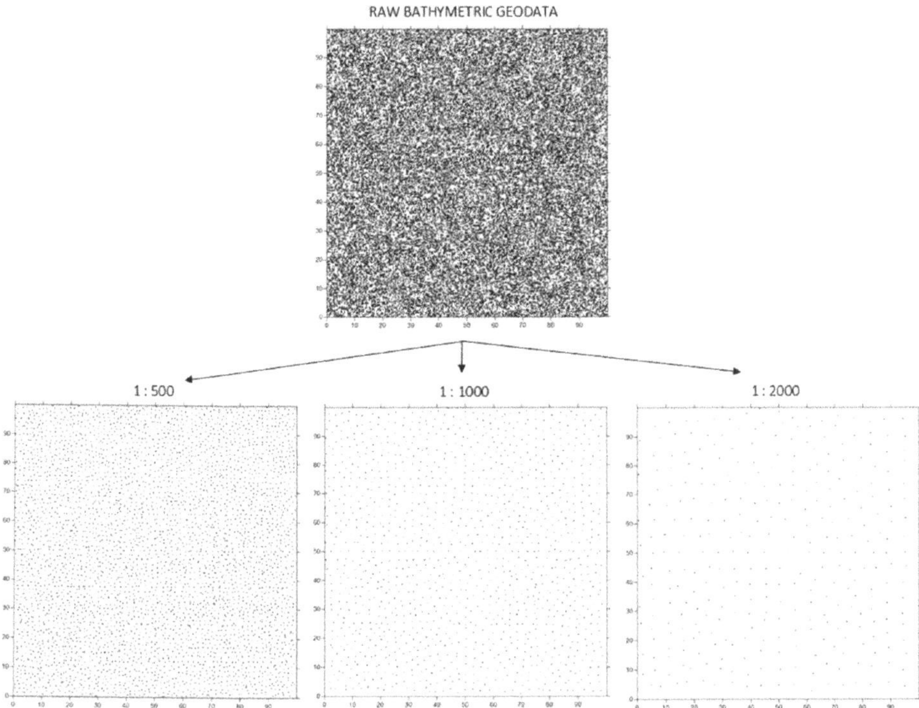

Figure 12. The points reduced in this study with our new method based on test surface one.

3.2. *Real Data*

In the last step of our research, we verified the new reduction method on real datasets. The assessed data were gathered by the usage of a GeoSwatch Plus multibeam echo sounder on board a Hydrograf XXI laboratory. As the main evaluation criteria, it was assumed that bathymetric data after reduction should still be in the real position and have real value of depth. The areas used in the research differed from each other in terms of the shape of the bottom and the characteristics of the occurring shore. The reduction was carried out for the three examined scales: 1:500, 1:1000, and 1:2000. According to the assumptions of the method, appropriate parameters were adopted for each scale.

The first test set comes from the Babina area. The input data included 5,864,171 bathymetric points. The minimal depth was 0.3 m and maximum depth was 9.34 m. After the reduction for the first area, three sets of reduced bathymetric geodata were obtained, which contained the following number of points:

- for scale 1:500 [illegible]

For the result points, digital terrain model using triangulation with linear interpolation method was developed, which are visualized in Figure 13.

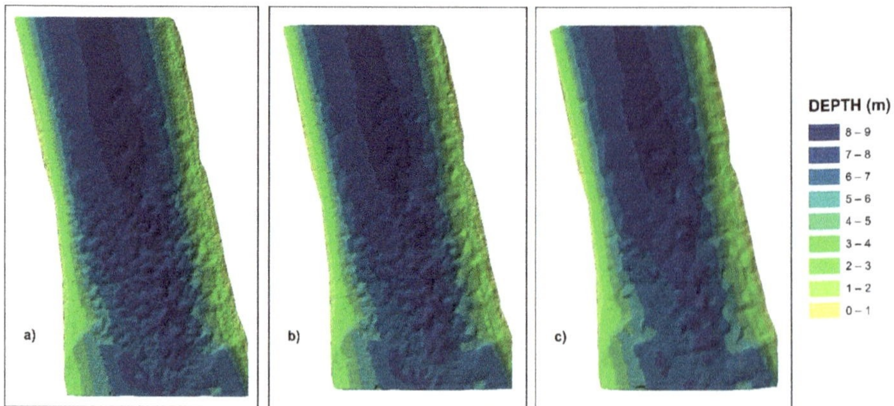

Figure 13. Surfaces from the Babina area for scale: (**a**) 1: 500, (**b**) 1:1000, and (**c**) 1:2000.

The second dataset was collected in the Debicki canal. The tested collection had 27,362,303 points and the minimal depth was 0.04 m and the maximum depth was 14.42 m. In this case, the following output datasets were obtained:

- for scale 1:500—54,346 points XYZ with minimum depth 0.3 m;
- for scale 1:1000—18,184 points XYZ with minimum depth 0.3 m;
- for scale 1:1000—7466 points XYZ with minimum depth 0.3 m.

The collections obtained during the reduction process were used to create digital terrain models, which are illustrated in Figure 14. As in the previous case, triangulation with linear interpolation method was used.

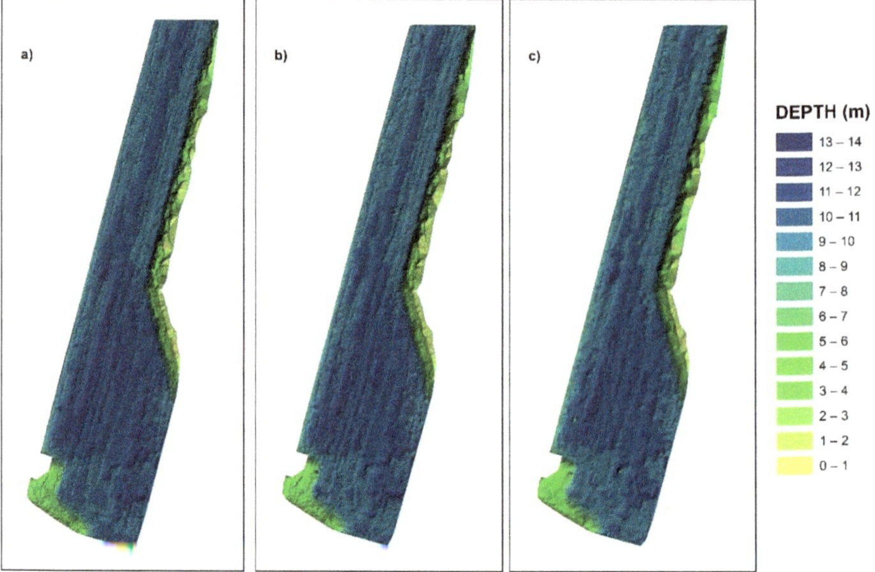

Figure 14. Surfaces from the Debicki canal for scale: (**a**) 1:500, (**b**) 1:1000, and (**c**) 1:2000.

The use of the method presented made it possible to retain 100% of the characteristics of the bottom surfaces. The geodata were commonly characterized by a lower rate of surface projection failure as real depth was preserved without interpolation. Reduction also made it possible to minimize the dataset from which results could be accurately obtained and opened a range of future processing possibilities.

4. Conclusions

Our research thesis was related to the reduction of bathymetric geodata, an issue that has a direct relationship to research in coastal zones. Our main goal was, therefore, to develop a method to reduce geodata while maintaining real measurement values. The method proposed here to reduce geodata enabled a selection of depth points to maintain the necessary accuracy of surface mapping. We tested our new method with a range of generated datasets that exhibited generally different characteristics related to the shape of the sea bottom across special areas. In order to properly evaluate the performance of our method, we defined a series of criteria and compared our results with those obtained from other methods. A comprehensive analysis of our results enabled us to present an objective evaluation of our new method. The reduction method was tested on real data. We chose coastal areas with a natural and constructed coastline and with different depth values.

The novel reduction method developed in this study allows real depth values to be maintained at their exact measurement locations. At the same time, the parameters of this method depend on the scope of input data and the depth of the water area being tested. This means that the set of input geodata is significantly reduced depending on the scale of the bathymetric map as a result of faster analysis and subsequent processing possibilities. The data we obtained was further characterized by a smaller surface mapping error, associated with maintaining real depth values without interpolation. Our new method also processes data via a novel artificial neural network approach; the bathymetric geodata obtained as a result of this reduction process can be implemented during the creation of a map or digital seabed model. We showed that a map in which XYZ points were reduced using our novel method does possess the necessary surface mapping accuracy. The results of this analysis showed that our new method for the reduction of bathymetric geodata can be utilized for the development of hydrographic products that require real data. Our research can be the basis for methods related to the creation of large numerical geographical surface models, as well as in the processing, management, and visualization of large geodata sets that comprise measurements obtained by modern remote sensing instruments.

Supplementary Materials: S1 File: Test dataset number one, S2 File: Test dataset number two, S3 File: Test dataset number three, S4 File: Test dataset number four, S1 Script: Script used to generate test surface number one, S2 Script: Script used to generate test surface number two, S3 Script: Script used to generate test surface number three, S4 Script: Script used to generate test surface number four, S1 Table: The number of subsets depending on the data range and the bathymetric chart scale, S2 Table: The results—statistical parameters.

Author Contributions: Conceptualization, M.W.-S., A.S. and J.L.; methodology, M.W.-S., A.S. and J.L.; software, M.W.-S.; validation, M.W.-S.; formal analysis, M.W.-S.; investigation, M.W.-S.; resources, M.W.-S.; data curation, M.W.-S.; writing—original draft preparation, M.W.-S.; writing—review and editing, M.W.-S., A.S. and J.L.; visualization, M.W.-S.

Funding: This research received no external funding.

Acknowledgments: This research outcome was achieved under the grant No 1/S/IG/16 financed from a subsidy of the Ministry of Science and Higher [illegible]

1. Brown, M.E.; Kraus, N.C. *Tips for Developing Bathymetry Grids for Coastal Modeling System Applications*; Army Engineer Research and Development Center, Coastal and Hydraulics Laboratory: Vicksburg, MS, USA, 2007.
2. Mishra, P.; Panda, U.S.; Pradhan, U.C.; Kumar, S.; Naik, S.; Begum, M.; Ishwarya, J. Coastal Water Quality Monitoring and Modelling Off Chennai City. *Procedia Eng.* **2015**, *116*, 955–962. [CrossRef]
3. Stansby, P.K. Coastal hydrodynamics—Present and future. *J. Hydraul. Res.* **2014**, *51*, 341–350. [CrossRef]
4. Bottelier, P.; Haagmans, R.; Kinneging, N. Fast Reduction of High Density Multibeam Echosounder Data for Near Real-Time Applications. *Hydrogr. J.* **2000**, *98*, 23–28.

5. Burroughes, J.; George, K.; Abbot, V. Interpolation of hydrographic survey data. *Hydrogr. J.* **2001**, *99*, 21–23.
6. Hansen, R.E.; Callow, H.J.; Sabo, T.O.; Synnes, S.A.V. Challenges in Seafloor Imaging and Mapping with Synthetic Aperture Sonar. *IEEE Trans. Geosci. Remote Sens.* **2011**, *49*, 3677–3687. [CrossRef]
7. Jong, C.D.; Lachapelle, G.; Skone, S.; Elema, I.A. *Hydrography*, 2nd ed.; DUP Blue Print: Delft, The Netherlands, 2010.
8. Moszynski, M.; Chybicki, A.; Kulawiak, M.; Lubniewski, Z. A novel method for archiving multibeam sonar data with emphasis on efficient record size reduction and storage. *Pol. Marit. Res.* **2013**, *20*, 77–86. [CrossRef]
9. Rezvani, M.; Hadi, S.; Abbas, A.; Alireza, A. Robust Automatic Reduction of Multibeam Bathymetric Data Based on M-estimators. *Mar. Geod.* **2015**, *38*, 327–344. [CrossRef]
10. Calder, B.R.; Mayer, L.A. Automatic processing of high-rate, high-density multibeam echosounder data. *Geochem. Geophys. Geosyst.* **2003**, *4*, 1048. [CrossRef]
11. Yang, F.; Li, J.; Han, L.; Liu, Z. The filtering and compressing of outer beams to multibeam bathymetric data. *Mar. Geophys. Res.* **2013**, *34*, 17–24. [CrossRef]
12. Kazimierski, W.; Wlodarczyk-Sielicka, M. Technology of Spatial Data Geometrical Simplification in Maritime Mobile Information System for Coastal Waters. *Pol. Marit. Res.* **2016**, *23*, 3–12. [CrossRef]
13. Specht, C.; Switalski, E.; Specht, M. Application of an Autonomous/Unmanned Survey Vessel (ASV/USV) in bathymetric measurements. *Pol. Marit. Res.* **2017**, *24*, 36–44. [CrossRef]
14. Kulawik, M.; Lubniewski, Z. Processing of LiDAR and multibeam sonar point cloud data for 3D surface and object shape reconstruction. In Proceedings of the 2016 Baltic Geodetic Congress (BGC Geomatics), Gdańsk, Poland, 2–4 June 2016. [CrossRef]
15. Blaszczak-Bak, W. New Optimum Dataset method in LiDAR processing. *Acta Geodyn. Geomater.* **2016**, *13*, 379–386. [CrossRef]
16. Blaszczak-Bak, W.; Sobieraj-Zlobinska, A.; Kowalik, M. The OptD-multi method in LiDAR processing. *Meas. Sci. Technol.* **2017**, *28*, 075009. [CrossRef]
17. Blaszczak-Bak, W.; Koppanyi, Z.; Toth, C. Reduction Method for Mobile Laser Scanning Data. *ISPRS Int. J. Geo Inf.* **2018**, *7*, 285. [CrossRef]
18. Holland, M.; Hoggarth, A.; Nicholson, J. Hydrographic processing considerations in the "Big Data" age: An overview of technology trends in ocean and coastal surveys. *Earth Environ. Sci.* **2016**, *34*, 012016. [CrossRef]
19. International Hydrographic Organization (IHO). *Transfer Standard for Digital Hydrographic Data*, 3rd ed.; Special Publication No. 57; International Hydrographic Organization: Monte Carlo, Monaco, 2002.
20. International Hydrographic Organization (IHO). *Standards for Hydrographic Surveys*, 5th ed.; Special Publication No. 44; International Hydrographic Organization: Monte Carlo, Monaco, 2008.
21. Wlodarczyk-Sielicka, M.; Stateczny, A. Clustering bathymetric data for electronic navigational charts. *J. Navig.* **2016**, *69*, 1143–1153. [CrossRef]
22. Lenk, U.; Kruse, I. Multibeam data processing. *Hydrogr. J.* **2001**, *102*, 9–14.
23. Maleika, W.; Palczynski, M.; Frejlichowski, D. Interpolation Methods and the Accuracy of Bathymetric Seabed Models Based on Multibeam Echosounder Data. *Lect. Notes Artif. Intell.* **2012**, *7198*, 466–475.
24. Cao, J.; Cui, H.; Shi, H.; Jiao, L. Big Data: A Parallel Particle Swarm Optimization-Back-Propagation Neural Network Algorithm Based on MapReduce. *PLoS ONE* **2016**, *11*, e0157551. [CrossRef]
25. Liu, S.; Wang, L.; Liu, H.; Su, H.; Li, X.; Zheng, W. Deriving Bathymetry from Optical Images with a Localized Neural Network Algorithm. *IEEE Trans. Geosci. Remote Sens.* **2018**, *56*, 5334–5342. [CrossRef]
26. Lubczonek, J. Hybrid neural model of the sea bottom surface, Artificial Intelligence and Soft Computing-ICAISC. *Lect. Notes Comput. Sci.* **2004**, *3070*, 1154–1160.
27. Sibaja-Cordero, J.A.; Troncoso, J.S.; Benavides-Varela, C.; Cortés, J. Distribution of shallow water soft and hard bottom seabeds in the Isla del Coco National Park, Pacific Costa Rica. *Rev. Biol. Trop.* **2012**, *60*, 53–66.
28. Kogut, T.; Niemeyer, J.; Bujakiewicz, A. Neural networks for the generation of sea bed models using airborne lidar bathymetry data. *Geod. Cartogr.* **2016**, *65*, 41–53. [CrossRef]
29. Huang, S.Y.; Liu, C.L.; Ren, H. Costal Bathymetry Estimation from Multispectral Image with Back Propagation Neural Network. *Int. Arch. Photogramm. Remote Sens. Spat. Inf. Sci.* **2016**, *41*, 1123–1125. [CrossRef]
30. Li, Z. *Algorithmic Foundation of Multi-Scale Spatial Representation*; CRC Press: Boca Raton, FL, USA, 2007.
31. Habib ur Rehman, M.; Chang, V.; Batool, A.; Wah, T.Y. Big data reduction framework for value creation in sustainable enterprises. *Int. J. Inf. Manag.* **2016**, *36*, 917–928. [CrossRef]

32. Habib ur Rehman, M.; Jayaraman, P.; Malik, S.; Khan, A.; Medhat Gaber, M. RedEdge: A Novel Architecture for Big Data Processing in Mobile Edge Computing Environments. *J. Sens. Actuator Netw.* **2017**, *6*, 17. [CrossRef]
33. Zhong, R.Y.; Newman, S.T.; Huang, G.Q.; Lan, S. Big Data for supply chain management in the service and manufacturing sectors: Challenges, opportunities, and future perspectives. *Comput. Ind. Eng.* **2016**, *101*, 572–591. [CrossRef]
34. Aykut, N.O.; Akpinar, B.; Aydin, O. Hydrographic data modeling methods for determining precise seafloor topography. *Comput. Geosci.* **2013**, *17*, 661–669. [CrossRef]
35. Kohonen, T. Self-organized formation of topologically correct feature maps. *Biol. Cybern.* **1982**, *43*, 59–69. [CrossRef]
36. Kohonen, T. The self-organizing map. *Proc. IEEE* **1990**, *78*, 1464–1480. [CrossRef]
37. Tang, X. Fuzzy clustering based self-organizing neural network for real time evaluation of wind music. *Cogn. Syst. Res.* **2018**, *52*, 359–364. [CrossRef]
38. Osowski, S. *Artificial Neural Networks for Information Processing*; Warsaw University of Technology Publishing House: Warszawa, Poland, 2000. (In Polish)
39. Wlodarczyk-Sielicka, M.; Lubczonek, J.; Stateczny, A. Comparison of Selected Clustering Algorithms of Raw Data Obtained by Interferometric Methods Using Artificial Neural Networks. In Proceedings of the 2016 17th International Radar Symposium (IRS), Krakow, Poland, 10–12 May 2016. [CrossRef]
40. Wlodarczyk-Sielicka, M. Importance of neighborhood parameters during clustering of bathymetric data using neural network. In *International Conference on Information and Software Technologies*; Dregvaite, G., Damasevicius, R., Eds.; Springer: Cham, Switzerland, 2016; pp. 441–452. [CrossRef]
41. Wlodarczyk-Sielicka, M.; Stateczny, A. Selection of SOM Parameters for the Needs of Clusterisation of Data Obtained by Interferometric Methods. In Proceedings of the 2015 16th International Radar Symposium (IRS), Dresden, Germany, 24–26 June 2015; pp. 1129–1134. [CrossRef]
42. Wlodarczyk-Sielicka, M.; Lubczonek, J. The Use of an Artificial Neural Network to Process Hydrographic Big Data during Surface Modeling. *Computers* **2019**, *8*, 26. [CrossRef]
43. *Caris, Bathy DataBASE Manager/Editor Reference Guide*, 2011.

Review

Advances in Remote Sensing Technology, Machine Learning and Deep Learning for Marine Oil Spill Detection, Prediction and Vulnerability Assessment

Shamsudeen Temitope Yekeen and Abdul-Lateef Balogun *

Geospatial Analysis and Modelling (GAM) Research Laboratory, Department of Civil and Environmental Engineering, Universiti Teknologi PETRONAS (UTP), Seri Iskandar 32610, Perak, Malaysia; Shamsudeen_18001318@utp.edu.my

* Correspondence: alateef.babatunde@utp.edu.my

Received: 28 August 2020; Accepted: 21 September 2020; Published: 18 October 2020

Abstract: Although advancements in remote sensing technology have facilitated quick capture and identification of the source and location of oil spills in water bodies, the presence of other biogenic elements (lookalikes) with similar visual attributes hinder rapid detection and prompt decision making for emergency response. To date, different methods have been applied to distinguish oil spills from lookalikes with limited success. In addition, accurately modeling the trajectory of oil spills remains a challenge. Thus, we aim to provide further insights on the multi-faceted problem by undertaking a holistic review of past and current approaches to marine oil spill disaster reduction as well as explore the potentials of emerging digital trends in minimizing oil spill hazards. The scope of previous reviews is extended by covering the inter-related dimensions of detection, discrimination, and trajectory prediction of oil spills for vulnerability assessment. Findings show that both optical and microwave airborne and satellite remote sensors are used for oil spill monitoring with microwave sensors being more widely used due to their ability to operate under any weather condition. However, the accuracy of both sensors is affected by the presence of biogenic elements, leading to false positive depiction of oil spills. Statistical image segmentation has been widely used to discriminate lookalikes from oil spills with varying levels of accuracy but the emergence of digitalization technologies in the fourth industrial revolution (IR 4.0) is enabling the use of Machine learning (ML) and deep learning (DL) models, which are more promising than the statistical methods. The Support Vector Machine (SVM) and Artificial Neural Network (ANN) are the most used machine learning algorithms for oil spill detection, although the restriction of ML models to feed forward image classification without support for the end-to-end trainable framework limits its accuracy. On the other hand, deep learning models' strong feature extraction and autonomous learning capability enhance their detection accuracy. Also, mathematical models based on lagrangian method have improved oil spill trajectory prediction with higher real time accuracy than the conventional worst case, average and survey-based approaches. However, these newer models are unable to quantify oil droplets and uncertainty in vulnerability prediction. Considering that there is yet no single best remote sensing technique

Keywords: oil spill; remote sensing; review; machine learning; deep learning; trajectory modeling; vulnerability assessment

1. Introduction

Oil spills are a global phenomenon that have been increasing with the rise in oil consumption [1]. Rapid world population growth [2], industrialization, and modern transportation have increased the demand for oil, which has escalated the occurrence of oil spills. Examples of oil spill incidents, which cut across diverse geographical locations, include the Canada Atlantic Empress between (2,100,000–2,400,000 Barrels), South African ship tank fire of Castillo De Bellver (1,850,000 Barrels), Mexico Ixtoc (3,300,000 Barrels), and Persian Gulf Iran–Iraq War (1,900,000 Barrels) [3]. Pipeline leakage, accidents arising from system failure, vandalism, human error [4], shipwreck, and collision are the major causes of oil spills. These result in severe ecological and economic disasters [5] which are exemplified by the cost of crude oil loss, cleaning costs, impact research funding, and rehabilitation costs. Lynch [6] reported the loss of 4.9 million barrels of crude oil and $68 billion incurred for environmental cleaning after the BP Deepwater Horizon oil spill at the Gulf of Mexico. Over $17 billion worth of natural resources were damaged [7], loss of environmental resources estimated at $37 billion was recorded [8], and tens of billions of USD were charged as fine and research funding following the disaster [9,10]. On the other hand, ecological problems include water pollution [11], shoreline, and beach contamination, which accounted for the loss of 5268 organisms after the Puerto Rico 1994 oil spill [12]. Marine invertebrate habitat degradation, oiling and smothering of individuals, interruption of food web and toxicity [13,14], sub-lethal, and mortality of marine birds [15–17] have also been documented. Further, loss of marine mammals and vegetation have been reported [18–20]. The severity of these hazards is aggravated by the slow response to oil spill disasters. Rapid response to oil spills prevents indiscriminate spread and minimizes likely consequences [21–24].

Attending to oil spills rapidly requires accurate identification of the location and extent of the spills, and several studies have been carried out to accomplish this [25–27]. Before the 1960s, traditional on-site monitoring methods were prominent in identifying oil spills [28]. However, this approach posed risks ranging from direct contact with oil to other site hazards. In addition, it was impossible to measure the extent of oil spills using this technique [25]. Ocean surveillance systems, comprising aircraft and coastguard forces, were later introduced to provide reliable and accurate monitoring. Although highly effective, its application is limited due to the high cost of acquiring data for larger areas [29,30]. Nonetheless, simple still and video photography were still common in the past although limited by short distance cover [31]. Hence, the need for larger area coverage technology such as remote sensing which is widely used today [32].

Remote sensing is the ability to acquire data from an object without physical contact [33,34]. The principle behind oil spill remote sensing monitoring was first established by Estes and Senger [35], stating the significance of aerial data for its application. Remote sensing uses sensors (optical and microwave) for data acquisition. Optical sensors utilize visible length and infrared rays while microwave sensors use longer wavelengths that receive microwaves. Both sensors have been applied to oil spill detection, but microwave sensors enjoy broader usage due to the ability to capture data at any time of the day and under any weather condition without being affected by cloud cover [25,30,36,37]. This explains the widespread adoption of Radar microwave technology for oil spill detection [28,38–41]. In identifying oil spills in water bodies, the satellites emit microwave and detect the reflected wave in return. This process reduces the energy returned to the satellites, causing it to appear as a black spot in the imagery [42] because of the visco elastic features of oil which suppress the wave growth and increase the wave dissipation [43]. However, the dark spots that appear on the SAR imageries from radar sensors are not exclusively oil spills. They could also be indicators of low wind areas, natural films, wind front areas, wind shadow at areas close to island, rain cells, current shear zones, ship wake [44], grease ice, algae blooms, and weed beds [40,45], which are termed lookalikes

To eliminate false positive identification, different methods have been used to distinguish oil spills from lookalikes. Fiscella et al. [46], Guo and Zhang [47], and Frate et al. [48] used texture and gray attribute and frequency domain, which considers geometric attribute and shapes in its classification. The threshold segmentation method that entails forecolor and background color distinction for dark spot detection and the use of threshold values to segregate oil spill from lookalikes has also been implemented [49]. However, challenges remain in clearly distinguishing oil spills from lookalikes because most of the existing methods such as traditional machine learning models, statistical models, and clustering models cannot adequately perform this classification.

Further, predicting the trajectory of oil spill is valuable in classifying the vulnerability of surrounding areas and prompt and accurate delineation of vulnerable areas is essential for decision making in disaster risk reduction. Several mathematical models based on Lagrangian particle works have been developed for this purpose but the ability to provide clear risk information with respect to the potential flow of the spilled oil is limited [50–53]. Recent advances in spatial data science, remote sensing technology and digitalization, particularly novel machine learning and deep learning algorithms, offer new opportunities to improve existing processes and address the challenges of oil spill disaster. The potential benefits of these emerging technologies have been widely projected, yet the available evidence remains limited. Many existing review studies do not include an updated review of various automated techniques, particularly novel digitalization approaches, for discriminating, detecting, and classifying oil spills from false positive elements in remote sensing imageries. Further, reviews of trajectory modeling, which is an important component of oil spill decision support systems, are limited.

Therefore, this paper aims to address this gap by undertaking a state-of-the-art review, covering the broad range of remote sensing oil spill detection, classification, and vulnerability assessment. We highlight the conceptual principles of the different remote sensing technologies for oil spill detection and the capability of novel digitalization tools for effective oil spill detection and trajectory modeling. The paper also identifies the various challenges confronting oil spill management. Consequently, the review distills what has and what has not been achieved, providing evidence-based basis for future research initiatives which is essential to addressing existing gaps in the literature in order to enhance oil spill disaster response and management utilizing remote sensing, machine learning, and deep learning algorithms.

2. Remote Sensing in Oil Spill Management

The emergence of numerous sensors over the years has improved the application of remote sensing for oil spill management, as illustrated in Figure 1 [25,30,31,54,55].

Figure 1. Remote sensing for oil spill management.

Generally, the appearance of oil slicks in water bodies varies from sensors. For example, oil spill detection in water bodies in SAR is based on the principle that oil on water surfaces decrease backscattering, thereby forming a dark spot that is different from the brightness of other surrounding areas [37]. This differs from other sensors (e.g., optical sensors) because the reflectance, absorbance, and contrast between oil and water change based on prevailing weather, oil type, and weathering reaction [25,56,57]. Remote sensing instruments for oil spills include video and photography, thermal infrared imaging, spectrometers, airborne optical, microwave, laser fluorosensors, and satellite or airborne optical and microwave sensors [58]. While spaceborne remote sensing has a higher efficiency for large area coverage, airborne surveillance is less efficient because of the limited area coverage. Comparatively, the latter is more suitable and efficient for identification of source, type, thickness, and extent of oil spills with higher spatial resolution [10,59–61] unlike the former with a lower spatial and temporal resolution but useful for monitoring remote areas. In the following sections, different oil spill detection sensors are reviewed, stating the underlying principles, advantages, and limitations of each sensor. Case studies that provide empirical perspectives on the different sensors are also examined.

2.1. Optical Remote Sensing

Optical sensors depend on externally emitted electromagnetic radiation (e.g., sun) due to its inability to support self-reflection [60]. It uses the light absorption theory, scattering, and reflection because of the different physical and chemical changes that occur after oil spill that affect its visualization as the thickness changes [62,63], and the prevailing atmospheric weather conditions [64,65]. Optical remote sensing for oil spill detection can be airborne or satellite based, and the thickness of the oil floating on water continues to change therein until it reaches 0.1 mm or less [66].

2.1.1. Optical Airborne Remote Sensing

Visible Spectrum Optical Remote Sensing

Historically, since the 1970s, visible scanners and thermal infrared have been widely used in the visible region of the spectrum due to their availability and easy identification of oil spills [58,67]. Charge-coupled device (CCD) was later developed because of its ability to provide more sensitive and selective oil spill information in the water [68,69]. For oil spill detection, the presence of moderate surface reflectance, difference, and absence of absorption in the visible region is an indicator of oil on water [25]. The reflection ranges between 480-570nm [70] with a strong reflectance and transmittance wavelengths. Generally, contrasts in oil on water surfaces are affected in two ways which can either be positive or negative depending on the viewing geometry and wind [71]. First, the sun-glint effect enhances the contrast of the non-observable oil due to wave-damping effect [72]. The second effect is the difference between the optical properties of oil and water, which are mostly characterized by high absorption in blue wavelengths [73]. Different studies have used the combination of sun glint and optical property differences for the characterization and classification of oil spills and waterbodies [56,60,74].

To improve the contrast of oil in video cameras, filters are used but their success has been limited by lack of positive discrimination. The improvement in the sensor technology led to the development of hyperspectral sensors such as the Airborne Visible/Infrared Imaging Spectrometer (AVIRIS) and Airborne Imaging Spectrometer for Applications (AISA). Hyperspectral imaging, with its high spectral resolution and large amount of data information, has become a leading technology in remote sensing applications [75]. It can acquire very narrow bands at several wavelengths in the visible, near-infrared, mid-infrared, and thermal infrared bands. Advancements in hyperspectral technology has enabled its application to oil spills detection too, showing great potential in quantitatively monitoring oil film types, areas, and oil spill volumes thereby making up for the deficiency of the existing sensors. This sensor has been applied in different oil spill disasters (Figure 2) such as the Deepwater Horizon oil spill [25,60] and Exxon Verdex Oil spill [37].

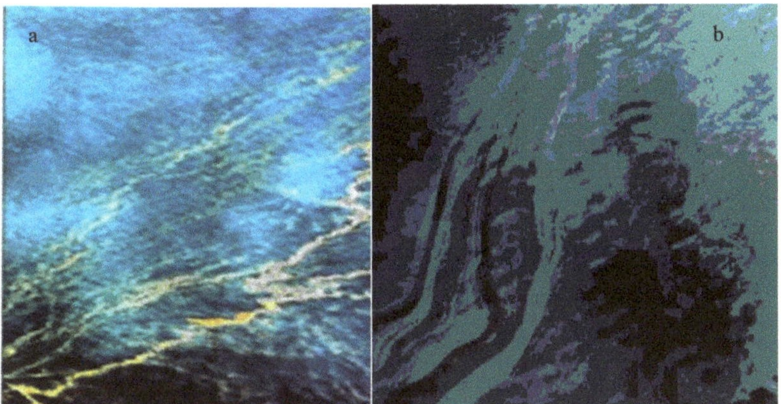

Figure 2. Examples of oil spill visible spectrum images: (**a**) Deepwater Horizon oil spill; and (**b**) Exxon Verdex Oil spill.

Infrared (IR) Remote Sensing

The thickness of oil increases the level of solar radiation absorption and re-emits thermal energy in form of radiation within the long-wave region of 8–14μm [31,58]. The appearance of oil varies in infrared images. For example, thick oil appears hot; moderately thick appears cool; and thin slick is mostly glossy or not seen because the minimum detectable layer is 2–70 μm. Differences between oil and water are distinct in a thermal infrared region since oil has a lower emissivity than water, making it a much different spectral signature than water [76]. This enhances the suitability of infrared remote sensing for the identification of oil thickness level as well as in situ identification of the oil spill origin [77]. A draw-back of this technique is that elements such as seaweed, shoreline, and other lookalikes have similar radiation with oil, giving room for false positive results. Studies (e.g., [77–82]) have used infrared airborne images for the detection of oil spills. Although oil spills were identified from the acquired imageries, the presence of lookalikes was a major limitation. However, statistical and machine learning models have been incorporated for the classification of the oil spill and lookalikes for better decision making, as illustrated in Figure 3 from the study in [83].

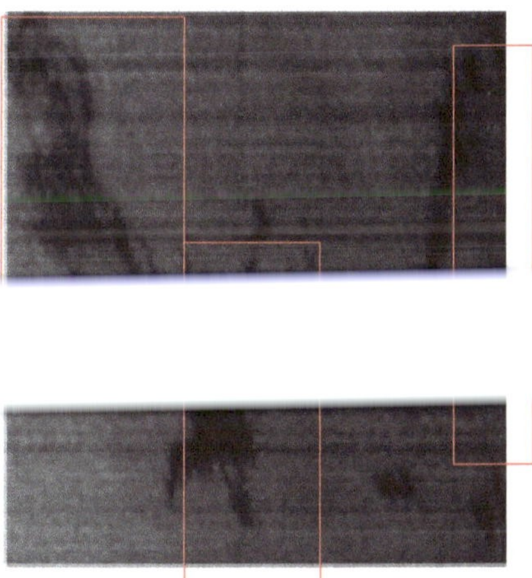

Figure 3. Example of Airborne IR imagery with oil slick in red box from the China Victory oil filed oil spill in 1998.

Near-Infrared (NIR) Remote Sensing

NIR airborne data have been widely applied in the detection of oil spills [79,84–86] and classification of oil slicks [87,88]. Spectral appearance of slicks in NIR are displayed differently in all ranges, with the highest variance in the 2–2.5 μm reflectance region, which makes the identification of oil spills challenging. The major spectral features that indicate the presence of oil slicks are the fundamental C-H stretching and bending vibration bands [79]. The near 1720–1730 and 1750–1760 nm with the addition of band near 2370 nm and second overtones near 1190–1210 nm are features capable of indicating oil slick in NIR [60,89,90]. Hörig et al. [91] observed the differences in the oil spill affected and the non-affected sand areas at 1730–2310 nm spectral feature in HyMap data. Similarly, Kühn et al. [92] developed an oil index using 1705, 1729, and 1741 nm radiance from HyMap.

Ultraviolet (UV) Sensor

The use of ultraviolet scanners for oil spill detection is based on the principle that the presence of oil in water bodies increase the reflectivity of water surfaces [93]. Being an optical sensor scanner, the presence of sunlight reflection at 0.32–0.38 μm region indicates the presence of oil in water. Due to the high reflectance of oil in water when using UV, very thin oil slick has reflectance which can be mapped when the sheens/slick is less than 0.1 μm [25,31,55,58,94]. However, when the micron is higher than 10 UV, scanners are unable to detect oil thickness [95] and can make a false detection in the presence of wind sheen, sun glints, and seaweed. The interference in UV and infrared images are different but the combination of the two gives an improved result for oil spill detection [58].

2.1.2. Optical Satellite Remote Sensing

The use of optical satellites for oil spill detection in place of airborne sensors has been limited due to the timing and frequency of the overpass [25,60]. For example, during the Exxon Valdex spill, within a month, only a day satellite imagery was acquired for the monitoring of oil [25]. Optical satellites are affected by cloud cover, bad weather, and absence of sunlight [96–98]. In addition, the processing of datasets from optical satellites requires more time, which could delay emergency response to oil spills.

With advances in remote sensing technology, satellites such as medium resolution satellites (MODIS) were utilized for oil spill detection to reduce the long-time observation interval to about three days [25]. Although other high spatial resolution optical sensors (e.g., SPOT and IKONOS) are also available, these are expensive, have few spectral channels and low temporal resolution, and are unable to provide daily observation data [3,36,37,60,99]. MODIS satellite has a moderate resolution band of 250–1000 m, and it is usually used during the day due to its ability to utilize illumination from sunlight [73]. However, the lower resolution makes it difficult to identify 200 m long oil slicks due to its coarse nature [73]. It is affected by cloud cover but its ability to provide multiple wavelengths gives more information to discriminate slicks caused by algae [30,100]. In addition, the availability of sun glint affects the appearance of oil spills. Hu et al. [101] examined the potential of MODIS in the identification of oil spills in marine environment by comparing its result with SAR imagery. Oil slick was identified on a 500-m resolution band blue and green at 469–555 nm and at the short-wave bands of 1240 and 2130 nm, highlighting the high potential of MODIS. However, the presence of freshwater runoff and surfactants from phytoplankton blooms affects the ability to distinguish between oil spills and lookalikes. Similarly, the study of Srivastava and Singh [102] used 250- and 500-m resolution MODIS imagery, revealing that oil spills are better identified from the shorter wavelengths of the visible channel and seen at bands 3, 2, and 1 at 645, 555, and 469 nm, respectively. In addition, Sun et al. [73] used MODIS for the detection of large size oil slick (1000 m) long at the early period of the oil spill near the damaged platform at the Gulf of Mexico, suggesting that MODIS is better for large size oil spill detection.

Other multispectral optical satellites such as thematic images (30 m) of Landsat satellites, which perform observations every 16 days, are also being used to monitor marine oil spills. The National Oceanic and Atmospheric Administration's (NOAA) weather satellite Advanced Very-High-Resolution Radiometer (AVHRR) image, which has a high temporal resolution, can be monitored four times a day in a certain sea area [60]. However, its spatial resolution (1.1 km) is low. For oil spill detection in water, the middle infrared (MIR) and thermal infrared (TIR) can indicate the presence of oil spill because the presence of oil in water shows a distinctly different spectral signature [103]. Oil spills are detected by identifying the difference in the thermal contrast between the emissivity of the oil slick and the background. The period of image acquisition in the day has a direct influence on the capability of the sensor's oil spill detection because it has higher brightness temperature during the day than at night. This is because of the heat capacities of water and oil as well as change in surface tension that reduces the heat transfer across the air-sea interface [104]. Consequently, the appearance of water containing spilled oil is brighter during the day than ordinary water with a reduction between 2 K (in AVHRR channel 3) and 3 K (in AVHRR channel 5). To achieve a desirable output, studies have combined the two satellites to complement each other. Nonetheless, very thin sheens of oil slicks at 20 µm are not detectable because they are affected by sea surface temperature [60]. Ning et al. [32], Su et al. [105], and Nie and Zhang [106] adopted this sensor for oil spill detection. Examples of applied optical remote sensing imagery are indicated in Figure 4 as adopted from [60].

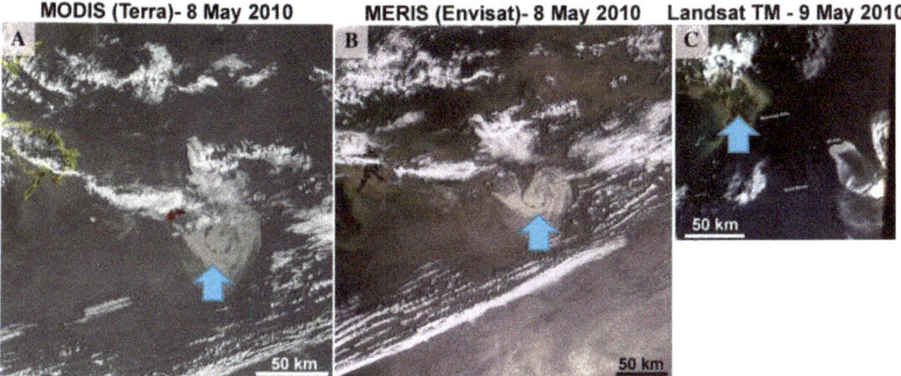

Figure 4. (**A**) MODIS; (**B**) MERIS; and (**C**) Landsat 5 TM showing the site of DWH oil spill in blue arrow.

In 2015, the European Space Agency (ESA) lunched the multispectral Sentinel-2 satellite to aid the operational activities of the Copernicus Program. The satellite comprises 13 spectral bands. The near infrared bands have a spatial resolution of 10 m while the spatial resolutions of the red-edge/short wave infrared band and the atmospheric correction band are 20 and 60 m, respectively [107,108] (see Figure 5). Nezhad et al. [108] indicated that oil slick can be detected at the wavelengths of 400 and 1400 µm. N

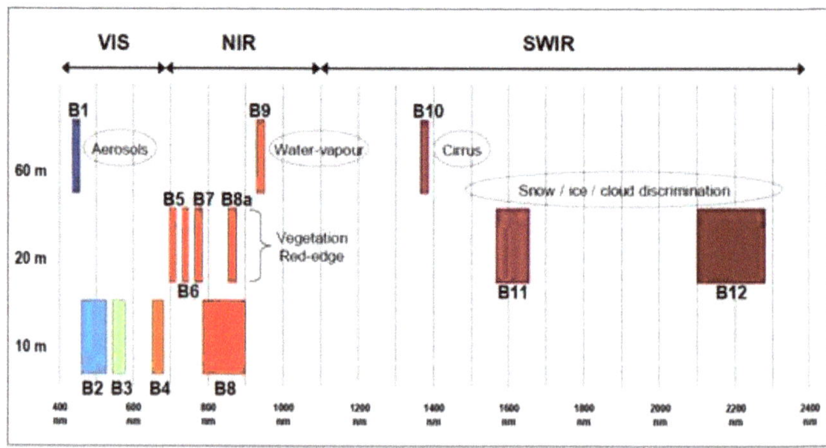

Figure 5. Sentinel 2A and 2B bands.

2.2. *Microwave Airborne and Satellite Remote Sensing*

The use of microwave sensors for marine pollution monitoring, especially oil spill detection, has received considerable attention due to its all-weather and all-day capability [30]. The presence of oil spills in water is measured through surface reflectance. Since the emissivity of water and oil are different at 0.4 and 0.8 μm, respectively, active remote sensing can determine this difference for detection of oil spill. Synthetic Aperture Radar (SAR) and Side-Looking Airborne Radar (SLAR) are examples of active sensors used in oil spill detection. SAR (<1 m) has a higher spatial resolution than SLAR (<10 m), is less expensive, and is often used for airborne remote sensing. Synthetic aperture technology is based on the principle of Doppler Effect, relying on short antennas to achieve high spatial resolution [76]. Its underlying concept is premised on the notion that oil in water reduce short gravity and capillary waves to shorter wavelengths which leads to a reduction in radar backscatter. This can mainly be caused by Bragg scattering from surface wave with similar wavelengths [111]. Over the years, different radar satellites have been launched with various configurations, frequency, and polarization that enable the performance of vital functions, including marine oil spill monitoring. Table 1 highlights the different radar satellites in use. Similar to passive sensors, the presence of biogenic elements such as seaweed and high waves indicate false positives of oil spill presence (lookalikes). In addition, SAR comprises both single and full polarization. The single polarimetric SAR has a large swath coverage and requires auxiliary information with large number of data samples for oil spill detection. The fully polarimetric SAR usually contain extra information for the measured scattering matrices, enabling the easy discrimination of oil spills from other ocean features [25,60].

Table 1. Past and current SAR Satellite Sensors.

Satellite	Launch Year	Frequency (GHz)	Band	Operator
SEASAT	1978	1.27 GHz	L	National Aeronautics and Space Administration (NASA)
ERS-1	1991	5.5 GHz	C	European Remote Sensing Satellite
ERS-2	1995	5.5 GHz	C	European Remote Sensing Satellite
ENVISAT-ASAR	2003	5.30	C	European Remote Sensing Satellite
TerraSAR-X	2005	9.65	X	European Remote Sensing Satellite
ALOS-PALSAR	2006	1.27	L	European Remote Sensing Satellite
RADARSAT-2	2007	5.40	X	German Earth observation satellite

Table 1. *Cont.*

Satellite	Launch Year	Frequency (GHz)	Band	Operator
Tandem-X	2010	9.65	X	German Earth observation satellite
Cosmos Skymed-1	2007	9.65	X	Italian Space Agency
Cosmos Skymed-2	2010	9.65	X	Italian Space Agency
TecSAR	2008	9.59	X	Israel Aerospace Industries
Kompsat-5	2013	9.66	X	Korean Space Agency
Sentinel-1a and -1b	2013/2016	5.405	C	European Space Agency
RADARSAT-Constellation (3 satellites)	2018	5.405	C	Canadian Space Agency

Several studies have used SAR images for oil spill detection. Ivanov and Zatyagalova [94] used SAR satellite remote sensing data with the aid of GIS to identify oil spills in the sea of Okhotsk, the Caspian Sea, the Black Sea, and the Gulf of Thailand and concluded that these tools are capable of providing information to aid the understanding of oil spillage in the marine environment. Kostianoy et al. [112], in a study on satellite remote sensing of oil spill pollution in the southeastern Baltic Sea, identified oil spill area of about 20.33 km^2 using SAR Imagery of ERS-2. Gallego et al. [113], in a study on segmentation of oil spills on side-looking airborne radar imagery with auto encoders, indicated that SLAR satellite images acquired from the placing of 2 SAR antennas on an aircraft are capable of detecting oil spill. Other studies (e.g., [29,61,114–117]) have also recorded successes in the application of microwave remote sensing for oil spills detection. However, similar to other sensors, the similarity in the visual appearance of oil slick with other lookalikes limits its application, as illustrated in Figure 6.

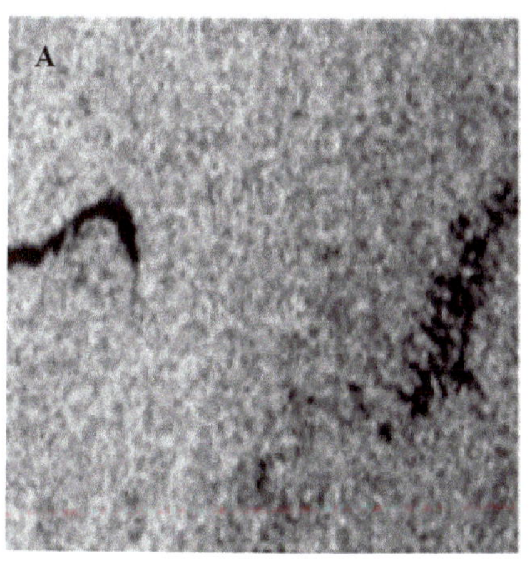

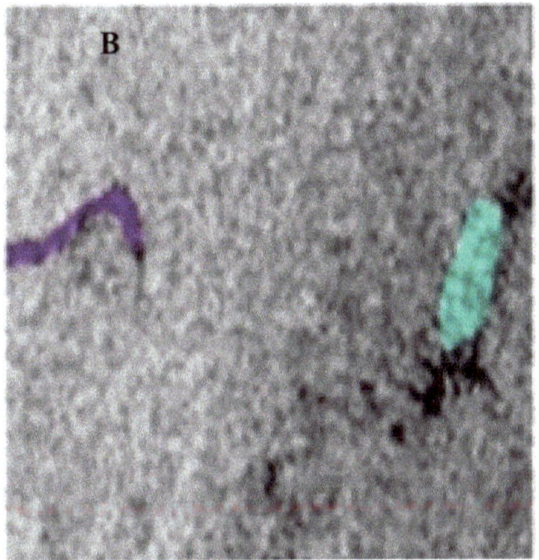

Different methods exist for oil spill detection, but, due to variations in oil spill incidents, there is no single most appropriate method [5]. To address the complexity in oil spill detection because of lookalikes in satellite imageries, different techniques have emerged. These methods, which include the use of graph representative [118], statistical based model [119], and clustering model [120], are based on threshold value and generally comprise of three stages: (1) dark spot oil spill region detection and identification from background; (2) feature extraction from identified dark spots; and (3) classification of features to oil slick and lookalikes [30,45]. Recently, the ubiquity of digitalization tools such as machine learning models (e.g., traditional classifiers, neural network, auto encoders, and deep learning convolutional neural network) is being leveraged to improve the oil spill detection and classification process. Details of these emerging digital trends are presented in the following section.

3.1. *Image Segmentation Technique*

Image segmentation is the process of dividing images into different regions of objects under consideration or interest [121]. The method converts images into highly contrasting levels, making it easier to examine and analyze isolated regions. It also requires lesser computation than other methods [122]. The thresholding and segmentation methods are used for the identification of oil spills represented as dark spots in satellite images due to the low level of backscattering.

3.1.1. Dark Spot Detection

This stage of oil spill detection is fundamental because identifying the dark spot is essential for obtaining perfect feature extraction for classification [37]. Overtime, several methods have been adopted for the dark spot detection. Manual cropping of dark spot formation and the use of threshold-based techniques were applied in early segmentation studies wherein bimodal N * N pixels (in which N = 25 pixels) histogram was used [123–125]. Vyas et al. [126] adopted a two-step methodological approach for spot detection and extraction using global threshold for pre-processing and pixel sampling. Adaptive thresholding using a multiscale pyramid technique and a clustering step to distinguish oil from its surrounding has also been implemented [127,128]. Hysteresis thresholding developed by Canny [129] was used by the authors of Kanaa et al. [130–134]. Chen et al. [135] used energy minimization function for the preservation of oil spill segmentation edges which gives a better performance than DRLSE method in the exact and precise oil spill region contours. Chen et al. [136] used a pixel segmentation technique for oil spill detection and classical classification method for the elimination of false positives in detection. Alattas [122] used Gamma distribution based on minimum cross-entropy thresholding as a dichotomous class pixel (also known as bi-modal images) for the detection of oil spill. Other studies (e.g., [135]) have used a combination of threshold algorithm and canny algorithm in the identification of oil spill region edge with a considerable high edge detection accuracy. Mira et al. [137] used two mathematical segmentation approaches (graph technique and j-image representation technique). They established that the use of graph segmentation has high performance in the detection of oil spills but produces high percentage of false positive detection. Notably, the J-image segmentation generated a lower percentage of false positives despite its overall lower performance.

Although improvements in this area have been recorded, the absence of proper procedure on how to acquire the features is a major setback. Other detection methods include the use of statistical approaches such as wavelets [138] and QinetiQ's algorithm [139]. Li et al. [140] used statistical K-distribution method with the aid of histogram values to identify areas of oil spill from dark spots in a satellite imagery. Despite the successes of these methods in identifying and preserving oil slick shape for feature extraction, their applications are limited when oil slicks are thin. In addition, these approaches are unable to distinguish between dark spots that are not oil slick since all dark spots are classified as oil slick.

3.1.2. Dark Spot Feature Extraction

The segmentation of the dark spots is followed by feature generation through the quantification of relevant attributes such as shape, size, backscattering, attenuation, and texture boundary [141]. Extracted features are grouped into three forms: geometric/shape feature, backscatter feature, and gradient feature [30,141]. Presently, there is no consensus on feature extraction procedures, evinced by the diverse feature selection approaches adopted by different researchers. Table 2 presents the commonly used feature extraction parameters. Some studies have used two features [142], and other studies adopted between eleven to fourteen features [46,48,143]. While Liu et al. [44] indicated that five features give better accuracy for oil spill and lookalikes selection, Topouzelis et al. [99] indicated that ten features are more appropriate. In addition, some other studies have successfully adopted the extraction of similar attributes for the classification of oil spills and lookalikes [141,144–146].

Table 2. Different Oil Spill Detection Features Feature classification.

N	Feature Category	Feature	Code
1		Area	A
2		Perimeter	P
3		slick complexity	C
4	Geometric/shape feature	perimeter to area ratio	P/A
5		shape factor I	SP1
6		slick width	SW
7		Spreading	S
8		Dark spot mean	DSMe
9		dark spot standard deviation	DSSd
		background mean	BMe
1	Backscatter feature	backgrounds standard deviation	BSd
1		dark spot power to mean ratio	OPm/Bpm
1		mean contrast	ConMe
1		max contrast	ConMax
1		Gradient mean	Gme
1		gradient standard deviation	Gsd
1	Gradient feature	gradient max	GMax
1		gradient min	GMin
1		gradient power to mean ratio	Gpm

Since Dark spots in satellite imageries could be oil slicks or other elements (e.g., sea grass, internal waves, low wind speed area, and other biogenic films), image classification is essential to distinguish oil spills from these other elements (lookalikes). Over time, several models have been developed to accomplish this. Solberg et al. [130] used a statistical model to automatically distinguish oil spill and lookalikes from an adaptive threshold dark spot segmentation of dual polarization ENVISAT SAR satellite imagery, which established the reliability of using statistical methods for automatic oil spill detection. Utilizing polarization and geometric and texture feature fusion, Wu et al. [147] classified oil spill and lookalikes from dark objects observed on a polarized SAR image with a 95% accuracy. In addition, fuzzy logic was adopted by Liu et al. [44] using eight features selected through ANOVA for the classification of oil spill and lookalikes. An accuracy of 80.9% was obtained. The relatively lower accuracy is attributable to the small number of SAR images used twenty-six (26)

[illegible] comparison of the methods accuracies is challenging due to variations in the number and type of data used. Other studies (e.g., [149–152]) have also adopted object-based image analysis (OBIA) models in combination with fuzzy logic for marine oil spill detection with reported high accuracy. It is noteworthy that the increasing popularity of digitalization tools such as novel machine learning and deep learning algorithms in the fourth industrial revolution (IR 4.0) offer potentials to automate oil spill detection and enhance classification processes with higher accuracy.

3.2. Machine Learning for Oil Spill Detection

Machine learning is one of the major digitalization tools in the fourth industrial revolution [153]. It is an aspect of Artificial Intelligence (AI) that evolved from pattern recognition and computational learning theory, depending on pattern and inference for execution of tasks [154,155]. Machine learning models are increasingly being used to solve a wide range of complex problems, including oil spill detection on satellite images for prompt decision making [115]. Different from the image segmentation statistical method, ML models utilize traditional classifiers following a five-step approach compromising of: (1) dark spot identification; (2) feature extraction; (3) feature selection; (4) model training and validation; and (5) feature classification. To date, various traditional machine learning models have been applied for the detection of oil spills from satellite imageries. These models are used for pertinent tasks such as segmentation, feature extraction and classification of oil spill and lookalike. This include Support Vector Machine, Decision Tree, Random Forest, Boosted Decision Tree, and Artificial Neural Network.

3.2.1. Support Vector Machine (SVM)

SVM model was developed to identify a linear class boundary. It is underpinned by the structural risk minimization principle and Vapnik–Chervonenkis dimension principle that enables it to achieve a high accuracy with limited samples. The SVM model has been applied to oil spill detection due to its capability to handle high data dimensionality. For instance, Li and Zhang [156] used SVM model for the detection of oil slick and lookalike, indicating that lower sample data size enables higher accuracy when compared with ANN-based method. The study further concluded that utilizing a large sample data size reduces model training time. In addition, Mera et al. [115] used five different feature selection machine learning algorithms to detect oil spill from a satellite imagery, with the SVM's recursive feature elimination achieving the highest accuracy and Cohen's kappa coefficient of 87.1% and 74.06%, respectively. Wan and Cheng [157] implemented a three-step approach of pre-processing, dark spot detection, and analysis with the aid of object based classification and SVM model to classify oil spill dark spots in a SAR image and analyze the changes in the slick on the sea. Analysis of these studies indicate that the SVM can achieve optimum classification despite using limited training samples.

3.2.2. Decision Tree (DT)

Decision Tree (DT) is a classification and regression model that comprises several trees. Decision trees are efficient in multi-feature extraction and robust de-noising and removing outliers without overfitting [158]. DT has been well adopted for marine oil spill detection. This is evident in the study of Topouzelis and Psyllos [159] which indicated the ability of decision forest model to identify elements with higher importance in the detection of oil slick and lookalike, attaining a cumulative accuracy of 84.4% from nine features out of the twenty-five features assessed. Equally, Li et al. [160] used DT on spectroscopic images for oil spill detection with a high accuracy. In addition, Singha et al. [158] reported a higher automated accuracy for decision tree in the selection of features and classification of oil spill and lookalike. The higher accuracy of decision tree is based on its repeated division of a set of training data into smaller subsets based on the principle of testing one or more of the feature values. Unlike other machine learning models, DT does not depend on the assumption of a variable specific distribution or the independence of the variables from one another.

3.2.3. Random Forest (RF)

Random forest (RF) is an ensemble classification algorithm that uses a majority vote to predict classes based on the partition of data from multiple decision tree. This model was developed by Breiman [161]. RF generates multiple trees through the random sub-setting of a predefined number of variables to subdivide each node of the decision trees, and by bagging. RF algorithm for marine oil spill detection is based on multi-feature technique which has been proven to have a robust performance to noise and outliers without overfitting [162,163]. Tong et al. [164] implemented a three-step approach for the detection and classification of marine oil spill and lookalike, accomplishing an accuracy of 92.99% and 82.25% for Radarsat-2 and UAVSAR polarimetric SAR datasets, respectively. However, a higher percentage of the available studies on RF focus more on oil spill vegetation classification with limitation in marine oil spill detection [96,97,153].

3.2.4. Artificial Neural Network (ANN)

ANN is mostly applied to marine oil spill detection due to the ability to represent each morphological feature with input neurons, resulting in a reduced false positive detection. Ma et al. [165] applied ANN's five step process for the classification of features after using principal component analysis (PCA) for dimensional reduction. Singha et al. [141] and Singha et al. [146] combined two different ANN models, using the first for the segmentation of oil spill spot and the second for classification of oil slick and lookalikes. While the former recorded 95.2% accuracy, the latter achieved 91.6% accuracy for oil spill detection. The variation in accuracies was due to the difference in the number of training images used. Following the three-stage oil spill detection approach, Topouzelis et al. [99] used two neural networks and ten feature selection attributes for the detection of dark spots and classification of oil spill and lookalikes. Accuracies of 94% for dark spot formation and 89% for oil spill and lookalikes classification were obtained. Insufficient data were a major limitation in this study. Comparative analyses of different methods have been documented in other studies. Karathanassi et al. [148] used neural network (NN) and a statistical correlation-based algorithm, Mahalanobis classification, for the classification of oil spill and lookalikes. The NN algorithm outperformed the statistical algorithm with a difference of 35%, 20%, 15%, and 30% for lookalikes, oil spill, land area, and sea surface, respectively. It is noteworthy that the above-mentioned studies have used either optical or microwave satellite sensor imageries. There is limited documentation of the use of other types of imageries, particularly airborne like Unmannered Ariel vehicle (UAV) and Sideward Looking Airborne Radar (SLAR) for oil spill detection.

Further advances in computer vision techniques have enabled the use of other AI algorithms (e.g., deep learning), which are more effective in individual class discrimination. Deep learning has begun to gain traction in oil spill detection tasks because of its higher accuracy and superior automation features.

3.3. Deep Learning

Deep learning models are based on the use of neural networks that contain units with certain activities and parameters [166,167]. The deep learning neuron comprises of several layers that transfer the input data to outputs with high level of progressive learning. Deep learning models include both supervised (e.g., convolutional neural network and recurrent neural network) and unsupervised algorithms (e.g., autoencoder and deep belief network) [168]. The recent application of deep learning models for marine oil spill detection is due to its increased success in various image classification tasks. Considering the complexity of oil spill and lookalike classification, DCNN is becoming widely used because of its ability to extract suitable features for classification. Similarly, Object Based Image Analysis (OBIA) is becoming more prominent since it can overcome the limitations of the pixel-based image classification models that assume that [illegible]

for image segmentation and the suitable feature for classification remains a major challenge.

The multi-layer neuron present in the deep learning algorithm enables it to describe complex functions with higher accuracy than the shallow networks in the traditional machine learning models. Thus far, different deep learning models have been applied to marine oil spill detection with varying accuracy outcomes. For example, the gradient boosting model, which is embedded with a multi-layer network, enhances oil spill detection accuracy, especially in SAR images [150,170]. Other deep network models have been applied to oil spill detection using polarimetric synthetic aperture radar images.

Chen et al. [171] used two deep learning algorithms (stacked auto encoder and deep belief network) to identify oil spills and categorize them into minerals, emulsion, and biological slicks during the Norwegian oil-on-water program of 2011. The study revealed that deep network has higher accuracy for oil spill detection than traditional ANN and SVM, irrespective of limited availability of data. Similarly, using unsupervised classification, Gallego et al. [113] utilized deep neural auto encoders to identify the oil spill point on a SLAR satellite image acquired from the placing of 2 SAR antennas on an aircraft and achieved 100% overall accuracy and an harmonic mean (F_1) of 93.01% at the pixel level. Jiao et al. [5] used deep convolutional neural network on a non-satellite image from unmanned aerial vehicle (UAV) for the detection of oil spill and Otsu algorithm for the filtering of the detection, resulting in 57.2% reduction in cost in comparison to using traditional manual inspection.

In addition, there are attempts to use object detection techniques for processing images [5]. This dates to the introduction of the adaboost algorithm as a boosted cascade method for simple feature detection in 2001 [172]. Although the method has been applied to different areas such as agriculture, transportation, and medicine [173–176], its application to oil spill detection is limited. In one of the few documented studies, Krestenitis et al. [45] used semantic segmentation detection and deep convolutional neural network models for oil spill detection in a SAR imagery, concluding that the DeepLabv2 model out-performed other models in accomplishing the task. However, more recent instance segmentation models (e.g., Mask R-CNN) and semantic segmentation models (e.g., YOLO V2, V3, and V4) have performed better in object recognition, detection, and segmentation [173–176].

4. Oil Spill Trajectory Modeling for Vulnerability Assessment

Classification of vulnerable oil spill zones is important to improve decision support systems, reduce oil spill cleaning cost, and identify environmentally sensitive areas [15]. In the past, many oil spill vulnerability assessments were based on worst case, average and survey-based approaches. Castanedo et al. [177] considered socioeconomic, physical, and biological features of the Cantabrain coast to classify the area's vulnerability to oil spill impacts into high, moderate, and low vulnerability levels. Depellegrin and Pereira [178] adopted similar procedures to classify vulnerable locations of 237km of the shoreline using physical and biological properties of the area, with consideration of shoreline sinuosity, orientation, and wave exposure. Azevedo et al. [179] developed a web-based GIS for the prediction of oil spill vulnerability based on the physical, socioeconomical, biological, and global vulnerability index of the intertidal area of the water body. Major shortcomings of these approaches are the absence of definite risk level of the coastal environment to different types of spills, such as pipeline leakage and spill intensity, and the possibility of uncertainty and subjectivity in experts' opinions.

To overcome the limitations of the worst case, average and survey-based methods, and improve the accuracy and spontaneity of the prediction of oil spill movement pattern as a means of identifying vulnerable coastal and marine ecosystems, several mathematical models have been developed to date [180,181]. These models form the basis for developing computer programs such as ADIOS 2, Oil Spill Contingency and Response (OSCAR), General National Oceanic and Atmospheric Administration Operational Oil Modeling Environment (GNOME), Medslik-II, and Spill Impact Model Application Package (SIMAP) [182,183]. More recently, the semi-implicit cross-scale hydro-science integrated system model (SCHISM) which uses water surface elevations and currents for oil spill trajectory modeling has been developed [183]. After the occurrence of oil spill, several physical and chemical transformation processes take place. The models are structured upon Lagrangian methods that use random walk and random flight method [184,185] for surface and sub-surface modeling of these processes, including oil transportation. Trajectory models are either in 2D on the ocean surface which depends on the prevailing environmental factors such as wind and wave or 3D which enable the sub-surface weathering with the capability of ocean current. While the spilled oil is treated as a series of Lagrangian elements that are tracked based on time and space when the oil is at the surface, it is treated as droplets that are tracked based on size when the oil is at the subsurface [185]. Oil spill spread, evaporation, entrainment, emulsification, dissolution, biodegradation, photo-oxidation, and sedimentation are the principal trajectory processes that interact with each other for vulnerability assessment [15,185].

The hydrodynamics, metrological, environmental features, and oil properties play important roles in modeling the trajectory. The spreading equation for predicting oil spill trajectory was first developed by Fay [186], forming the basis of most spreading algorithms such as the thick slick of [187,188]. Other models, as documented by Elliott et al. [189], Johansen [190], and Galt and Overstreet [191], have been developed since then. These are based on the principle that the spread of oil depends on the oil viscosity and the breaking energy. The Fay [186] algorithm is appropriate for immediate oil spill spread modeling. Analytical models such as ADIOS and the recent OILTRANS model of Berry et al. [192] have also been used at various times.

Integration of the existing oil spill trajectory models have been found in the study of Gług and Wąs [182] that combined Lagrangian discrete particle algorithm with cellular automata for the modeling of oil spill spread using ocean wind, current, heterogeneous state of space and instability of time as conditioning parameters. With the aid of MEDSLIK II, Yu et al. [193] developed an oil spill risk assessment map of Chinese Bohai Sea using locally available oceanographic data for the prediction of oil spill trajectory which indicated five different coastal zones that were highly vulnerable to oil spill. Similarly, Chiu et al. [183] predicted oil spill trajectories using SCHISM and X-band Radar and established that schism model can perform accurate oil spill trajectory simulation using wind, surface elevations and ocean surface current as metrological parameters. Amir-Heidari and Raie [194] used wind (advection) and ocean current as metrological parameters for the creation of an active and passive decision support system (DSS) based on consequence modeling of GNOME for response planning for accidental oil spills in the Persian Gulf. To address the peculiarities of regions where snow and ice cover large sections of the water body, Nordam et al. [51] adapted the OSCAR model by using sea-ice velocity to predict the trajectory of oil spill. An evaluation of the three case studies revealed that the use of sea-ice velocity, ocean wind, wave, and current is not always feasible for predicting trajectory, especially when there is constraint by land. In addition, the quantification of oil droplet size distribution has not really been accurate in the present studies thereby affecting the oil spill transportation prediction. The applicability of these models is limited when the vertical and horizontal spreading are different, spatially, and temporally. In addition, modeling the dispersion above the edge of linear features like coastlines where the oil spill is concentrated at a point remains problematic. Further, not all the models enable integration with Geographical Information System (GIS) framework to aid better visual output and geo-statistical analysis.

5. Discussion

5.1. *Remote Sensing for Oil Spill Detection*

This study has described different remote sensing techniques used for oil spill monitoring. The remote sensing applications were divided into optical airborne, optical remote sensing satellite, microwave airborne, and microwave satellite. The distinctive difference between the optical and microwave remote sensing is the ability of the latter to work under any weather condition and period of

[illegible] similar visual representation as oil spills, leading to false positive detection of oil spills [15]. However, optical remote sensing imageries contain diagnostic spectral information that indicate the spectral band and value, distinguishing oil spills [197,198]. In addition, the use of airborne remote sensing is less efficient compared to the satellite observations in terms of area coverage. Thus, the airborne are more suitable and efficient for rapid identification of source, type, thickness, and extent of oil spills [199]. There are differences in the appearance of oil spills across the different sensors. For example, the presence of oil spill in optical airborne remote sensing at the visible area indicates high surface reflectance difference. It also indicates the absence of absorption with reflectance ranging from 480 to 570 nm. The appearance of oil in IR airborne sensors is based on the difference in oil and water that creates a distinct thermal infrared region due to the lower emissivity from oil than water. The appearance of oil in a NIR is based on the fundamental C-H stretching and bending vibration bands while UV depends on the sun reflection value to indicate the presence of oil slick. In the optical satellite, oil slick is identified on 500-m resolution blue and green bands at 469–555-nm and short-wave bands at 1240 and 2130 nm of MODIS. For the multispectral satellite, oil spills are detected by identifying the difference in the thermal contrast between the emissivity of the oil slick and the background. Oil spills are seen as dark spots in both the airborne and satellite microwave sensors which are measured through surface reflectance. However, the absence of specialized sensors for marine oil spill detection is evident in the literature, which indicates the need for further research in this area.

In determining the suitability of sensors for oil spill monitoring, the following criteria should be considered: resolution, cost, data collection and processing duration, spatial resolution, operational period, altitude, and weather dependency. As ascertained by the authors of Topouzelis [37,200], the width window of a typical oil spill does not exceed 10 m, which indicates that a sensor with 10-m resolution is more appropriate for oil spill detection. Table 3 characterizes different sensors for oil spill applications in literature. Analysis of Table 3 shows that operational ability at any time of the day is a crucial factor in the selection of sensors. While other sensors are affected by weather and cloud cover, radar sensors can be used at any time and under any weather condition [28,37,42,44,113,122,133]. This supports real time capturing of oil spill data and enables early decision making before the weathering process commences. Due to the urgent need for oil spill data to support rapid mapping and cleanup activities [37,54,200], and considering the higher spatial and temporal resolutions of airborne remote sensing in comparison to satellite remote sensing, optical airborne sensors are still being adopted [25,31,36,55,76,201]. However, for a synoptic capture of large areas of oil spills, satellite sensors are more appropriate [25,37,60].

Table 3. Summary comparison of the different oil spill sensors.

Sensors	Spatial Resolution	24-h Operation Ability	False Positive Effect	Altitude	Weather Operation
Visible	high	No	Affected by elements such as seaweed, darker shoreline)	Below or Above 500 m	Affected by cloudy and non-clear weather
Infrared	High	Yes	Affected by seaweed and shoreline	Below or above 250 m	Affected by heavy fog and cloudy sky
Near-Infrared	High	Yes	Affected by seaweed and shoreline	Below or above 250 m	Affected by heavy fog and cloudy sky
Ultraviolent	High	No	Sun glint, wind sheen, and seaweed cause lookalikes.	Below or above 250 m.	Requires a clearer atmosphere
Optical Satellite	Medium	Yes	Sun glint, wind sheen, and seaweed cause lookalikes.	700–900 km	Affected by heavy cloud cover
Radar sensor	High	Yes	Affected by several elements such as high wave, seaweed, grease, etc.	Airborne (10–12 km) and Satellite (700–900 km)	It can work under any weather condition, but wind speed contributes to the detection of oil spill.

3.2. Automatic Oil Spill Detection

The automated detection of marine oil spill has been conducted on different remote sensing images. For optimal performance, existing models (statistical, machine learning, and deep learning) are mostly applied on SAR microwave imageries more than other types of remote sensing imageries because of its high resolution and easy depiction of dark spots. The variation in the sizes of oil spills compound the selection of an appropriate oil spill detection method. However, over time, as illustrated in Table 4, there has been steady development in the detection methods starting with the segmentation process which comprises of three stages: darks spot identification, feature extraction, and feature classification [30,45]. For the dark spot detection, statistical methods that include hysteresis thresholding, adaptive thresholding, wavelets, and QinetiQ's algorithm are mostly used although limited by the inability to discriminate between oil spill and lookalikes because they both form dark spots. Further, these methods are unable to detect oil spills when the slick is thin. The extraction of features after dark spot identification is based on the geometric, backscattering, and gradient features. The absence of clear guidelines and inconsistencies in the number of features extracted affects the outcome. While some studies have used just two features, some have used fourteen [46,48,143]. Some other studies have used similar extracted geometric, backscatter, and gradient features [141,144–146] and have been able to classify between oil spill and lookalikes based on that.

Table 4. Different Automated oil spill Detection Methods.

Author	Task	Algorithm	Method	Accuracy
[164]	Oil spill detection from Radarsat-2 and UAVSAR polarimetric SAR images	Random Forest classifier	Machine learning	Overall accuracy of 92.99% and 82.25% were achieved from the two datasets respectively.
[113]	Segmentation of oil spills on side-looking airborne radar imagery	Deep Neural autoencoders	Deep learning	F1 score accuracy of 93.1%
[171]	Oil spill detection using Polarimetric Synthetic Aperture Radar Images	Deep Learning Algorithms (Stacked Auto Encoder and Deep Belief Network)	Deep learning	Above 80.0% ROC accuracy
[203]	Observation of oil spills through Landsat thermal infrared imagery	Ocean surface temperature.	Traditional segmentation method	Reported a high accuracy (accuracy percentage was not documented)
[156]	Oil spill detection based on morphological attributes	SVM	Machine learning	Reported a high accuracy (accuracy percentage was not documented)
[126]	Using feature extraction and threshold-based segmentation for oil spill detection on SAR images	Spot Extraction and Global Threshold	Traditional segmentation method	Reported a high accuracy (accuracy percentage was not documented)
[204]	Satellite SAR oil spill detection using wind history information	Wind history	Traditional segmentation method	Reported a high accuracy (accuracy percentage was not documented)
[45]	Identification of marine oil spill from SAR images	Semantic segmentation algorithms (UNet, LinkNet, PSPNet, DeepLabv2, DeepLabv2 (msc), DeepLabv3+)	Deep learning	DeepLab3+ had the highest mIoU accuracy 65.06%
[149]	dynamics in optical MODIS images	Object-based images analysis	Object-based images analysis	Highest user accuracy at 94.2% and producer accuracy at 73.5%.
[150]	Coast, Ship and oil spill detection from side-looking airborne radar images	Two-stage CNN	Deep learning	98.34% overall accuracy was achieved.
[151]	Detection and object-based classification of oil spill and lookalike	Object-based fuzzy classification	Object-based images analysis	83% overall accuracy for oil spills and 77% for lookalikes

Table 4. *Cont.*

Author	Task	Algorithm	Method	Accuracy
[152]	Oil spill detection from SAR images	Object-based images analysis and Fuzzy logic	Object-based images analysis	97.34% oil spill probability accuracy
[205]	Oil spill detection from spaceborne SAR images	Deep convolutional neural networks algorithm	Deep learning	An overall accuracy, recall and precision value of 94.1%, 83.51 and 85.70% were achieved respectively.
[206]	Oil spill detection from Quad-Polarimetric SAR images	CNN	Deep learning	Mean Intersection over Union (MIoU) accuracy of 90.5% was achieved.
[169,202]	Detection of oil spill, lookalike, ship, and land area	Mask R-CNN	Deep leaning instance segmentation	Overall accuracy of 96.6%

Feature classification, the last stage in the automatic oil spill detection process entails discriminating lookalikes from oil spill based on the extracted features. Statistical threshold method and fuzzy logic have been implemented for this purpose. A major drawback in current approaches is the inconsistency in the type and number of data which makes it difficult to undertake accuracy assessments of these methods. Digitalization, one of the megatrends in the fourth industrial revolution, has facilitated the use of machine learning models for oil spill management via the following techniques: dark spot identification; feature extraction and section; model training; model validation and feature classification. To date, different traditional machine learning models (e.g., SVM, DT, RF, and ANN) have been applied to marine oil spill detection with high accuracy. Out of the multitude, SVM and ANN have been mostly applied. SVM produces higher accuracy despite using fewer data while ANN can treat each oil spill feature as an independent neuron, reducing false positive error. Thus far, the different machine learning models applied have been limited to feed forward image classification. Support for the end-to-end trainable framework is generally lacking, limiting the accuracy of the models.

More recently, deep learning models, buoyed by advances in computer vision are being utilized for instance and semantic segmentation due to their strong feature extraction and autonomous learning capability. Unlike machine learning models, deep learning algorithms possess a high number of hidden layers that enable better data abstraction, prediction, generalization, and transferability. Evaluation of various deep learning architecture including deep belief network, auto-encoder, convolutional neural network (CNN), and recurrent neural network (RNN) indicates that CNN has a higher accuracy. This is because it supports both pixel-level classification and detects object categories independently. CNN's higher accuracy in object recognition, detection, and segmentation as well as its ability to localize the instance under consideration which enables oil spill detection in complex scenarios (e.g., where oil spill and lookalike overlay each other) have been highlighted [169,202]. The CNN could be applied for both semantic and instance segmentation. However, limitations exist in the application of instance segmentation algorithms to oil spill detection. Nonetheless, recent studies suggest that considering other features in the sea (e.g., ship, water body, and land area) can enhance detection accuracy [169,202].

5.3. Oil Spill Trajectory Modeling for Vulnerability Assessment

The traditional use of experts' opinion affects oil spill vulnerability assessment because it lacks real-time predictive capabilities and it has limited applicability to different scenarios of oil spill incidents such as pipeline leakage, vandalism, and tank trucks. The transformation processes that occur immediately after oil spill is also not considered, limiting the accuracy of the method. Modern mathematical models, which are based on Lagrangian methods, enable both surface and sub-surface oil spill tracking and accurate and real-time prediction of oil spill movement for area vulnerability assessment.

Despite the successes of these mathematical models, accurate quantification of the size of oil droplets remains a challenge, limiting the accurate calculation of the uncertainty. Current models also use wind speed and ocean current without taking into consideration the tidal current exemplified by the rise and fall of sea level which affects the modeling of the oil spill trajectory. Further, the current models prevent oil spill dispersal above the edge of costal lines and frameworks for robust integration of trajectory models with geospatial technologies for better output, visualization, and geo-statistical analysis are still lacking. However, GNOME and MEDSLIK are exceptions since they work well with GIS for better visualization of oil spill pattern.

6. Lessons Learned

6.1. *Remote Sensing*

(1) Remote sensing for oil spill monitoring and management can be divided into two broad categories (optical (active) and microwave (passive) sensors), which can be further classified into four subcategories: (optical and passive airborne; optical and passive satellites).

(2) Appearance and thickness of oil spills in optical airborne remote sensing vary across sensors. For example, in the visible sensors, the presence of oil spill indicates high surface reflectance difference. In addition, absence of absorption in the visible region indicates the presence of oil on water with reflectance ranging 480–570 nm.

(3) The appearance of oil in IN passive airborne sensor is based on the difference in oil and water that makes a distinct thermal infrared region due to the lower emissivity from oil than water.

(4) The appearance of oil in a NIR is based on the fundamental C-H stretching and bending vibration bands while UV depends on the sun's reflection value to indicate the presence of oil slick.

(5) Improvements in the visible sensor hyperspectral remote sensing led to the emergence of AVIRIS and AISA, which have high signal to noise ratio and good spectral resolution.

(6) The quantity of oil slick in water can be best measured with airborne NIR.

(7) Passive satellite sensors for oil spill monitoring are mostly affected by cloud cover, bad weather, absence of sunlight, and limited ability to differentiate between lookalikes and oil slick.

(8) Presently, active sensors are the most widely used remote sensing technology for oil spill detection due to its ability to operate under any weather condition. They detect oil spills from wind speeds within 2–10 ms^{-1}. However, false positive appearance of lookalikes affects the reliability of these sensors.

(9) To date, there is no single best remote sensing technique that can unambiguously and reliably detect oil spills without lookalikes.

(10) Developing remote sensing technology that can detect oil spills without the appearance of lookalikes is a vital field of research that is worth exploring.

6.2. *Automatic Oil Spill Detection*

(1) The deficiency of remote sensing technology in distinguishing oil slick from lookalikes necessitated [illegible]

[illegible] machine learning models have been mostly applied for the classification of marine oil spill and lookalike.

(4) The limitation of machine learning models to feed forward image classification with no support for the end-to-end trainable framework affects its accuracy.

(5) Deep learning models' strong feature extraction and autonomous learning capability enhance their performance and facilitates high accuracy in marine oil spill detection.

(6) The application of instance segmentation models for rapid detection, recognition, and segmentation of oil spills from lookalikes are still limited.
(7) The inclusion of other elements in the sea enhances model detection accuracy
(8) The absence of verified database of oil spill images affects automated detection accuracy since present modeling approaches depend on either data augmentation or transfer learning on existing models to enhance the accuracy.

6.3. Oil Spill Trajectory Modeling for Vulnerability Assessment

(1) The need for accurate and timely mapping of vulnerable locations inspired the development of different oil spill trajectory models.
(2) Oil spill trajectory models use Lagrangian particles to indicate oil spill in water surfaces and sub-surfaces.
(3) The available oil spill trajectory models comprise a series of algorithms which make it impossible for individual fate of Lagrangian particles to be processed independently.
(4) Existing models do not support the quantification of uncertainty in the vulnerability prediction
(5) Some of the existing models are limited in their integration with GIS, which hinders visualization of oil spill movement.

Although other reviews have been done on remote sensing for marine oil spill monitoring and management, such as those by Brekke and Solberg [30], Fingas and Brown [76], Robbe et al. [207], most only consider satellite remote sensing without assessing airborne remote sensing. Fingas and Brown [25], Topouzelis [37], Jha et al. [58], Leifer et al. [60], Fingas and Brown [208], and Fingas and Brown [209] reviewed both airborne and satellite sensors for marine oil spill monitoring but the present review presents a more comprehensive comparison of both sensors. In addition, existing studies do not include an updated review of various automated techniques, particularly emerging Artificial intelligence approaches for discriminating, detecting, and classifying oil spills from false positive elements in remote sensing imageries, which is implemented in this study. Further, reviews of trajectory modeling, which is an important component of oil spill decision support systems, are limited. Based on the foregoing, this is the first attempt in the literature, in the authors' estimation, that undertakes a comprehensive review covering the broad spectrum of marine oil spill detection using remote sensing, machine learning, and deep learning, in addition to oil spill trajectory modeling and vulnerability assessment in the same study for marine oil spill management.

7. Conclusions and Future Outlook

Oil spills significantly affect the marine and terrestrial ecosystems and the rapid identification of spilled oil as well as accurate prediction of its trajectory to identify vulnerable locations is essential for disaster risk reduction and management. This review evaluates various remote sensing technologies that are used for the identification of oil spills in the marine environment. To understand the limitations of remote sensing devices, particularly the false positive visual representation of lookalikes in the imageries, several automatic oil detection models, including segmentation thresholding models, machine learning classifiers, and other emerging computer vision models, are systematically reviewed. The paper also provides a comprehensive overview of the opportunities and challenges of RS, ML, and DL in oil spill management in addition to trajectory prediction models. Based on the review, it is concluded that there is no single best remote sensing technique for unambiguous and accurate detection of oil spills, although active sensors are more widely used due to their ability to operate under any weather condition. However, the sensors are also affected by the false positive appearance of lookalikes. Different image segmentation techniques have been developed to overcome the limitations of remote sensing by automating the oil spill detection process, with SVM and ANN being the most utilized machine learning algorithms for classifying oil spills and lookalikes. Nonetheless, the accuracy of machine learning is limited by their restriction to feed forward image classification with no support for the end-to-end trainable framework. In contrast, deep learning models' strong feature extraction and autonomous learning features enhance their accuracy in marine oil spill detection. For trajectory modeling, the vulnerability prediction models use tracking Lagrangian elements to predict surface and sub-surface oil spills. However, the models are unable to independently process individual fate of Lagrangian particles (oil splots). In addition, the available models do not support the quantification of uncertainty in the oil spill environment and effective integration with GIS platforms for accurate visualization of oil spill movement remains a challenge.

In the future, it is necessary to advance research to gain deeper insights on RS technology that can accurately detect and discriminate oil spills from lookalikes and improve oil spill classification leveraging machine learning and deep learning algorithms. In addition, the use of image fusion methods that can enable multi sensor images to provide better information to aid oil spill detection should be explored. Existing models are usually developed for unique applications to a particle image type. This prevents the general application of such models to different types of remote sensing images. With the aid of deep learning algorithms, it is imperative to develop a universal model that can detect marine oil spills from different remote sensing images. In addition, developing a repository for the storage of oil spill remote sensing images is essential to reduce the dependence of oil spill detection models on transfer learning which is currently necessitated by the limited availability of oil spill remote sensing images. Research to address uncertainties in Lagrangian models through the integration of oil droplet quantification with tidal current, taking into consideration spatiotemporal differences, should be pursued. It is equally important to develop a framework to enable better integration of trajectory models with GIS. Further, specialized sensors for oil spill pollution monitoring are essential to address the limitations of false positive identification, weather effects, and limited area coverage, in addition to the incorporation of emerging geospatial computer vision models for the detection of oil spill. To overcome limitations of existing generalized oil spill trajectory models, it is imperative to develop specialized models that are applicable to diverse oil spill scenarios, taking into consideration the peculiarity of each situation. The envisioned trajectory model should be compatible and easily integrated with geospatial models for optimal performance.

Author Contributions: Conceptualization, A.-L.B. and S.T.Y.; methodology, S.T.Y.; formal analysis, S.T.Y.; investigation, S.T.Y.; resources, A.-L.B.; data curation, S.T.Y.; writing—original draft preparation, S.T.Y.; writing—review and editing, A.-L.B. and S.T.Y.; visualization, S.T.Y.; supervision, A.-L.B.; project administration, A.-L.B.; and funding acquisition, A.-L.B. All authors have read and agreed to the published version of the manuscript.

Funding: This research was funded by the University Teknologi PETRONAS (UTP) Y-UTP Research Project Grant (015LC0-003).

Acknowledgments: The authors acknowledge the financial support from the University Teknologi PETRONAS (UTP) Y-UTP Research Project Grant (015LC0-003) and the valuable comments of the anonymous reviewers.

Conflicts of Interest: The authors declare no conflict of interest.

References

1. Chen, J.; Zhang, W.; Wan, Z.; Li, S.; Huang, T.; Fei, Y. Oil spills from global tankers: Status review and future

Springer International Publishing: Cham, Switzerland, 2019; pp. 1–13.

3. Michel, J.; Fingas, M. Oil spills: Causes, consequences, prevention, and countermeasures. In *Fossil Fuels*; Research Planning, Inc.: Columbia, SC, USA, 2015.
4. Pelta, R.; Carmon, N.; Ben-Dor, E. A machine learning approach to detect crude oil contamination in a real scenario using hyperspectral remote sensing. *Int. J. Appl. Earth Obs. Geoinf.* **2019**, *82*, 101901. [CrossRef]
5. Jiao, Z.; Jia, G.; Cai, Y. A new approach to oil spill detection that combines deep learning with unmanned aerial vehicles. *Comput. Ind. Eng.* **2019**, *135*, 1300–1311. [CrossRef]

6. Lynch, L.E. Statement by Attorney General Loretta E. Lynch on the Agreement in Principle with BP to Settle Civil Claims for the Deepwater Horizon Oil Spill. 31 March 2015. Available online: https://www.justice.gov/opa/pr/statement-attorney-general-loretta-e-lynch-agreement-principle-bp-settle-civil-claims (accessed on 29 November 2019).
7. Bishop, R.C.; Boyle, K.J.; Carson, R.T.; Chapman, D.; Hanemann, W.M.; Kanninen, B.; Kopp, R.J.; Krosnick, J.A.; List, J.; Meade, N. Putting a value on injuries to natural assets: The BP oil spill. *J. Sci.* **2017**, *356*, 253–254. [CrossRef]
8. Smith, L.C.; Smith, M.; Ashcroft, P. Analysis of environmental and economic damages from British Petroleum's Deepwater Horizon oil spill. *Alban. Law Rev.* **2011**, *4*, 563–585. [CrossRef]
9. Murphy, D.; Gemmell, B.; Vaccari, L.; Li, C.; Bacosa, H.P.; Evans, M.; Gemmell, C.; Harvey, T.; Jalali, M.; Niepa, T.H. An in-depth survey of the oil spill literature since 1968: Long term trends and changes since Deepwater Horizon. *Mar. Pollut. Bull.* **2016**, *113*, 371–379. [CrossRef]
10. Cornwall, W. Deepwater Horizon: After the oil. *Science* **2015**, *348*, 22–29. [CrossRef]
11. Nwachukwu, A.N.; Osuagwu, J.C. Effects of Oil Spillage on Groundwater Quality In Nigeria. *Am. J. Eng. Res. AJER* **2014**, *3*, 271–274.
12. Mignucci-Giannoni, A. Assessment and rehabilitation of wildlife affected by an oil spill in Puerto Rico. *Environ. Pollut.* **1999**, *104*, 323–333. [CrossRef]
13. Fingas, M. *The Basics of Oil Spill Cleanup*; CRC Press: Boca Raton, FL, USA, 2012.
14. National Research Council. *Oil in the Sea III: Inputs, Fates, and Effects*; National Academies Press (US): Washington, DC, USA, 2003.
15. Li, P.; Cai, Q.; Lin, W.; Chen, B.; Zhang, B. Offshore oil spill response practices and emerging challenges. *Mar. Pollut. Bull.* **2016**, *110*, 6–27. [CrossRef]
16. Westerholm, D.A.; Rauch, S.D., III; Kennedy, D.M.; Basta, D.J. Deepwater Horizon oil spill: Final programmatic damage assessment and restoration plan and final programmatic environmental impact statement. In *Natural Resources Science Plan 2011–2015*; Springer: Berlin/Heidelberg, Germany, 2011.
17. Piatt, J.F.; Lensink, C.J.; Butler, W.; Nysewander, D.R. Immediate Impact of the 'Exxon Valdez' Oil Spill on Marine Birds. *Auk* **1990**, *107*, 387–397. [CrossRef]
18. Nevalainen, M.; Helle, I.; Vanhatalo, J.P. Estimating the acute impacts of Arctic marine oil spills using expert elicitation. *Mar. Pollut. Bull.* **2018**, *131*, 782–792. [CrossRef] [PubMed]
19. Prabowo, A.R.; Bae, D.M. Environmental risk of maritime territory subjected to accidental phenomena: Correlation of oil spill and ship grounding in the Exxon Valdez's case. *Results Eng.* **2019**, *4*, 100035. [CrossRef]
20. Amir-Heidari, P.; Arneborg, L.; Lindgren, J.F.; Lindhe, A.; Rosén, L.; Raie, M.; Axell, L.; Hassellöv, I.-M. A state-of-the-art model for spatial and stochastic oil spill risk assessment: A case study of oil spill from a shipwreck. *Environ. Int.* **2019**, *126*, 309–320. [CrossRef] [PubMed]
21. Grubesic, T.H.; Nelson, J.R.; Wei, R. A strategic planning approach for protecting environmentally sensitive coastlines from oil spills: Allocating response resources on a limited budget. *Mar. Policy* **2019**, *108*, 103549. [CrossRef]
22. Fan, C.; Hsu, C.-J.; Lin, J.-Y.; Kuan, Y.-K.; Yang, C.-C.; Liu, J.-H.; Yeh, J.-H. Taiwan's legal framework for marine pollution control and responses to marine oil spills and its implementation on T.S. Taipei cargo shipwreck salvage. *Mar. Pollut. Bull.* **2018**, *136*, 84–91. [CrossRef]
23. Bullock, R.J.; Perkins, R.A.; Aggarwal, S. In-situ burning with chemical herders for Arctic oil spill response: Meta-analysis and review. *Sci. Total. Environ.* **2019**, *675*, 705–716. [CrossRef]
24. Sardi, S.S.; Qurban, M.A.; Li, W.; Kadinjappalli, K.P.; Manikandan, K.P.; Hariri, M.M.; Al-Tawabini, B.S.; Khalil, A.B.; El-Askary, H. Assessment of areas environmentally sensitive to oil spills in the western Arabian Gulf, Saudi Arabia, for planning and undertaking an effective response. *Mar. Pollut. Bull.* **2019**, *150*, 110588. [CrossRef]
25. Fingas, M.; Brown, C.E. Review of oil spill remote sensing. *Mar. Pollut. Bull.* **2014**, *83*, 9–23. [CrossRef]
26. Ko, T.-J.; Hwang, J.-H.; Davis, D.; Shawkat, M.S.; Han, S.S.; Rodriguez, K.L.; Oh, K.H.; Lee, W.H.; Jung, Y. Superhydrophobic MoS2-based multifunctional sponge for recovery and detection of spilled oil. *Curr. Appl. Phys.* **2019**, *20*, 344–351. [CrossRef]
27. Marghany, M. Chapter 13—Quantum immune fast spectral clustering for automatic detection of oil spill. In *Synthetic Aperture Radar Imaging Mechanism for Oil Spills*; Marghany, M., Ed.; Gulf Professional Publishing: Houston, TX, USA, 2020; pp. 243–267.

28. Fan, J.; Zhang, F.; Zhao, D.; Wang, J. Oil Spill Monitoring Based on SAR Remote Sensing Imagery. *Aquat. Proced.* **2015**, *3*, 112–118. [CrossRef]
29. Fustes, D.; Cantorna, D.; Dafonte, C.; Arcay, B.; Iglesias, A.; Manteiga, M. A cloud-integrated web platform for marine monitoring using GIS and remote sensing. Application to oil spill detection through SAR images. *Futur. Gener. Comput. Syst.* **2014**, *34*, 155–160. [CrossRef]
30. Brekke, C.; Solberg, A.H. Oil spill detection by satellite remote sensing. *Remote Sens. Environ.* **2005**, *95*, 1–13. [CrossRef]
31. Fingas, M.; Brown, C.E. Chapter 5—Oil Spill Remote Sensing. In *Oil Spill Science and Technology*, 2nd ed.; Fingas, M., Ed.; Gulf Professional Publishing: Boston, MA, USA, 2017; pp. 305–385.
32. Ning, J.L.; Chen, Z.L.; Wang, C.Y.; Xie, W.J. Analysis of Marine Oil Spill Pollution Monitoring Based on Satellite Remote Sensing Technology. In *IOP Conference Series: Materials Science and Engineering*; IOP Publishing: Bristol, UK, 2018; Volume 392, p. 042045.
33. Schott, J.R. *Remote Sensing: The Image Chain Approach*; Oxford University Press on Demand: Oxford, UK, 2007.
34. Campbell, J.B.; Wynne, R.H. *Introduction to Remote Sensing*; Guilford Press: New York, NY, USA, 2011.
35. Estes, J.; Senger, L. The multispectral concept as applied to marine oil spills. *Remote Sens. Environ.* **1971**, *2*, 141–163. [CrossRef]
36. Fingas, M. *Review of Oil Spill Remote Sensing Technologies*; Spill Science: Edmonton, AB, Canada, 2015; p. 9.
37. Topouzelis, K. Oil Spill Detection by SAR Images: Dark Formation Detection, Feature Extraction and Classification Algorithms. *Sensors* **2008**, *8*, 6642–6659. [CrossRef]
38. Migliaccio, M.; Nunziata, F.; Gambardella, A. Polarimetric signature for oil spill observation. In Proceedings of the 2008 IEEE/OES US/EU-Baltic International Symposium, Tallinn, Estonia, 27–29 May 2008; pp. 1–5.
39. Migliaccio, M.; Nunziata, F.; Buono, A. SAR Polarimetry for Effective Sea Oil Slick Observation. In Proceedings of the 2018 IEEE/OES Baltic International Symposium (BALTIC), Klaipeda, Lithuania, 12–15 June 2018; pp. 1–5.
40. Solberg, A.H.S. Remote sensing of ocean oil-spill pollution. *Proc. IEEE* **2012**, *100*, 2931–2945. [CrossRef]
41. Genovez, P.; Ebecken, N.F.F.; Freitas, C.; Bentz, C.; Freitas, R. Intelligent hybrid system for dark spot detection using SAR data. *Expert Syst. Appl.* **2017**, *81*, 384–397. [CrossRef]
42. Cantorna, D.; Dafonte, C.; Iglesias, A.; Arcay, B. Oil spill segmentation in SAR images using convolutional neural networks. A comparative analysis with clustering and logistic regression algorithms. *Appl. Soft Comput.* **2019**, *84*, 105716. [CrossRef]
43. Alpers, W.; Hühnerfuss, H. Radar signatures of oil films floating on the sea surface and the Marangoni effect. *J. Geophys. Res. Space Phys.* **1988**, *93*, 3642–3648. [CrossRef]
44. PLiu, P.; Zhao, C.; Li, X.; He, M.; Pichel, W.G. Identification of ocean oil spills in SAR imagery based on fuzzy logic algorithm. *Int. J.Remote Sens.* **2010**, *31*, 4819–4833.
45. Krestenitis, M.; Orfanidis, G.; Ioannidis, K.; Avgerinakis, K.; Vrochidis, S.; Kompatsiaris, I. Oil Spill Identification from Satellite Images Using Deep Neural Networks. *Remote Sens.* **2019**, *11*, 1762.
46. Fiscella, B.; Giancaspro, A.; Nirchio, F.; Pavese, P.; Trivero, P. Oil spill detection using marine SAR images. *Int. J. Remote Sens.* **2000**, *21*, 3561–3566.
47. Guo, Y.; Zhang, H.Z. Oil spill detection using synthetic aperture radar images and feature selection in shape space. *Int. J. Appl. Earth Obs. Geoinformation* **2014**, *30*, 146–157.
48. Frate, F.D.; Petrocchi, A.; Lichtenegger, J.; Calabresi, G. Neural networks for oil spill detection using ERS-SAR data. *IEEE Trans. Geosci. Remote Sens.* **2000**, *38*, 2282–2287.

[illegible] Wang, G.; Yin, L.; Zeng, K.; Zhang, Y.; Zhang, M.; Xiao, B.; Jiang, S.; Chen, H.; Chen, G. Modelling oil trajectories and potentially contaminated areas from the Sanchi oil spill. *Sci. Total Environ.* **2019**, *685*, 856–866.
51. Nordam, T.; Beegle-Krause, C.; Skancke, J.; Nepstad, R.; Reed, M. Improving oil spill trajectory modelling in the Arctic. *Mar. Pollut. Bull.* **2019**, *140*, 65–74. [PubMed]
52. Bozkurtoğlu, Ş.N.E. Modeling oil spill trajectory in Bosphorus for contingency planning. *Mar. Pollut. Bull.* **2017**, *123*, 57–72.

53. Abascal, A.J.; Sanchez, J.; Chiri, H.; Ferrer, M.I.; Cárdenas, M.; Gallego, A.; Castanedo, S.; Medina, R.; Alonso-Martirena, A.; Berx, B.; et al. Operational oil spill trajectory modelling using HF radar currents: A northwest European continental shelf case study. *Mar. Pollut. Bull.* **2017**, *119*, 336–350. [PubMed]
54. Goodman, R. Overview and future trends in oil spill remote sensing. *Spill Sci. Technol. Bull.* **1994**, *1*, 11–21.
55. Fingas, M. *Oil Spill Science and Technology*; Gulf Professional Publishing: Houston, TX, USA, 2016.
56. Bulgarelli, B.; Djavidnia, S. On MODIS Retrieval of Oil Spill Spectral Properties in the Marine Environment. *IEEE Geosci. Remote Sens. Lett.* **2012**, *9*, 398–402.
57. Andreou, C.; Karathanassi, V. Endmember detection in marine environment with oil spill event. In Proceedings of the Image and Signal Processing for Remote Sensing XVII, Prague, Czech Republic, 26 October 2011; International Society for Optics and Photonics: Bellingham, WA, USA, 2011; Volume 8180, p. 81800P.
58. Jha, M.N.; Levy, J.; Gao, Y. Advances in Remote Sensing for Oil Spill Disaster Management: State-of-the-Art Sensors Technology for Oil Spill Surveillance. *Sensors* **2008**, *8*, 236–255. [CrossRef] [PubMed]
59. Trieschmann, O.; Hunsaenger, T.; Tufte, L.; Barjenbruch, U. Data assimilation of an airborne multiple-remote-sensor system and of satellite images for the North Sea and Baltic Sea. *Remote Sensing Ocean Sea Ice* **2003**, *5233*, 51–61.
60. Leifer, I.; Lehr, W.J.; Simecek-Beatty, D.; Bradley, E.; Clark, R.; Dennison, P.E.; Hu, Y.; Matheson, S.; Jones, C.E.; Holt, B.; et al. State of the art satellite and airborne marine oil spill remote sensing: Application to the BP Deepwater Horizon oil spill. *Remote Sens. Environ.* **2012**, *124*, 185–209. [CrossRef]
61. Garcia-Pineda, O.; Staples, G.; Jones, C.E.; Hu, C.; Holt, B.; Kourafalou, V.; Graettinger, G.; Dipinto, L.; Ramirez, E.; Streett, D.; et al. Classification of oil spill by thicknesses using multiple remote sensors. *Remote Sens. Environ.* **2020**, *236*, 111421. [CrossRef]
62. Zhong, Z.; You, F. Oil spill response planning with consideration of physicochemical evolution of the oil slick: A multiobjective optimization approach. *Comput. Chem. Eng.* **2011**, *35*, 1614–1630. (In English) [CrossRef]
63. Lu, Y.; Li, X.; Tian, Q.; Zheng, G.; Sun, S.; Liu, Y.; Yang, Q. Progress in Marine Oil Spill Optical Remote Sensing: Detected Targets, Spectral Response Characteristics, and Theories. *Mar. Geodesy* **2013**, *36*, 334–346. [CrossRef]
64. Shi, L.; Ivanov, A.Y.; He, M.; Zhao, C. Oil spill mapping in the western part of the East China Sea using synthetic aperture radar imagery. *Int. J. Remote Sens.* **2008**, *29*, 6315–6329. [CrossRef]
65. Yin, Q.-Z.; Li, K.; Zhou, C.; Liu, C.; Chu, X.-M.; Zheng, J.; Yin, Q.-Z. Research on Oil Spill Monitoring Experiments Based on OFD-1 Oil Film Detector. In Proceedings of the 2012 2nd International Conference on Remote Sensing, Environment and Transportation Engineering; Institute of Electrical and Electronics Engineers (IEEE), Nanjing, China, 1–3 June 2012; pp. 1–4.
66. Lu, Y.; Chen, J.; Bao, Y.; Han, W.; Li, X.; Tian, Q.; Zhang, X. Using HJ-1 satellite CCD data for remote sensing analysis and information extraction in oil spill scenarios. *J. Sci. Sin. Inf.* **2011**, *41*, 193–201.
67. Wismann, V.; Gade, M.; Alpers, W.; Huhnerfuss, H. Radar signatures of marine mineral oil spills measured by an airborne multi-frequency radar. *Int. J. Remote Sens.* **1998**, *19*, 3607–3623. [CrossRef]
68. Wang, D.; Gong, F.; Pan, D.; Hao, Z.; Zhu, Q. Introduction to the airborne marine surveillance platform and its application to water quality monitoring in China. *Acta Oceanol. Sin.* **2010**, *29*, 33–39. [CrossRef]
69. Svejkovský, J.; Hess, M.; Muskat, J.; Nedwed, T.J.; McCall, J.; Garcia, O. Characterization of surface oil thickness distribution patterns observed during the Deepwater Horizon (MC-252) oil spill with aerial and satellite remote sensing. *Mar. Pollut. Bull.* **2016**, *110*, 162–176. [CrossRef]
70. Boochs, F.; Kupfer, G.; Dockter, K.; Kühbauch, W. Shape of the red edge as vitality indicator for plants. *Int. J. Remote Sens.* **1990**, *11*, 1741–1753. [CrossRef]
71. Lu, Y.; Sun, S.; Zhang, M.; Murch, B.; Hu, C. Refinement of the critical angle calculation for the contrast reversal of oil slicks under sunglint. *J. Geophys. Res. Oceans* **2016**, *121*, 148–161. [CrossRef]
72. Sun, S.; Hu, C.; Feng, L.; Swayze, G.A.; Holmes, J.; Graettinger, G.; Macdonald, I.; Garcia, O.; Leifer, I. Oil slick morphology derived from AVIRIS measurements of the Deepwater Horizon oil spill: Implications for spatial resolution requirements of remote sensors. *Mar. Pollut. Bull.* **2016**, *103*, 276–285. [CrossRef] [PubMed]
73. Sun, S.; Hu, C.; Garcia-Pineda, O.; Kourafalou, V.H.; Le Hénaff, M.; Androulidakis, Y. Remote sensing assessment of oil spills near a damaged platform in the Gulf of Mexico. *Mar. Pollut. Bull.* **2018**, *136*, 141–151. [CrossRef]
74. Lu, Y.; Tian, Q.; Wang, J.; Zheng, G.; Li, X. Determining oil slick thickness using hyperspectral remote sensing in the Bohai Sea of China. *Int. J. Digit. Earth* **2013**, *6*, 76–93. [CrossRef]

75. González, C.; Sánchez, S.; Paz, A.; Resano, J.; Mozos, D.; Plaza, A. Use of FPGA or GPU-based architectures for remotely sensed hyperspectral image processing. *Integration* **2013**, *46*, 89–103. [CrossRef]
76. Fingas, M.; Brown, C.E. A Review of Oil Spill Remote Sensing. *Sensors* **2017**, *18*, 91. [CrossRef]
77. Brown, C.W.; Lynch, P.F.; Ahmadjian, M. Applications of Infrared Spectroscopy in Petroleum Analysis and Oil Spill Identification. *Appl. Spectrosc. Rev.* **1975**, *9*, 223–248. [CrossRef]
78. Fortes, F.; Ctvrtníckovà, T.; Mateo, M.; Cabalín, L.; Nicolas, G.; Laserna, J. Spectrochemical study for the in situ detection of oil spill residues using laser-induced breakdown spectroscopy. *Anal. Chim. Acta* **2010**, *683*, 52–57. [CrossRef]
79. Salisbury, J.W.; D'Aria, D.M.; Sabins, F.F. Thermal infrared remote sensing of crude oil slicks. *Remote Sens. Environ.* **1993**, *45*, 225–231.
80. Pavlova, A.; Papazova, D. Oil-Spill Identification by Gas Chromatography-Mass Spectrometry. *J. Chromatogr. Sci.* **2003**, *41*, 271–273. [PubMed]
81. Reddy, C.M.; Quinn, J.G. GC-MS analysis of total petroleum hydrocarbons and polycyclic aromatic hydrocarbons in seawater samples after the North Cape oil spill. *Mar. Pollut. Bull.* **1999**, *38*, 126–135.
82. Khanna, S.; Santos, M.J.; Ustin, S.L.; Koltunov, A.; Kokaly, R.F.; Roberts, D.A. Detection of Salt Marsh Vegetation Stress and Recovery after the Deepwater Horizon Oil Spill in Barataria Bay, Gulf of Mexico Using AVIRIS Data. *PLoS ONE* **2013**, *8*, e78989.
83. Jing, Y.; An, J.; Liu, Z. A Novel Edge Detection Algorithm Based on Global Minimization Active Contour Model for Oil Slick Infrared Aerial Image. *IEEE Trans. Geosci. Remote Sens.* **2011**, *49*, 2005–2013.
84. Howari, F.M. Investigation of Hydrocarbon Pollution in the Vicinity of United Arab Emirates Coasts Using Visible and Near Infrared Remote Sensing Data. *J. Coast. Res.* **2004**, *204*, 1089–1095.
85. Brown, C.W.; Alberts, J.J. Fiber Optic Sensor for Petroleum. Google Patents. U.S. Patent No 6,144,026, 20 November 2001.
86. Lu, Y.; Tian, Q.; Wang, J.; Wang, X.; Qi, X. Experimental study on spectral responses of offshore oil slick. *Sci. Bull.* **2008**, *53*, 3937–3941.
87. De Carolis, G.; Adamo, M.; Pasquariello, G. Thickness estimation of marine oil slicks with near-infrared MERIS and MODIS imagery: The Lebanon oil spill case study. In Proceedings of the 2012 IEEE International Geoscience and Remote Sensing Symposium, Munich, Germany, 22–27 July 2012; pp. 3002–3005.
88. Clark, R.N.; Swayze, G.A.; Leifer, I.; Livo, K.E.; Lundeen, S.; Eastwood, M.; Green, R.O.; Kokaly, R.F.; Hoefen, T.; Sarture, C.; et al. *A Method for Qualitative Mapping of Thick Oil Using Imaging Spectroscopy*; United States Geological Survey: Reston, VA, USA, 2010.
89. Clark, R.N. Spectroscopy of rocks and minerals, and principles of spectroscopy. *Manual Remote Sens.* **1999**, *3*, 2.
90. Lammoglia, T.; de Souza Filho, C.R. Spectroscopic characterization of oils yielded from Brazilian offshore basins: Potential applications of remote sensing. *Remote Sens. Environ.* **2011**, *115*, 2525–2535.
91. Hörig, B.; Kühn, F.; Oschütz, F.; Lehmann, F. HyMap hyperspectral remote sensing to detect hydrocarbons. *Int. J. Remote Sens.* **2001**, *22*, 1413–1422.
92. Kühn, F.; Oppermann, K.; Hörig, B. Hydrocarbon Index—An algorithm for hyperspectral detection of hydrocarbons. *Int. J. Remote Sens.* **2004**, *25*, 2467–2473. [CrossRef]
93. Alpers, W. Remote sensing of oil spills. In *Maritime Disaster Management Symposium*; Citeseer: New York, NY, USA, 2002; pp. 19–23.
94. Ivanov, A.Y.; Zatyagalova, V.V. A GI[illegible]

[illegible] M.S.; Kaduk, J.D.; Jarvis, C.H. Mapping terrestrial oil spill impact using machine learning random forest and Landsat 8 OLI imagery: A case site within the Niger Delta region of Nigeria. *Environ. Sci. Pollut. Res.* **2019**, *26*, 3621–3635. [CrossRef] [PubMed]
97. Ozigis, M.S.; Kaduk, J.D.; Jarvis, C.H.; da Conceição Bispo, P.; Balzter, H. Detection of oil pollution impacts on vegetation using multifrequency SAR, multispectral images with fuzzy forest and random forest methods. *Environ. Pollut.* **2019**, *256*, 113360. [CrossRef] [PubMed]

98. Balogun, A.L.; Yekeen, S.T.; Pradhan, B.; Althuwaynee, O.F. Spatio-Temporal Analysis of Oil Spill Impact and Recovery Pattern of Coastal Vegetation and Wetland Using Multispectral Satellite Landsat 8-OLI Imagery and Machine Learning Models. *Remote Sens.* **2020**, *12*, 1225. [CrossRef]
99. Topouzelis, K.; Karathanassi, V.; Pavlakis, P.; Rokos, D. Detection and discrimination between oil spills and look-alike phenomena through neural networks. *ISPRS J. Photogr. Remote Sens.* **2007**, *62*, 264–270. [CrossRef]
100. Hu, C.; Müller-Karger, F.E.; Taylor, C.; Myhre, D.; Murch, B.; Odriozola, A.L.; Godoy, G. MODIS detects oil spills in Lake Maracaibo, Venezuela. *Eos Trans. Am. Geophys. Union* **2003**, *84*, 313–319.
101. Hu, C.; Li, X.; Pichel, W.G.; Muller-Karger, F.E. Detection of natural oil slicks in the NW Gulf of Mexico using MODIS imagery. *Geophys. Res. Lett.* **2009**, *36*, L01604. [CrossRef]
102. Srivastava, H.; Singh, T.P. Assessment and development of algorithms to detection of oil spills using MODIS data. *J. Ind. Soc. Remote Sens.* **2010**, *38*, 161–167. [CrossRef]
103. Casciello, D.; Lacava, T.; Pergola, N.; Tramutoli, V. Robust Satellite Techniques for oil spill detection and monitoring using AVHRR thermal infrared bands. *Int. J. Remote Sens.* **2011**, *32*, 4107–4129. [CrossRef]
104. Lo, C.P. *Applied Remote Sensing*; Burnt Mill; Longman: Harlow, UK; New York, NY, USA, 1986.
105. Su, W.; Su, F.; Zhou, C.; Du, Y. Optical Satellite Remote Sensing Capabilities Analysis of the Marine Oil Spill. *J. GeoInf. Sci.* **2012**, *14*, 523–530. [CrossRef]
106. Nie, W.; Zhang, X. Detecting Marine Oil Spill Pollution Based on Borda Count Method of Ocean Water Surface Image. In Proceedings of the 2012 2nd International Conference on Remote Sensing, Environment and Transportation Engineering, Nanjing, China, 1–3 June 2012; pp. 1–4.
107. Castro Gomez, M.G. Joint Use of Sentinel-1 and Sentinel-2 for Land Cover Classification: A Machine Learning Approach. Master's Thesis, Lund University, Lund, Switzerland, 2017.
108. Nezhad, M.M.; Groppi, D.; Laneve, G.; Marzialetti, P.; Piras, G. Oil spill detection analyzing "Sentinel 2" satellite images: A Persian Gulf case study. In Proceedings of the 3rd World Congress on Civil, Structural, and Environmental Engineering, Budapest, Hungary, 8–10 April 2018; pp. 1–8.
109. Kolokoussis, P.; Karathanassi, V. Oil Spill Detection and Mapping Using Sentinel 2 Imagery. *J. Mar. Sci. Eng.* **2018**, *6*, 4. [CrossRef]
110. Setiani, P.; Ramdani, F. Oil spill mapping using multi-sensor Sentinel data in Balikpapan Bay, Indonesia. In Proceedings of the 2018 4th International Symposium on Geoinformatics (ISyG), Malang, Indonesia, 10–12 November 2018; pp. 1–4.
111. Espedal, H. Detection of oil spill and natural film in the marine environment by spaceborne SAR. In Proceedings of the IEEE 1999 International Geoscience and Remote Sensing Symposium. IGARSS'99 (Cat. No.99CH36293), Hamburg, Germany, 28 June–2 July 1999; Volume 3, pp. 1478–1480.
112. Kostianoy, A.G.; Lebedev, S.A.; Litovchenko, K.T.; Stanichny, S.V.; Pichuzhkina, O.E. Satellite remote sensing of oil spill pollution in the southeastern Baltic Sea. *Gayana Concepción* **2004**, *68*, 327–2332. [CrossRef]
113. Gallego, A.-J.; Gil, P.; Pertusa, A.; Fisher, R.B. Segmentation of Oil Spills on Side-Looking Airborne Radar Imagery with Autoencoders. *Sensors* **2018**, *18*, 797. [CrossRef]
114. Chang, L.; Tang, Z.; Chang, S.; Chang, Y.-L. A region-based GLRT detection of oil spills in SAR images. *Pattern Recognit. Lett.* **2008**, *29*, 1915–1923. [CrossRef]
115. Mera, D.; Bolón-Canedo, V.; Cotos, J.M.; Alonso-Betanzos, A. On the use of feature selection to improve the detection of sea oil spills in SAR images. *Comput. Geosci.* **2017**, *100*, 166–178. [CrossRef]
116. Yu, F.; Sun, W.; Li, J.; Zhao, Y.; Zhang, Y.; Chen, G. An improved Otsu method for oil spill detection from SAR images. *Oceanologia* **2017**, *59*, 311–317. [CrossRef]
117. Chaturvedi, S.K.; Banerjee, S.; Lele, S. An assessment of oil spill detection using Sentinel 1 SAR-C images. *J. Ocean Eng. Sci.* **2020**, *5*, 116–135. [CrossRef]
118. Xu, L.; Shafiee, M.J.; Wong, A.; Clausi, D.A. Fully Connected Continuous Conditional Random Field with Stochastic Cliques for Dark-Spot Detection In SAR Imagery. *IEEE J. Sel. Top. Appl. Earth Obs. Remote Sens.* **2016**, *9*, 2882–2890. [CrossRef]
119. Ma, Y.L.; Jiao, L.C. Sar image segmentation based on watershed and spectral clustering. *J. Infared. Millim. Waves* **2008**, *6*, 013.
120. Gao, G.; Liu, L.; Zhao, L.; Shi, G.; Kuang, G. An Adaptive and Fast CFAR Algorithm Based on Automatic Censoring for Target Detection in High-Resolution SAR Images. *IEEE Trans. Geosci. Remote Sens.* **2008**, *47*, [illegible]. [CrossRef]
121. Gonzalez, R.C.; Woods, R.E. *Digital Image Processing*; Prentice Hall: Upper Saddle River, NJ, USA, 2002.

122. Alattas, R. Oil spill detection in SAR images using minimum cross-entropy thresholding. In Proceedings of the 2014 7th International Congress on Image and Signal Processing, Dalian, China, 14–16 October 2014; pp. 709–713.
123. Skøelv, Å.; Wahl, T. *Oil Spill Detection Using Satellite Based SAR*; Phase 1B competition report; Norwegian Defence Research Establishment: Kjeller, Norway, 1993.
124. Vachon, P.W.; Thomas, S.J.; Cranton, J.A.; Bjerkelund, C.; Dobson, F.W.; Olsen, R.B. Monitoring the coastal zone with the RADARSAT satellite. In *Oceanology International*; Brighton, UK, 10–13 March 1998; Volume 98, pp. 29–38.
125. Manore, M.J.; Vachon, P.W.; Bjerkelund, C.; Edel, H.R.; Ramsay, B. Operational use of Radarsat SAR in the coastal zone—The Canadian experience. *Inf. Sustain.* **1998**, *1998*, 115–118.
126. Vyas, G.; Bhan, A.; Gupta, D. Detection of oil spills using feature extraction and threshold based segmentation techniques. In Proceedings of the 2015 2nd International Conference on Signal Processing and Integrated Networks (SPIN), Noida, India, 19–20 February 2015.
127. Solberg, A.S.; Storvik, G.; Solberg, R.; Volden, E. Automatic detection of oil spills in ERS SAR images. *IEEE Trans. Geosci. Remote Sens.* **1999**, *37*, 1916–1924. [CrossRef]
128. Solberg, A.H.; Dokken, S.T.; Solberg, R. Automatic detection of oil spills in Envisat, Radarsat and ERS SAR images. In Proceedings of the IGARSS 2003, 2003 IEEE International Geoscience and Remote Sensing Symposium, Proceedings (IEEE Cat. No. 03CH37477), Toulouse, France, 21–25 July 2003; Volume 4, pp. 2747–2749.
129. Canny, J. A computational approach to edge detection. *IEEE Trans. Pattern Anal. Mach. Intel.* **1986**, *6*, 679–698. [CrossRef]
130. Kanaa, T.; Tonye, E.; Mercier, G.; Onana, V.; Ngono, J.; Frison, P.-L.; Rudant, J.-P.; Garello, R. Detection of oil slick signatures in SAR images by fusion of hysteresis thresholding responses. In Proceedings of the IGARSS 2003, 2003 IEEE International Geoscience and Remote Sensing Symposium, Proceedings (IEEE Cat. No. 03CH37477), Toulouse, France, 21–25 July 2003; Volume 4, pp. 2750–2752.
131. Pelizzari, S.; Bioucas-Dias, J. Oil spill segmentation of SAR images via graph cuts. In Proceedings of the 2007 IEEE International Geoscience and Remote Sensing Symposium, Barcelona, Spain, 23–28 July 2007; pp. 1318–1321.
132. Huang, B.; Li, H.; Huang, X. A level set method for oil slick segmentation in SAR images. *Remote Sens.* **2005**, *26*, 1145–1156. [CrossRef]
133. Hu, G.; Xiao, X. Edge detection of oil spill using SAR image. In Proceedings of the 2013 Cross Strait Quad-Regional Radio Science and Wireless Technology Conference, Chengdu, China, 21–25 July 2013; pp. 466–469.
134. Vyas, K.; Shah, P.; Patel, U.; Zaveri, T. Oil spill detection from SAR image data for remote monitoring of marine pollution using light weight imageJ implementation. In Proceedings of the 2015 5th Nirma University International Conference on Engineering (NUiCONE), Ahmedabad, India, 26–28 November 2015; pp. 1–6.
135. Chen, F.; Yu, X.; Jiang, X.; Ren, P. Level sets with self-guided filtering for marine oil spill segmentation. In Proceedings of the 2017 IEEE International Geoscience and Remote Sensing Symposium (IGARSS), Fort Worth, TX, USA, 23–28 July 2017; pp. 1772–1775.
136. Chen, Z.; Wang, C.; Teng, X.; Cao, L.; & Li, J. Oil spill detection based on a superpixel segmentation method for SAR image. In Proceedings of the 2014 IEEE Geoscience and Remote Sensing Symposium, Quebec City, QC, Canada, 13–18 July 2014; pp. 1725–1728.

[illegible] Liu, A.K. Towards an automated ocean feature detection, extraction and classification scheme for SAR imagery. *Int. J. Remote Sens.* **2003**, *24*, 935–951. [CrossRef]
139. Barni, M.; Betti, M.; Mecocci, A. A fuzzy approach to oil spill detection an SAR images. In Proceedings of the 1995 International Geoscience and Remote Sensing Symposium, IGARSS'95. Quantitative Remote Sensing for Science and Applications, Firenze, Italy, 10–14 July 1995; Volume 1, pp. 157–159.
140. Li, H.Z.; Wang, C.; Zhang, H.; Wu, F.; Li, J. Oil slick spot detection using K-distribution model of the sea background. In Proceedings of the 2009 IEEE International Geoscience and Remote Sensing Symposium, Cape Town, South Africa, 12–17 July 2009; Volume 4, pp. IV-470–IV-473.

141. Singha, S.; Bellerby, T.J.; Trieschmann, O. Satellite Oil Spill Detection Using Artificial Neural Networks. *IEEE J. Sel. Top. Appl. Earth Obs. Remote Sens.* **2013**, *6*, 2355–2363. [CrossRef]
142. Gasull, A.; Fàbregas, X.; Jiménez, J.; Marqués, F.; Moreno, V.; Herrero, M.A. Oil spills detection in SAR images using mathematical morphology. In Proceedings of the 2002 11th European Signal Processing Conference, Toulouse, France, 3–6 September 2002; pp. 1–4.
143. Solberg, A.H.; Brekke, C.; Husoy, P.O. Oil Spill Detection in Radarsat and Envisat SAR Images. *IEEE Trans. Geosci. Remote Sens.* **2007**, *45*, 746–755. [CrossRef]
144. Montali, A.; Giacinto, G.; Migliaccio, M.; Gambardella, A. Supervised pattern classification techniques for oil spill classification in SAR images: Preliminary results. In Proceedings of the SEASAR2006 Workshop, ESAESRIN, Frascati, Italy, 24–26 January 2006; pp. 23–26.
145. Migliaccio, M.; Tranfaglia, M. Oil spill observation by SAR: A review. In Proceedings of the 2004 USA-Baltic Internation Symposium, Klaipeda, Lithuania, 15–17 June 2004; pp. 1–6.
146. Singha, S.; Bellerby, T.J.; Trieschmann, O. Detection and classification of oil spill and look-alike spots from SAR imagery using an Artificial Neural Network. In Proceedings of the 2012 IEEE International Geoscience and Remote Sensing Symposium, Munich, Germany, 22–27 July 2012; pp. 5630–5633.
147. Wu, D.; Guo, H.; An, J. Research on Multi-Feature Fusion for Discriminating Oil Spill and Look-Alike Spots. In Proceedings of the 2017 4th International Conference on Information Science and Control Engineering (ICISCE), Changsha, China, 21–23 July 2017; pp. 607–610.
148. Karathanassi, V.; Topouzelis, K.; Pavlakis, P.; Rokos, D. An object-oriented methodology to detect oil spills. *Int. J. Remote Sens.* **2006**, *27*, 5235–5251. [CrossRef]
149. Maianti, P.; Rusmini, M.; Tortini, R.; Via, G.D.; Frassy, F.; Marchesi, A.; Nodari, F.R.; Gianinetto, M. Monitoring large oil slick dynamics with moderate resolution multispectral satellite data. *Nat. Hazards* **2014**, *73*, 473–492. [CrossRef]
150. Nieto-Hidalgo, M.; Gallego, A.-J.; Gil, P.; Pertusa, A. Two-Stage Convolutional Neural Network for Ship and Spill Detection Using SLAR Images. *IEEE Trans. Geosci. Remote Sens.* **2018**, *56*, 5217–5230. [CrossRef]
151. Akar, S.; Süzen, M.L.; Kaymakçi, N. Detection and object-based classification of offshore oil slicks using ENVISAT-ASAR images. *Environ. Monit. Assess.* **2011**, *183*, 409–423. [CrossRef] [PubMed]
152. Su, T.F.; Li, H.Y.; Liu, T.X. Sea Oil Spill Detection Method Using SAR Imagery Combined with Object-Based Image Analysis and Fuzzy Logic. *Adv. Mater. Res.* **2014**, *1065*, 3192–3200. [CrossRef]
153. Balogun, A.-L.; Marks, D.; Sharma, R.; Shekhar, H.; Balmes, C.; Maheng, D.; Arshad, A.; Salehi, P. Assessing the Potentials of Digitalization as a Tool for Climate Change Adaptation and Sustainable Development in Urban Centres. *Sustain. Cities Soc.* **2020**, *53*, 101888. [CrossRef]
154. Yalcin, A.; Reis, S.; Aydinoglu, A.; Yomralioglu, T. A GIS-based comparative study of frequency ratio, analytical hierarchy process, bivariate statistics and logistics regression methods for landslide susceptibility mapping in Trabzon, NE Turkey. *Catena* **2011**, *85*, 274–287. [CrossRef]
155. Marwan, M.; Ali, K.; Ouahmane, H. Security Enhancement in Healthcare Cloud using Machine Learning. *Procedia Comput. Sci.* **2018**, *127*, 388–397. [CrossRef]
156. Li, Y.; Zhang, Y. Synthetic aperture radar oil spills detection based on morphological characteristics. *Geospatial Inf. Sci.* **2014**, *17*, 8–16. [CrossRef]
157. Wan, J.; Cheng, Y. Remote sensing monitoring of Gulf of Mexico oil spill using ENVISAT ASAR images. In Proceedings of the 2013 21st International Conference on Geoinformatics; Institute of Electrical and Electronics Engineers (IEEE), Kaifeng, China, 20–22 June 2013; pp. 1–5.
158. Singha, S.; Vespe, M.; Trieschmann, O. Automatic Synthetic Aperture Radar based oil spill detection and performance estimation via a semi-automatic operational service benchmark. *Mar. Pollut. Bull.* **2013**, *73*, 199–209. [CrossRef]
159. Topouzelis, K.; Psyllos, A. Oil spill feature selection and classification using decision tree forest on SAR image data. *ISPRS J. Photogramm. Remote Sens.* **2012**, *68*, 135–143. [CrossRef]
160. [illegible] Zhang, D.; [illegible] Oil Slick Index: An algorithm for crude oil spill detection with imaging spectroscopy. In Proceedings of the 2012 IEEE 2nd International Workshop on Earth Observation and Remote Sensing Applications, Shanghai, China, 8–11 June 2012; pp. 46–49.
161. Breiman, L. Random Forests. *Mach. Learn.* **2001**, *45*, 5–32. [CrossRef]

162. Mahdianpari, M.; Salehi, B.; Mohammadimanesh, F.; Motagh, M. Random forest wetland classification using ALOS-2 L-band, RADARSAT-2 C-band, and TerraSAR-X imagery. *ISPRS J. Photogramm. Remote Sens.* **2017**, *130*, 13–31. [CrossRef]
163. Wang, W.; Yang, X.; Li, X.; Chen, K.; Liu, G.; Li, Z.; Gade, M. A Fully Polarimetric SAR Imagery Classification Scheme for Mud and Sand Flats in Intertidal Zones. *IEEE Trans. Geosci. Remote Sens.* **2016**, *55*, 1–9. [CrossRef]
164. Tong, S.; Liu, X.; Chen, Q.-H.; Zhang, Z.; Xie, G. Multi-Feature Based Ocean Oil Spill Detection for Polarimetric SAR Data Using Random Forest and the Self-Similarity Parameter. *Remote Sens.* **2019**, *11*, 451. [CrossRef]
165. Ma, Y.; Zeng, K.; Zhao, C.; Ding, X.; He, M. Feature selection and classification of oil spills in SAR image based on statistics and artificial neural network. In Proceedings of the 2014 IEEE Geoscience and Remote Sensing Symposium, Quebec City, QC, Canada, 13–18 July 2014; pp. 569–571.
166. Litjens, G.; Kooi, T.; Bejnordi, B.E.; Setio, A.A.A.; Ciompi, F.; Ghafoorian, M.; Van Der Laak, J.A.; Van Ginneken, B.; Sánchez, C.I. A survey on deep learning in medical image analysis. *Med. Image Anal.* **2017**, *42*, 60–88. [CrossRef] [PubMed]
167. Schmidhuber, J. Deep learning in neural networks: An overview. *Neural Netw.* **2015**, *61*, 85–117. [CrossRef]
168. Ma, L.; Liu, Y.; Zhang, X.; Ye, Y.; Yin, G.; Johnson, B.A. Deep learning in remote sensing applications: A meta-analysis and review. *ISPRS J. Photogramm. Remote Sens.* **2019**, *152*, 166–177. [CrossRef]
169. Yekeen, S.T.; Balogun, A.; Yusof, K.B.W. A novel deep learning instance segmentation model for automated marine oil spill detection. *ISPRS J. Photogramm. Remote Sens.* **2020**, *167*, 190–200. [CrossRef]
170. Ramalho, G.L.B.; Medeiros, F. Using Boosting to Improve Oil Spill Detection in SAR Images. In Proceedings of the IEEE 18th International Conference on Pattern Recognition (ICPR'06), Hong Kong, China, 20–24 August 2006; Volume 2, pp. 1066–1069.
171. Chen, G.; Li, Y.; Sun, G.; Zhang, Y. Application of Deep Networks to Oil Spill Detection Using Polarimetric Synthetic Aperture Radar Images. *Appl. Sci.* **2017**, *7*, 968. [CrossRef]
172. Viola, P.; Jones, M.J.C. Rapid object detection using a boosted cascade of simple features. In Proceedings of the 2001 IEEE Computer Society Conference on Computer Vision and Pattern Recognition (CVPR 2001), Kauai, HI, USA, 8–14 December 2001; Volume 1, p. 3.
173. Gu, Q.; Sheng, L.; Zhang, T.; Lu, Y.; Zhang, Z.; Zheng, K.; Hu, H.; Zhou, H. Early detection of tomato spotted wilt virus infection in tobacco using the hyperspectral imaging technique and machine learning algorithms. *Comput. Electron. Agric.* **2019**, *167*, 105066. [CrossRef]
174. Barbat, M.M.; Wesche, C.; Werhli, A.V.; Mata, M.M. An adaptive machine learning approach to improve automatic iceberg detection from SAR images. *ISPRS J. Photogramm. Remote Sens.* **2019**, *156*, 247–259.
175. Barbedo, J.G.A. Detection of nutrition deficiencies in plants using proximal images and machine learning: A review. *Comput. Electron. Agric.* **2019**, *162*, 482–492.
176. Gunter, N.B.; Schwarz, C.G.; Graff-Radford, J.; Gunter, J.L.; Jones, D.T.; Graff-Radford, N.R.; Petersen, R.C.; Knopman, D.S.; Jack, C.R. Automated detection of imaging features of disproportionately enlarged subarachnoid space hydrocephalus using machine learning methods. *NeuroImage Clin.* **2019**, *21*, 101605. [PubMed]
177. Castanedo, S.; Juanes, J.A.; Medina, R.; Puente, A.; Fernandez, F.; Olabarrieta, M.; Pombo, C. Oil spill vulnerability assessment integrating physical, biological and socio-economical aspects: Application to the Cantabrian coast (Bay of Biscay, Spain). *J. Environ. Manag.* **2009**, *91*, 149–159.
178. DePellegrin, D.; Pereira, P. Assessing oil spill sensitivity in unsheltered coastal environments: A case study for Lithuanian-Russian coasts, South eastern Baltic Sea. *Mar. Pollut. Bull.* [illegible]
180. Kankara, R.S.; Arockiaraj, S.; Prabhu, K. Environmental sensitivity mapping and risk assessment for oil spill along the Chennai Coast in India. *Mar. Pollut. Bull.* **2016**, *106*, 95–103.
181. Guo, W.; Hao, Y.; Zhang, L.; Xu, T.; Ren, X.; Cao, F.; Wang, S. Development and application of an oil spill model with wave–current interactions in coastal areas. *Mar. Pollut. Bull.* **2014**, *84*, 213–224.
182. Gług, M.; Wąs, J. Modeling of oil spill spreading disasters using combination of Langrangian discrete particle algorithm with Cellular Automata approach. *Ocean Eng.* **2018**, *156*, 396–405.
183. Chiu, C.-M.; Huang, C.-J.; Wu, L.-C.; Zhang, Y.J.; Chuang, L.Z.-H.; Fan, Y.; Yu, H.-C. Forecasting of oil-spill trajectories by using SCHISM and X-band radar. *Mar. Pollut. Bull.* **2018**, *137*, 566–581.

184. Lynch, D.R.; Greenberg, D.A.; Bilgili, A.; McGillicuddy, D.J., Jr.; Manning, J.P.; Aretxabaleta, A.L. *Particles in the Coastal Ocean: Theory and Applications*; Cambridge University Press: Cambridge, UK, 2014.
185. Spaulding, M.L. State of the art review and future directions in oil spill modeling. *Mar. Pollut. Bull.* **2017**, *115*, 7–19. [CrossRef]
186. Fay, J.A. Physical processes in the spread of oil on a water surface. In *International Oil Spill Conference*; American Petroleum Institute: Washington, DC, USA, 1971; Volume 1971, pp. 463–467.
187. Mackay, D.; Buist, I.; Mascarenhas, R.; Paterson, S. Oil spill processes and models. In *Environment Canada Report EE-8*; Environmental Protection Service: Ottawa, ON, Canada, 1980.
188. Mackay, D.; Shiu, W.Y.; Hossain, K.; Stiver, W.; McCurdy, D. *Development and Calibration of an Oil Spill Behavior Model*; Toronto University Dept of Chemical Engineering and Applied Chemistry: Toronto, ON, Canada, 1982.
189. Elliott, A.; Hurford, N.; Penn, C. Shear diffusion and the spreading of oil slicks. *Mar. Pollut. Bull.* **1986**, *17*, 308–313. [CrossRef]
190. Johansen, O. The Halten Bank experiment-observations and model studies of drift and fate of oil in the marine environment. In Proceedings of the 11th Arctic Marine Oil Spill Program (AMOP) Techn. Seminar. Environment Canada, Ottawa, ON, Canada, 7–9 June 1984; pp. 18–36.
191. Galt, J.A.; Overstreet, R. Development of Spreading Algorithms for the ROC. In *Response Options Calculator*; Genwest: Edmonds, WA, USA, 2009.
192. Berry, A.; Dabrowski, T.; Lyons, K. The oil spill model OILTRANS and its application to the Celtic Sea. *Mar. Pollut. Bull.* **2012**, *64*, 2489–2501. [CrossRef] [PubMed]
193. Yu, F.; Xue, S.; Zhao, Y.; Chen, G. Risk assessment of oil spills in the Chinese Bohai Sea for prevention and readiness. *Mar. Pollut. Bull.* **2018**, *135*, 915–922. [CrossRef] [PubMed]
194. Amir-Heidari, P.; Raie, M. Response planning for accidental oil spills in Persian Gulf: A decision support system (DSS) based on consequence modeling. *Mar. Pollut. Bull.* **2019**, *140*, 116–128. [CrossRef]
195. Jones, A.; Logan, G.; Kennard, J.; Rollet, N. Reassessing potential origins of synthetic aperture radar (sar) slicks from the timor sea region of the north west shelf on the basis of field and ancillary data. *Appea J.* **2005**, *45*, 311–332. [CrossRef]
196. Jones, A.; Thankappan, M.; Logan, G.A.; Kennard, J.M.; Smith, C.J.; Williams, A.K.; Lawrence, G.M. Coral spawn and bathymetric slicks in Synthetic Aperture Radar (SAR) data from the Timor Sea, north-west Australia. *Int. J. Remote Sens.* **2006**, *27*, 2063–2069. [CrossRef]
197. Thankappan, M.; Rollet, N.; Smith, C.J.; Jones, A.; Logan, G.; Kennard, J. Assessment of SAR ocean features using optical and marine survey data. In Proceedings of the Envisat Symposium, Montreux, Switzerland, 23–27 April 2007; Volume 2327.
198. Wettle, M.; Daniel, P.J.; Logan, G.A.; Thankappan, M. Assessing the effect of hydrocarbon oil type and thickness on a remote sensing signal: A sensitivity study based on the optical properties of two different oil types and the HYMAP and Quickbird sensors. *Remote Sens. Environ.* **2009**, *113*, 2000–2010. [CrossRef]
199. Lotliker, A.A.; Mupparthy, R.S.; Tummala, S.K.; Nayak, S.R. Evaluation of high resolution MODIS-Aqua data for oil spill monitoring. In *Remote Sensing of Inland, Coastal, and Oceanic Waters*; International Society for Optics and Photonics: Bellingham, WA, USA, 2008; Volume 7150, p. 71500S.
200. Brown, C.E.; Fingas, M.F. New space-borne sensors for oil spill response. In *International Oil Spill Conference*; American Petroleum Institute: Washington, DC, USA, 2001; Volume 2001, pp. 911–916.
201. Fingas, M.; Brown, C.E. Review of oil spill remote sensing. *Spill Sci. Technol. Bull.* **1997**, *4*, 199–208. [CrossRef]
202. Yekeen, S.T.; Balogun, A.L. Automated Marine Oil Spill Detection Using Deep Learning Instance Segmentation Model. *Int. Arch. Photogramm. Remote Sens. Spatial Inf. Sci.* **2020**, *XLIII-B3-2020*, 1271–1276. [CrossRef]
203. Xing, Q.; Li, L.; Lou, M.; Bing, L.; Zhao, R.; Li, Z. Observation of Oil Spills through Landsat Thermal Infrared Imagery: A Case of Deepwater Horizon. *Aquat. Procedia* **2015**, *3*, 151–156. [CrossRef]
204. Espedal, H. Satellite SAR oil spill detection using wind history information. *Int. J. Remote Sens.* **1999**, *20*, 49–65. [CrossRef]
205. Zeng, K.; Wang, Y. A Deep Convolutional Neural Network for Oil Spill Detection from Spaceborne SAR Images. *Remote Sens.* **2020**, *12*, 1015. [CrossRef]

206. Zhang, J.; Feng, H.; Luo, Q.; Li, Y.; Wei, J.; Li, J. Oil Spill Detection in Quad-Polarimetric SAR Images Using an Advanced Convolutional Neural Network Based on SuperPixel Model. *Remote Sens.* **2020**, *12*, 944. [CrossRef]
207. Robbe, N.; Hengstermann, T. Remote sensing of marine oil spills from airborne platforms using multi-sensor systems. *Des. Nat. III Comparing Des. Nat. Sci. Eng.* **2006**, *95*, 347–355.
208. Fingas, M.; Brown, C.E. Chapter 6—Oil Spill Remote Sensing: A Review. In *Oil Spill Science and Technology*; Fingas, M., Ed.; Gulf Professional Publishing: Boston, MA, USA, 2011; pp. 111–169.
209. Fingas, M.; Brown, C.E. Oil spill remote sensing. In *Earth System Monitoring*; Springer: New York, NY, USA, 2013; pp. 337–388.

Publisher's Note: MDPI stays neutral with regard to jurisdictional claims in published maps and institutional affiliations.

Article

Combining Satellite Imagery and Numerical Modelling to Study the Occurrence of Warm Upwellings in the Southern Baltic Sea in Winter

Halina Kowalewska-Kalkowska [1,*] and Marek Kowalewski [2,3]

1 Institute of Marine and Environmental Sciences, University of Szczecin, 70-383 Szczecin, Poland
2 Institute of Oceanography, University of Gdańsk, 81-378 Gdynia, Poland; marek.kowalewski@ug.edu.pl or ocemk@iopan.gda.pl
3 Institute of Oceanology of Polish Academy of Sciences, 81-712 Sopot, Poland
* Correspondence: halina.kowalewska@usz.edu.pl; Tel.: +48-9144-42-534

Received: 30 October 2019; Accepted: 9 December 2019; Published: 12 December 2019

Abstract: Coastal upwelling involves an upward movement of deeper, usually colder, water to the surface. Satellite sea surface temperature (SST) observations and simulations with a hydrodynamic model show, however, that the coastal upwelling in the Baltic Sea in winter can bring warmer water to the surface. In this study, the satellite SST data collected by the advanced very high resolution radiometer (AVHRR) and the moderate-resolution imaging spectroradiometer (MODIS), as well as simulations with the Parallel Model 3D (PM3D) were used to identify upwelling events in the southern Baltic Sea during the 2010–2017 winter seasons. The PM3D is a three-dimensional hydrodynamic model of the Baltic Sea developed at the Institute of Oceanography, University of Gdańsk, Poland, in which parallel calculations enable high-resolution modelling. A validation of the model results with in situ observations and satellite-derived SST data showed the PM3D to adequately represent thermal conditions in upwelling areas in winter (91.5% agreement). Analysis of the frequency of warm upwellings in 12 areas of the southern Baltic Sea showed a high variability in January and February. In those months, the upwelling was most frequent, both in satellite imagery and in model results, off the Hel Peninsula (38% and 43% frequency, respectively). Upwelling was also frequent off the Vistula Spit, west of the Island of Rügen, and off the eastern coast of Skåne, where the upwelling frequency estimated from satellite images exceeded 26%. As determined by the PM3D, the upwelling frequency off VS and R was at least 25%, while off the eastern coast of Skåne, it reached 17%. The faithful simulation of SST variability in the winters of 2010–2017 by the high-resolution model used was shown to be a reliable tool with which to identify warm upwellings in the southern Baltic Sea.

Keywords: warm upwelling; sea surface temperature; numerical modelling; winter; southern Baltic Sea

... water masses to the surface [1]. According to the Ekman theory, upwelling in the northern hemisphere can occur when the current moves along the shore situated to the left of the velocity vector. The upwelled water is frequently different in its physical and chemical properties from the surface water. Therefore, in a thermally stratified sea, such as the Baltic Sea, the occurrence of an upwelling can be inferred not only from in situ measurements, but also from satellite imagery [2,3]. Studies carried out in the summer showed upwelling to be an important process affecting water mixing and coastal weather [4–7]. The upwelled water is usually nutrient-rich; therefore, upwelling affects primary production and the phytoplankton biomass [6,8–11].

The Baltic Sea is a shallow intracontinental sea of the Atlantic Ocean (Figure 1). It is a semi-enclosed water body with a surface area of 392,978 km^2 and a 54 m mean depth [12]. The water temperature changes seasonally. During spring, the water column temperature profile shows a thermocline which separates the warm upper layer from the cooler intermediate water. This thermocline impedes vertical mixing within the upper layer until late autumn [13]. During that time, under an appropriate wind regime, the emergence of an upwelling may result in a rapid temperature drop by as much as more than 10 °C [10]. In winter, when the surface temperature decreases, the effects of warmer upwelled water are visible as the surface temperature increases. Off the southern coasts of the Gulf of Finland, the surface temperature may then raise to 3.5–4 °C [14].

The increasing use of satellite-based remote sensing, in progress since the early 1980s, has made it possible to detect upwellings in the Baltic Sea, not only from direct in situ measurements, but also through satellite-based records of the surface temperature [2,15–19]. At present, there is a variety of remote techniques with which to detect upwellings [11,20–24].

A thermal infrared remote sensing-based analysis of upwellings is possible only at the absence of cloud cover, cloudiness over the Baltic Sea being usually quite considerable. As shown by Finkensieper et al. [25], in 2010–2015, the annual mean frequency of cloud cover over the Baltic Sea varied from 62.0% to 89.5%. Although passive microwave devices can be used when the cloudiness limits the use of infrared techniques, their resolution is too low in comparison to upwelling sizes to detect them. Therefore, upwelling hydrodynamics in different regions of the Baltic Sea has been addressed by many numerical studies, beginning with the 1990s research on modelling the upwelling off the Island of Rügen [26,27]. Subsequently, other numerical models describing upwelling in response to atmospheric and hydrological forcing have been developed [3,5,28–32]. In recent years, many studies addressed upwellings by using both satellite-based remote sensing and numerical modelling [6,7,33,34].

The summer coastal upwelling sites and frequency in the southern Baltic Sea are relatively well known. When studying upwellings off the western coast of the Island of Rügen and off the Polish coast, Horstmann [15] demonstrated the appearance of upwelling with easterly and southeasterly winds. Gidhagen [16], who analysed upwellings off the Swedish coast, found them to occur most frequently with winds from the western sector. Bychkova and Viktorov [4] identified 14 upwelling zones in the Baltic Sea proper. For each zone, they reported basic characteristics such as the upwelling size and typical wind regime. A more comprehensive list of upwelling sites throughout the Baltic Sea, depending on the atmospheric circulation type, was given by Bychkova et al. [17]. Myrberg and Andrejev [29], using an upwelling index based on the numerical computation of vertical velocity, attempted to calculate the frequency of upwellings throughout the Baltic Sea. They demonstrated that in the southern Baltic Sea, the upwelling frequency in some areas exceeds 30%. Based on the analysis of simulated vertical velocity, Kowalewski and Ostrowski [34] determined the upwelling frequency in 12 southern Baltic areas. They determined the frequency for every month in a year and identified conditions favourable for upwelling generation. Krężel et al. [2] assessed the size of upwelling off the Hel Peninsula, Kołobrzeg, and Łeba within areas of 1400, 5000, and 3500 km^2, respectively. He pointed out that the temperature difference between surface and deep water often make it possible to trace the range and directions of the spreading upwelled water on satellite images for as long as ten days, sometimes even longer. Based on satellite imagery and numerical simulations, Lehmann et al. [6] showed that the upwelling frequency in May–September in some coastal areas of the Baltic Sea occasionally reached 40%. In the southern Baltic Sea, they observed upwellings to be particularly frequent off the coast of the Hanö Bay and off the southernmost coast of Sweden. Moreover, Lehmann et al. [6] demonstrated a positive trend of upwelling frequencies along the Swedish coast of the Baltic Sea and the Finnish coast of the Gulf of Finland as well as a negative trend along the Polish, Latvian and Estonian coasts. At the eastern coast of the southern Baltic Sea, several authors found that northerly winds make a significant contribution to upwelling generation [4,11,23,34].

While numerous studies have addressed problems associated with the detection of sites and frequency of summer upwellings in the Baltic Sea, there is a paucity of literature on upwelling sites and frequency in winter. This most likely stems from methodological differences. On the one hand, upwellings occur irregularly and their spatial and temporal scales are small; on the other, the dense cloud cover over the Baltic Sea in winter largely prevents the use of satellite information on the sea surface temperature (SST). The few studies addressing the problem relied primarily on results of in situ measurements. Svansson [35] documented the presence of winter upwelling off the east coast of Sweden (outside Västervik), whereas Suursaar [14] described the effect off the southern coast of the Gulf of Finland (near Sillamäe). Suursaar [14] observed, inter alia, that the occurrence of upwelling in winter produced a stronger record in water salinity and currents than in water temperature. The upwelling frequency of 12 areas of the southern Baltic in winter was determined based on numerical simulations of vertical velocities by Kowalewski and Ostrowski [34]. They found that in winter, as a result of the prevalence of westerly and southwesterly winds, upwellings off the east coast of Skåne are more frequent than downwelling. Along the Polish coastline, downwelling prevails. Off the Hel Peninsula, the frequencies of strong upwellings in January and February are 30% and 21%, respectively. Off the Hel Peninsula and off the Kołobrzeg, the probability was the highest when winds were from the southerly to northeasterly sectors. Off the Łeba, upwelling was generated mainly by the southeasterly to northeasterly winds. The only region on the southern coast of the Baltic with prevailing upwelling is the area off the Vistula Spit, in spite of prevailing westerly winds. In the eastern part of the southern Baltic, in winter, downwelling was more frequent than upwellings. It is noteworthy that the warm water transport by upwelling in winter and its effects on the marine environment have been studied so far primarily in Arctic areas [36–38]. In the opinion of Randelhoff and Sundfjord [39], in a strongly stratified sea such as the Beaufort Sea, should the upwelling frequency in winter increase in the future because of reduced sea ice cover, this can be an important factor contributing to the pre-bloom nutrient pool.

The present work was aimed at identifying sites and the frequency of upwelling in the southern Baltic Sea in winter based on the SST information acquired from the Advanced Very High-Resolution Radiometer (AVHRR) and the Moderate-Resolution Imaging Spectroradiometer (MODIS) and results of numerical simulations of a parallel, high spatial resolution version of the three-dimensional hydrodynamic model of the Baltic Sea (PM3D). The use of the PM3D was necessitated by a poor availability of satellite data due to a dense cloud cover observed above the Baltic Sea in winter. Upwelling events were identified in January and February of 2010–2017. Section 2 describes the area of interest, the data used for the upwelling detection, and the main features of the PM3D; it also specifies details of SST filtration and assimilation from the AVHRR and MODIS systems, from satellite maps available in the SatBałtyk System. This subsection closes with the description of validation methods. Section 3 presents results of validation of the simulated SST for 2010–2017 and analysis of concordance between numerical simulations and satellite observations in upwelling areas. The section analyses the winter upwelling frequency in the Baltic Sea and presents the representation of warm upwellings in different regions of the southern Baltic Sea. Discussion and concluding remarks bring the paper to a close.

2. Materials and Methods

[illegible] the PM3D and the available satellite imagery from the AVHRR and MODIS radiometers for 12 regions of the southern Baltic Sea (Table 1, Figure 1). The regions were identified based on the available literature [6,17,29,34] and on the analysis of upwelling generation sites indicated by the model.

Table 1. Areas of upwelling occurrence in the southern Baltic Sea, based on [6,17,29,34].

No.	Abbreviation	Area
1	CS	off the Curonian Spit
2	SP	off the Sambia Peninsula
3	VS	off the Vistula Spit
4	HP	off the Hel Peninsula
5	Ł	Polish coast off Łeba
6	K	Polish coast off Kołobrzeg
7	R	off the Island of Rügen
8	SWB	off southwestern Bornholm
9	NEB	off northeastern Bornholm
10	SS	off southern Skåne
11	ES	off eastern Skåne
12	B	off Blekinge

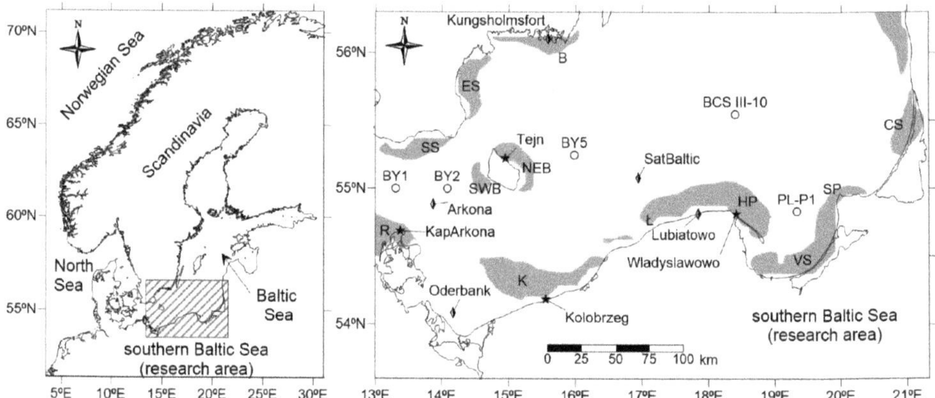

Figure 1. The upwelling regions, with the location of coastal stations (stars), coastal buoys (rhombs) and monitoring stations (circles). Upwelling regions are shown as grey areas; abbreviations as in Table 1.

2.2. Satellite Data

Upwelling detection was undertaken using satellite images registered by AVHRR/3 used by the NOAA (the National Oceanic and Atmospheric Administration) and METOP (Meteorological Operational Satellite) satellites as well as from MODIS mounted on the Terra and Aqua satellites. The data were acquired from the SatBałtyk System (http://satbaltyk.iopan.gda.pl) [40]. Until 2012, the AVHRR data were received at the University of Gdańsk via the HRPT Station (NOAA RAW High Resolution Picture Transmission data); then the data were downloaded from the EUMETCast system (NOAA RAW and MetOp Level-0) or from the EUMETSAT (European Organisation for the Exploitation of Meteorological Satellites) Archive (NOAA RAW and MetOp Level-1b). Instrumental correction and the calculation of the brightness temperature (AVHRR channels 4 and 5) were carried out in accordance with standard NOAA procedures using AAPP software [41]. Ice and cloud detection was performed using the MAIA 3 algorithm [42]. The cloud mask was extended further by a buffer of one pixel in width around clouds in order to hinder the influence of bad cloud masking. The bulk SST was computed according to the nonlinear split-window formula [43]. The AVHRR and MODIS data were provided several times a day at irregular intervals. They were georeferenced according to the procedure put forward by Kowalewski and Krężel [44] and reprojected into a 1 km grid in the Lambert Azimuthal Equal-Area projection used in the SatBałtyk system.

2.3. The Model

2.3.1. The Model Output

The three-dimensional hydrodynamic model of the Baltic Sea (M3D) has been used at the Institute of Oceanography, University of Gdańsk for more than twenty years. The model is based on the model of oceanic coastal circulation, known as the Princeton Ocean Model (POM) [45]. The POM was adapted to the Baltic Sea conditions by Kowalewski [46]. The operational version of the M3D became operational in 1999 [47]. The M3D was used then as a basis on which to build other modules, like the ProDeMo (Production and Destruction of Organic Matter Model) [48,49] or to model the sea ice thermodynamics and dynamics in the Baltic Sea [50]. Subsequent studies showed the M3D to be very useful for the analysis of physical processes in the coastal zone, including modelling thermal conditions at upwelling sites [34] as well as sea- and freshwater mixing during storm surges [51]. Within the framework of the project "The satellite monitoring of the Baltic Sea environment" (acronym: SatBałtyk) [52,53] dealing with remote sensing methods for monitoring of the Baltic Sea ecosystem, measures were taken to increase the M3D resolution to render it applicable to enhance SST maps in cloud-covered areas. As a result, a new version, the Parallel Model 3D (PM3D) was developed; parallel computations used by the model made it possible to increase the resolution to that achieved in satellite imagery [54]. This, in turn, facilitated development of an algorithm with which to complement information from the satellite SST imagery under cloudy conditions by using a mosaic of consecutive satellite maps together with numerical models as an additional source of information [40]. The PM3D assimilates the SST data retrieved from the AVHRR and MODIS radiometers, so its results do not differ significantly from those observed by satellite remote sensing. The PM3D has been already used to study the long-term variability of currents and circulation patterns in the Baltic Sea [55], as well as to storm surge prediction [54].

In the PM3D, calculations are conducted in parallel for two areas (computational domains) of different spatial resolution: 1 nautical mile (NM), i.e., 1.85 km for the Baltic Sea and the Skagerrak and 0.5 NM (about 0.9 km) for the southern part of the Baltic Sea. Simulation of a 24 h period is effected in 64 min [54]. Parallel calculations, which facilitate high-resolution SST modelling, have made it possible to use the PM3D while supplementing SST satellite maps in overcast areas without any loss of accuracy in the SatBałtyk System (http://satbaltyk.iopan.gda.pl/, [40]). A total of 18 layers in the sigma representation have been defined in the vertical plane [34].

The water exchange proceeds through the open boundary between the North Sea and the Skagerrak. The boundary involved a radiation marginal condition for flows, the sea level changes being introduced based on 1 h interval observation data for station Tregde (http://www.sjokart.no/en/sehavniva/). The model utilises meteorological data from the 4 km resolution operational weather Unified Model (UM) [56], the solar energy input is derived from the diagnostic SolRad model [57], and the monthly mean inflow of more than 150 river into the Baltic Sea is included (http://nest.su.se/bed/).

2.3.2. Filtration and Assimilation of SST Satellite Data from AVHRR and MODIS Radiometers

Data supplied by numerical models are always approximations. [illegible] by hydrodynamic models. The major advantage of such data sets is a possibility to record, with a high resolution, temperature fields over large areas; another advantage is the rapidity with which such data are available. The major disadvantage of remotely sensed SST, especially in the Baltic Sea region, is the low accuracy on the order of 1 °C [40,58]. Another problem stems from the fact that the measurement pertains only to a very thin surface water layer. Despite the employment of procedure of skin temperature correction for the calculation of bulk SST, under certain conditions, a difference between the satellite-derived and conventionally measured SST may be quite large, on the order of a few degrees. This is the case, for instance, on windless sunny days when the absence of mixing allows the surface to be heated quite strongly. This phenomenon is called a hotspot. On the other hand, the cloud cover precludes SST recording by radiometers operating in the infra-red range. When a part of the sky area is overcast, satellite imagery—to be applicable—requires masking the clouds, which is not straightforward. Very thin clouds and fogs are difficult to identify and—when left unmasked—may render the SST underestimated. All those drawbacks of the satellite SST estimation do not rule it out for the correction of results of modelling but should be taken into account when developing a data assimilation method.

The assimilation of oceanographic data is usually effected with methods adapted from meteorology. One of the first methods for the objective analysis, a simple but numerically very effective, was proposed by Cressman [59]. The statistical method of optimal interpolation [60] produces the best linear unbiased field estimator. There have been numerous algorithms with which to solve a simplified version of the Kalman filter, including the Ensemble Kalman filter [61,62], the Singular Evolutive Extended Kalman filter [63], and the Singular Evolutive Interpolated Kalman filter [63]. An alternative approach involves an iteration-based solution of the cost function utilising variance-based methods such as the 3DVar and 4DVar [64]. Many of those methods were applied to hydrodynamic models of the Baltic Sea to assimilate both the point source and the satellite data [65–69].

When assimilating satellite imagery, the problem of spatial interpolation is of secondary importance because the spatial resolution of the images is similar to or better than the resolution of the models. As the resolution is similar to that of the model, no advanced interpolation techniques are necessary and the SST images are directly reprojected to the model grid by bilinear interpolation. In addition, the satellite images are uniformly distributed over the rectangular spatial grid. Therefore, in this case, simple, numerically effective methods are successful, e.g., the Cressman method [58,70]. The basic problem in this case is, however, the low accuracy of the bulk surface temperature computed from the observed satellite skin temperature. Although the root-mean-square error (RMSE) is usually in the Baltic Sea on the order of 1 °C, in some cases, e.g., if the cloud has not been masked in a given pixel, it may be as high as 10 °C [58]. Therefore, an important part of the assimilation algorithm is an appropriate preliminary filtration of the data.

Thermal infrared satellite imagery contains information on the SST in cloudless areas only. In cloudy areas, the cloud radiation temperature is shown instead of information on the actual water temperature. Consequently, initial filtration of the satellite SST scenes is necessary to remove overcast fragments and those with skin temperature much higher than that of the surface water layer (hotspots). Cloudiness, which is usually seen on satellite images as a reduced temperature, was detected using the threshold technique. The operation proceeds in two stages. First, the SST in a pixel is compared with that shown in the preceding satellite image if that is cloudless. If the temperature dropped more than the threshold value, the pixel is assumed to be cloudy, i.e., the cloudiness condition is

$$\Delta T_{sat} > SST^{n}{}_{sat} - SST^{n-1}{}_{sat} \tag{1}$$

where ΔT_{sat} is the threshold difference between SST in two consecutive satellite images, above which the pixel is regarded as cloudy.

At the second stage, the difference between SST in the satellite image pixel and the corresponding node of the model grid is calculated. If the difference exceeds the appropriate threshold value, the pixel is regarded as cloudy, i.e., the cloudiness condition is

$$\Delta T_{sat\_modell} > SST_{sat} - SST_{model} \tag{2}$$

where ΔT_{sat_model} is the threshold difference between SST in the satellite image and provided by the model; above the difference, the pixel is regarded as cloudy.

At the next phase of filtration, pixels with excessive surficial temperature, the so-called hotspots, are detected. As for determining the cloudiness, the threshold technique based on the SST difference between the satellite image and the model was applied. The condition for classifying an SST as a hotspot is

$$\Delta T_{hotspot} > SST_{sat} - SST_{model} \tag{3}$$

where $\Delta T_{hotspot}$ is the threshold difference between SST in the satellite image and that provided by the model, above which the pixel is regarded as a hotspot.

As a result of filtration, all the pixels regarded as hotspots or cloudy are excluded from assimilation. For the remaining areas, assimilation corrections are determined for each model grid node from a difference (ΔT) between the satellite SST and the model value:

$$\Delta T = SST_{SAT} - SST_{Model} \tag{4}$$

Due to the presence of the surface mixed layer, temperature corrections are applied to the surface layers modelled, down to the depth R_z; it is assumed that the correction value will decrease linearly with depth following the function G(z). The temperature of the ith layer of the model is corrected in a single Δt calculation step as in

$$T_i^a = T_i^m + \frac{C_{assim} \Delta t \cdot G_i(z)}{R_t} \cdot \Delta T \tag{5}$$

$$G_i(z_i) = 1 - \frac{z_i}{R_z} \qquad \text{for} \qquad z_i - z < R_z \tag{6}$$

$$G_i(z_i) = 0 \qquad \text{for} \qquad z_i - z \geq R_z \tag{7}$$

where

C_{assim}, parameter (0 to 1) defining the degree of assimilation;
R_t, temporal range of assimilation;
R_Z, vertical range of assimilation;
Δt, calculation step of the model;
T^a, temperature calculated by the model after assimilation;
T^m, temperature calculated by the model prior to assimilation;
z_i, depth of the ith layer.

The degree of assimilation (C_{assim}) determines that part of the correction which will be applied during assimilation. To prevent rapid temperature changes, it is added gradually at each calculation step of the model from the moment of satellite observation until time R_t. The optimal values of $\Delta T_{\text{sat}} = -2$ °C, $\Delta T_{sat\_modell} = -2$ °C, $\Delta T_{hotspot} = 3$ °C, $C_{assim} = 0.5$, $R_Z = 5$ m and $R_t = 1$ h were selected by calibration. For these parameters, the RMSE of SST calculation by the PM3D with assimilation was 0.73 °C, compared to 0.89 °C without data assimilation [40].

2.4. Methods of Validation

The validation of the PM3D results with in situ measurements was based on sea surface temperature [illegible]

[illegible]

A validation of upwelling occurrence in the PM3D results with satellite-derived SST data was undertaken using 672 AVHRR scenes and 110 MODIS images taken in January and February of 2010–2017. The agreement in time and space of upwelling events generated by the PM3D was estimated for the 12 southern Baltic regions (Figure 1). The highest number of scenes making upwelling detection possible was obtained in 2011 and 2016, the lowest number being obtained in 2013 (Table 2). Individual days yielded up to 11 usable AVHRR images. Most often, one to three images were used (in 25% of the studied time span). AVHRR images were unsuitable for upwelling detection, mainly because of cloudiness, in 58% of the days. The MODIS data were amenable to upwelling detection during as little as 23% of the time span covered by the study. Most often, the image involved one scene recorded at noon (as few as three days yielded two images). On each satellite image (both AVHRR and MODIS), the areas not covered by clouds were examined.

Table 2. The number of available satellite sea surface temperature (SST) images in January and February (2010–2017).

Year	AVHRR	MODIS
2010	46	19
2011	145	8
2012	112	12
2013	9	3
2014	50	14
2015	45	16
2016	139	20
2017	126	18
Sum	672	110

To check the agreement, satellite-based SST distributions were compared with those generated by the PM3D. Upwelling detection on satellite images involved observing water of a temperature higher by 0.5 °C than that in the surrounding area. Because the presence of warmer water near the shore is not always caused by upwelling, during the analysis, the impact of two other factors on the increase in water temperature was accounted for in each area of examination. It was the spread of warmer river water in the sea and the heating of coastal waters caused by a positive heat exchange balance through the sea surface in shallow waters. If a case could not be classified unequivocally, it was excluded in the validation procedure. In the modelled situations, analysis of SST was supplemented by analysing the water temperature at a depth of 10 m, as well as the salinity and surface currents. The differences between the absolute values of SST in the satellite image and those provided by the model were of less importance because an upwelling was identified as a relative increase in water temperature in comparison to the surrounding waters.

3. Results

3.1. Model Validation

3.1.1. Validation with In Situ Measurements

Evaluation of the PM3D performance showed a good fit between the simulations and temperature readings from in situ measurements collected in 2010–2017. The best correlation was achieved for the open waters of the southern Baltic, where the coefficients of correlation were higher than 0.992 (Table 3). Correlation coefficients between the numerical and the observed readings as measured at the coastal stations were only slightly lower, ranging from 0.969 in Władysławowo to 0.987 in Tejn. For most of the stations, the simulated water temperatures were slightly lower than the measured values. Although the modelled mean values were lower by not more than 0.5 °C (the highest differences in mean values was found for Kołobrzeg and Kap Arkona), in rare cases, the modelled and the observed SSTs differed by a few degrees. On the other hand, RMSE ranged from 0.47 °C for the SatBaltic buoy to over 1 °C for coastal stations. The highest RMSE was that at Władysławowo (1.44 °C). The relatively large RMSE errors shown in Table 3 for coastal stations resulted mainly from their localization. Coastal stations are usually situated in harbour basins protected by a breakwater and therefore, the water temperature there may be different than that of open sea.

Table 3. Statistical indicators of differences between the modelled and in situ measurements of SST in the southern Baltic Sea.

Station	Bias [°C]	RMSE [°C]	Correlation Coefficient	Number of Records	Observation Period
Kołobrzeg	−0.51	1.33	0.973	30238	2014–2017
Władysławowo	0.00	1.44	0.969	30235	2014–2017
Lubiatowo	−0.44	0.57	0.993	2080	2015–2017
Oder Bank	−0.38	0.50	0.996	5885	2010–2017
Arkona	−0.16	0.63	0.994	8384	2010–2017
Kap Arkona	−0.52	1.17	0.978	29420	2010–2017
Tejn	−0.23	0.93	0.987	66307	2010–2017
Kungsholmsfort	−0.39	1.13	0.983	60165	2010–2017
SatBałtic	−0.09	0.47	0.997	2336	2013
PL-P1	−0.18	0.69	0.994	111	2010–2017
BCS III-10	0.04	0.68	0.995	97	2010–2015
BY1	0.23	0.70	0.994	45	2010–2012
BY2	−0.11	0.73	0.992	143	2010–2015
BY5	0.03	0.50	0.997	147	2010–2015

3.1.2. Validation with Satellite SST Data

Out of the 782 scenes analysed, a set of images without cloud cover was selected for each area (Table 4). The number of such situations ranged from 79 in SP to 104 in R. The agreement was achieved if both an image and the model showed the presence (or the absence) of an upwelling. The results obtained showed a good agreement (91.5%) between satellite images and modelled SST in the upwelling areas in winter. Most of the upwelling identified in the SST images was reflected by the model results. The closest agreement (above 95%) was typical of SS, SWB, CS and Ł, showing 14 upwelling events in the 344 consistent situations. In HP, the agreement was 89.8% (out of the 79 consistent situations, 32 were those of upwelling). Although the PM3D failed to identify upwelling in HP visible on the image on a single case only, the number of upwellings detected by the model was overestimated. In R, the area of the second most frequent upwellings, the agreement was 91.3% (25 upwelling events out of 95 consistent situations). The lowest agreement (77.8%) was recorded for VS, for which the highest tendency towards both over- and underestimation of upwelling events in the model results compared to the satellite observations was shown. Inconsistencies observed in VS may have been the result of its location near the mouth of the Vistula River, one of the biggest rivers of the Baltic Sea catchment area, which makes the interpretation of satellite scenes more difficult. Errors may also be produced by the model, in which monthly mean flows and temperatures of the Vistula River were assumed.

Table 4. A comparison of simultaneous occurrence of upwelling event in model simulation (M) and observed in satellite images (S) in individual areas of the Baltic Sea (January and February 2010–2017).

Area	Number of Cases	M(+), S(+)		M(+), S(−)		M(−), S(+)		M(−), S(−)		Consistent		Inconsistent	
[illegible]	[illegible]	[illegible]	[illegible]	11	[illegible]	[illegible]	10.0	54	60.0	70	77.8	20	22.2
HP	88	32	36.4	8	9.1	1	1.1	47	53.4	79	89.8	9	10.2
Ł	82	1	1.2	0	0.0	4	4.9	77	93.9	78	95.1	4	4.9
K	89	8	9.0	4	4.5	8	9.0	69	77.5	77	86.5	12	13.5
R	104	25	24.0	5	4.8	4	3.8	70	67.3	95	91.3	9	8.7
SWB	88	0	0.0	1	1.1	1	1.1	86	97.7	86	97.7	2	2.3
NEB	91	2	2.2	9	9.9	5	5.5	75	82.4	77	84.6	14	15.4
SS	102	1	1.0	0	0.0	1	1.0	100	98.0	101	99.0	1	1.0

Table 4. *Cont.*

Area	Number of Cases Analysed	M(+), S(+)		M(+), S(−)		M(−), S(+)		M(−), S(−)		Consistent Situations		Inconsistent Situations	
		No.	%	No.	%	No.	%	No.	%	No.	%	No.	%
ES	103	24	23.3	7	6.8	3	2.9	69	67.0	93	90.3	10	9.7
B	102	8	7.8	4	3.9	3	2.9	87	85.3	95	93.1	7	6.9

Note: M(+), S(+): both M and S showed the presence of an upwelling; M(+), S(−): M showed the presence of an upwelling, S – the absence of an upwelling; M(−), S(+): M showed the absence of an upwelling, S—the presence of an upwelling; M(−), S(−): both M and S showed the absence of an upwelling; situations where upwelling or its absence were observed in both image and model results were classified as consistent.

3.2. Frequency of Upwellings in the Southern Baltic Sea in January and February (2010–2017)

The analysis of the winter upwelling frequency in the southern Baltic Sea in 2010–2017 detected by model simulations and the available satellite imagery showed fairly large differences between the areas. The results obtained with both approaches are consistent. Upwelling was most frequent, both in the images and in the model outcomes, off the Hel Peninsula (HP), occurring with frequencies of 38% and 43%, respectively (Table 5). Upwelling was very frequent off the Vistula Spit (VS) and west of the Island of Rügen (R), the satellite SST imagery showing a frequency of 28%. As determined by the model, the upwelling frequency off VS and R was 29% and 25%, respectively. Off the eastern coast of Skåne (ES), the upwelling frequency was relatively high too (26% and 17% in satellite imagery and model simulations, respectively). Upwelling was less frequent at the Polish coast off Kołobrzeg (K) and the Curonian Spit (CS), 18% and 15%, respectively, as shown by satellite images. The PM3D-generated upwelling for the area showed a frequency lower by a few per cent. The frequency of upwelling off Blekinge (B) and the Sambia Peninsula (SP), as calculated with both methods, was about 10% and 8%, respectively. Off the southern coast of Skåne (SS) and off Łeba (Ł), winter upwelling was very rare, both in satellite images and in model simulations (two cases in SS and five and four cases in Ł). Winter upwelling was somewhat more frequent off the northeastern coast of Bornholm (NEB), with 8% and 9% shown by the SST images and model simulations, respectively. Winter upwelling was at its rarest off the southwestern coast of Bornholm (SWB) (one case in SST images and three cases generated by the PM3D). It is noteworthy that the satellite image-based analysis of frequencies may carry a substantial error due to the extensive cloud cover persisting over the Baltic Sea in winter (in 2013, upwellings could be detected in 12 images only; Table 2). The good agreement between the numerical simulations and observations allows one to regard the model as a reliable tool with which to identify upwelling in the southern Baltic Sea in winter.

Table 5. Upwelling events as recorded on satellite images and modelled by the Parallel Model 3D (PM3D).

Area	Satellite SST Images				PM3D		
	Available Dates [1]	%	Upwelling Events No.	%	Available Dates	Upwelling Events No.	%
CS	81	17.1	12	14.8	466	52	11.2
SP	79	16.7	6	7.6	466	35	7.5
VS	90	19.0	25	27.8	466	137	29.4
HP	88	18.6	33	37.5	466	202	43.3
Ł	82	17.3	5	6.1	466	4	0.9
K	89	18.8	16	18.0	466	40	8.6
R	104	21.9	29	27.9	466	114	24.5
SWB	88	18.6	1	1.1	466	3	0.6
NEB	91	[illegible]	7	7.7	466	43	9.2
SS	102	21.5	2	2.0	466	2	0.4
ES	103	21.7	27	26.2	466	79	17.0
B	102	21.5	11	10.8	466	46	9.9

[1] The availability of SST satellite images.

3.3. *Winter Upwellings as Represented by the PM3D*

3.3.1. Upwelling Event in February 2013

The PM3D generated an increased water temperature associated with a winter upwelling off the Hel Peninsula and off the Vistula Spit in late February to early March 2013. In HP, upwelling was detected with a surface current (0.5 ms^{-1}) directed NW, a current (0.4 ms^{-1}) directed west prevailing at that time in VS. In both areas, the upwelling situation persisted until 1 March and was most visible in simulations for 25 and 26 February (Figure 2). The difference in water temperature between the upwelling centre and the surrounding water was 2 and 1 °C in HP and VS, respectively. The strongest current was recorded on 24 February, 0.7 and 0.4 ms^{-1} in HP and VS, respectively. The currents enhancing upwelling generation persisted until 27 February when they switched direction to an opposite one, whereby the upwelling in both areas was being gradually extinguished in the subsequent days.

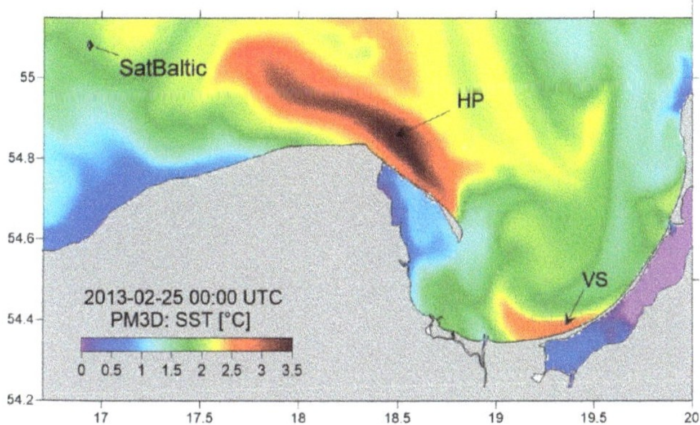

Figure 2. PM3D-generated sea surface temperature distribution on 25 February 2013, showing the presence of upwelling off the Hel Peninsula (HP) and off the Vistula Spit (VS). The location of the SatBaltic buoy used for the PM3D validation is indicated by a rhomb.

A comparison of the simulated SST with the only available SST image from MODIS of 1 March at 11:30 shows the temperature distributions in the area of investigation to be similar (Figure 3). The model reflects the SST increase in HP and VS produced by the upwelling. The simulated upwelling areas were similar in shape to those visible in the satellite image. It was only in HP that the simulated water temperature in the upwelling centre, reaching 3 °C, was 1 °C higher than that shown by the MODIS image. The difference between SST in the satellite image and provided by the model in VS did not exceed 0.5 °C.

A comparison of SSTs recorded by the SatBaltic buoy in the Słupsk Furrow and generated by the model showed a

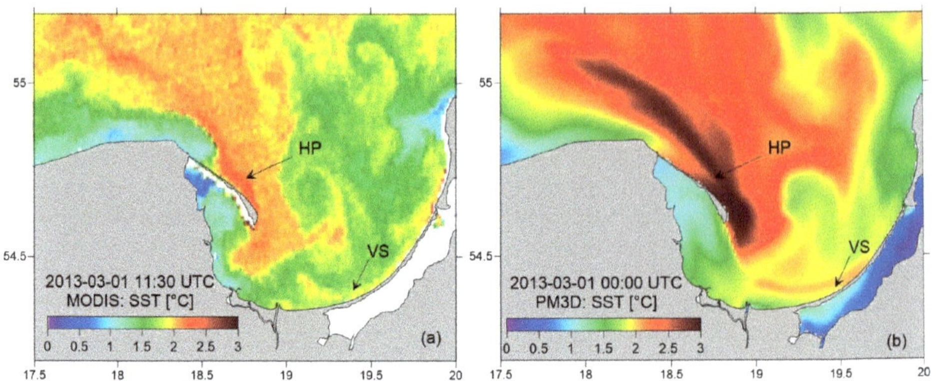

Figure 3. Upwelling off the Hel Peninsula (HP) and off the Vistula Spit (VS): a comparison of SST distributions as determined from SST for 1 March 2013 in (**a**) the MODIS (Moderate-Resolution Imaging Spectroradiometer) image with (**b**) the simulation generated by the PM3D. White patches in the satellite image indicate cloud-covered areas.

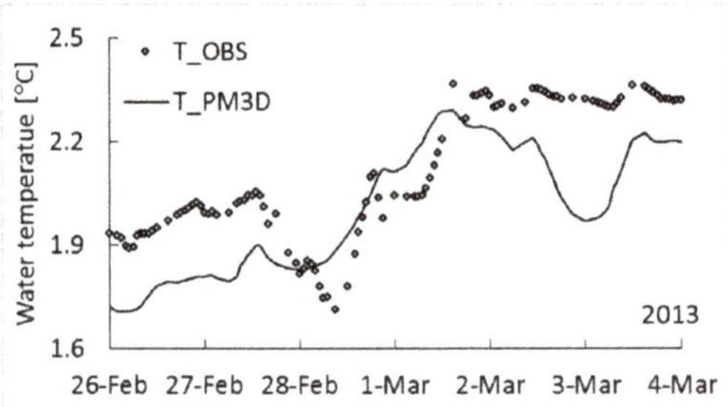

Figure 4. Water temperature fluctuations recorded by the SatBaltic buoy (T_OBS) and simulations produced by PM3D (T_PM3D) between 26 February and 3 March 2013.

3.3.2. Upwelling Event in February 2011

Another example of upwelling off the Hel Peninsula coast was the late February 2011 situation recorded by AVHRR. The model simulation showed the upwelling to occur in the area from 26 February until 2 March, with temperature difference between the centre and the surrounding water exceeding 2 °C. The upwelling was generated when the current, up to 0.6 ms^{-1}, was flowing NW. SST distributions in the area of investigation generated by the PM3D and shown by the AVHRR image from 28 February at 12:17 showed their high concordance (Figure 5). The upwelling area simulated by the PM3D was close in shape to that visible in the satellite image. The model-calculated SST in the upwelling centre, reaching 2.6 °C, was only 0.4 °C lower than that recorded by the satellite. The modelled SST of surrounding waters was lower than that recorded in the image by about 1 °C.

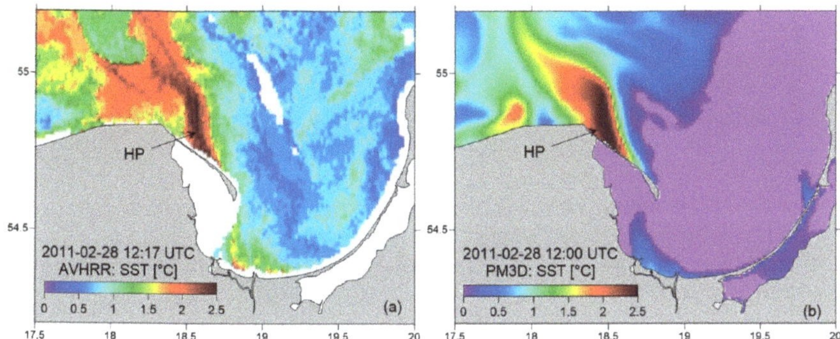

Figure 5. An upwelling off the Hel Peninsula (HP): a comparison of SST distributions as determined from SST for 28 February 2011 in (**a**) the AVHRR image with (**b**) the simulation generated by the PM3D. For explanation of white patches, see Figure 3.

3.3.3. Upwelling Event in January 2012

Between 26 February and 2 March 2012, the analysis of SST distributions generated by the model and shown by the available satellite imagery indicates the upwelling to be present in some areas of the southern Baltic Sea. Upwelling was detected in R, K, Ł, HP, VS, SP, CS and NEB. As shown in Figure 6, the AVHRR-generated SST distribution on 27 January at 00:33 points to the presence of warmer water in HP, VS, SP and CS. The water temperature in the upwelling centres in both areas HP and VS reached 4 °C. Differences between SST in the upwelling centre and in the surrounding water in those areas were in the order of 1 °C. Numerical simulations indicated the presence of upwelling-enhancing currents, i.e., directed to the west or northwest. The current velocity in HP and VS was 0.6 and 0.2 ms^{-1}, respectively. A comparison of the satellite-borne and PM3D-generated SSTs demonstrates a good representation of the actual thermal conditions by the model. The model generated upwelling areas in VS and HP, although the upwellings are more visible on the satellite image. Although in the HP upwelling centre, the modelled SST was calculated by the model accurately, in the VS upwelling centre, the numerical SST was lower by 0.5 °C. The upwelling in CS was partly obscured by clouds, but the analysis also indicated the simulated SST to be somewhat lower. Numerical simulations also show the presence of warmer water in SP, although the occurrence of upwelling could not be inferred with any certainty.

Besides the upwellings in areas of HP, VS, R, and NEB, the AVHRR image of 30 January at 12:12 showed upwelling to occur off Kołobrzeg and Łeba (Figure 7). Numerical simulations indicated westward currents of up to 0.6 and 0.2 ms^{-1} in K and Ł, respectively. The two cases of upwelling are well visible in the satellite image. However, the model failed to produce unequivocal confirmation of an upwelling event.

Upwelling off the islands of Rügen and Bornholm is clearly visible in the AVHRR image of 31 January at 02:05 (Figure 8). SST in R and NEB was up to 5 °C and 4 °C, respectively. At that time, the PM3D generated currents directed [illegible]

however, the modelled SST was 1–2 °C lower than that shown in the AVHRR image.

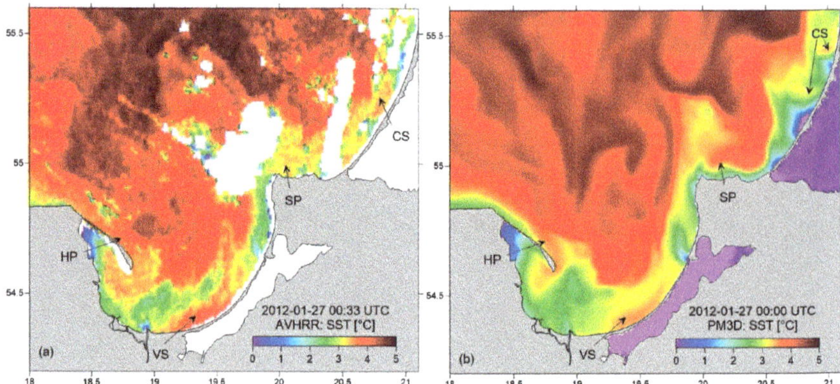

Figure 6. Upwelling off the Hel Peninsula (HP), off the Vistula Spit (VS) and off the Curonian Spit (CS): a comparison of SST distributions as determined from SST for 27 January 2012 in (**a**) the AVHRR image with (**b**) the simulation generated by the PM3D. The presence of upwelling off the Sambia Peninsula (SP) could not be classified unequivocally. For explanation of white patches, see Figure 3.

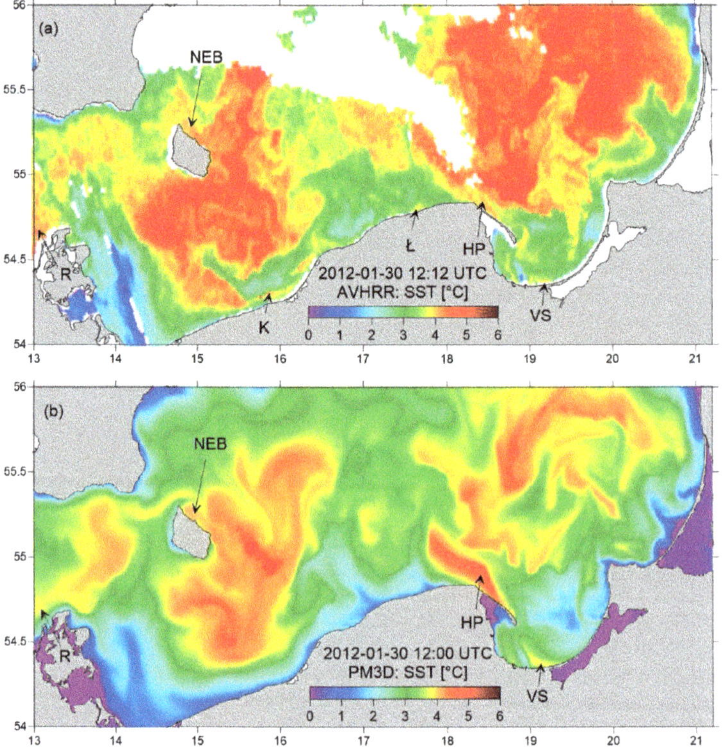

Figure 7. Upwelling at the southern Baltic coast: a comparison of SST distributions as determined from SST for 30 January 2012 in (**a**) the AVHRR image with (**b**) the simulation generated by the PM3D, abbreviations as in Table 1. For explanation of white patches, see Figure 3.

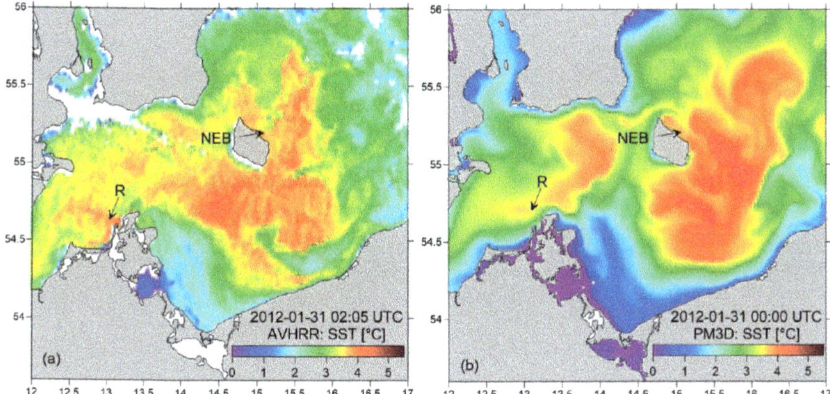

Figure 8. Upwelling off the islands of Rügen (R) and Bornholm (NEB): a comparison of SST distributions as determined from SST for 31 January 2012 in (**a**) the AVHRR image with (**b**) the simulation generated by the PM3D. For explanation of white patches, see Figure 3.

3.3.4. Upwelling Event in January 2016

The upwelling off the NW coast of the Island of Rügen, recorded by both AVHRR and MODIS, occurred in early January 2016. The SST distribution on 4 January 2016 (AVHRR: at 19:57; MODIS: at 12:25) involved the presence of warm water, reaching 7 °C in R (Figure 9). The difference in water temperature between the upwelling centre and the surrounding water was in the order of 4 °C. The comparison of the modelled and radiometer-recorded SST revealed that the upwelling area simulated by the PM3D was similar in shape to that visible in the satellite images. However, the SST model computed in the upwelling centre was by about 1 °C lower than that determined from the images and in the surrounding waters, it was lower by about 2 °C. It is worth pointing out that the upwelling was produced by westward 0.8 ms^{-1} currents, which subsequently veered to the southwest.

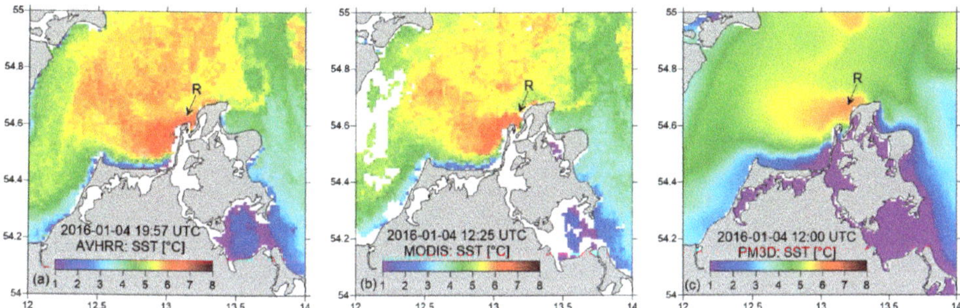

3.3.5. Upwelling Event in February 2017

An example of an upwelling generated by the PM3D off the southeastern and southern coasts of Sweden is that occurring from 21 until 25 February 2017, associated with an eastward current. On 23 February, the current velocity exceeded 0.7 ms^{-1}. Similar upwelling-enhancing currents were generated by the PM3D off Blekinge. A change in current direction and current weakening resulted in the cessation of the upwelling during the subsequent days. The SST distribution off the Swedish coast generated by the PM3D for 23 February 2017 at 18:00 shows a clearly warmer water, associated with the upwelling, in ES, SS and B (Figure 10). The water temperature in the centre of upwelling in SS was 3.2 °C, and in the areas of ES and B, it reached 3.6 °C. Differences between SST in the upwelling centres and beyond were in the order of 0.5 °C.

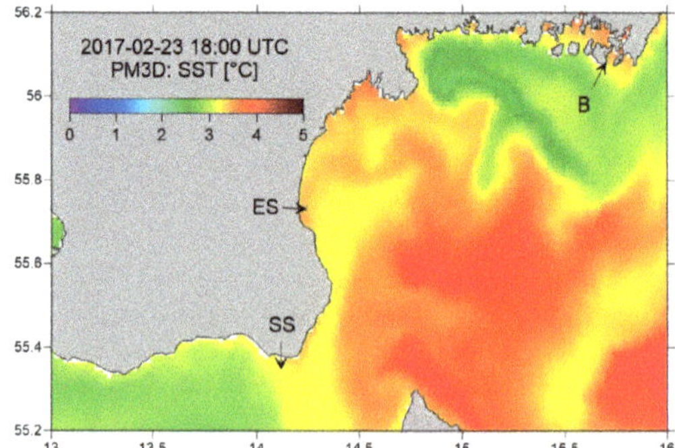

Figure 10. PM3D-generated sea surface temperature distribution on 23 February 2017 showing the presence of upwellings off southern Skåne (SS), off eastern Skåne (ES), and off Blekinge (B).

A comparison of SST simulated numerically by the PM3D with the SST AVHRR image available for 24 February at 19:30 showed the spatial SST distributions off the southern and southeastern Baltic coasts of Sweden to be similar (Figure 11). The model generated upwellings occurring in SS and SE, and predicted the upwelling cessation in B. The simulated water temperature in the upwelling centre off the southern Skåne was about 1 °C lower than that shown by the AVHRR image. On the contrary, it was 0.3 °C and 1 °C higher off eastern Skåne and off the Blekinge coast, respectively.

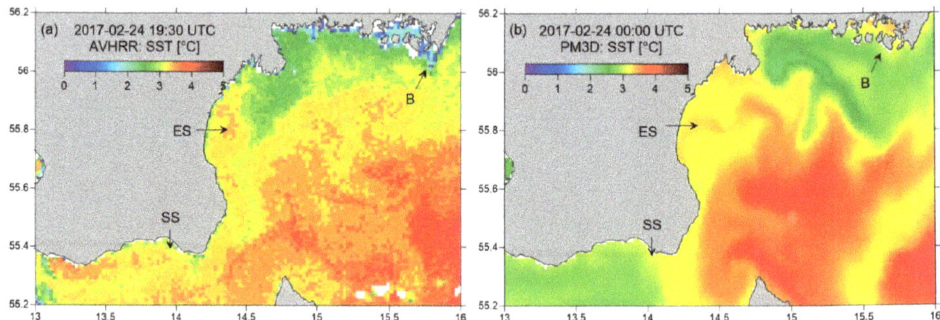

Figure 11. Upwelling at the Swedish coast: a comparison of SST distributions as determined from SST for 24 February 2017 in (**a**) the AVHRR image with (**b**) the simulation generated by the PM3D, abbreviations as in Table 1. For explanation of white patches, see Figure 3.

4. Discussion

This study, based on satellite-recorded SST and results of numerical simulations by a high-resolution model, allowed for the determination of the frequency throughout the southern Baltic Sea. The upwelling frequencies determined from satellite imagery in winter are hardly amenable to comparisons with those determined for the summer season [6,17,29], because computations were based on scarce (because of the cloud cover) data acquired at irregular time intervals. The results, including those produced by model simulations, allow the conclusion, however, that similarly to the summer, southern Baltic upwelling is most frequent off the Hel Peninsula, west of the Island of Rügen as well as off the southeast coast of Skåne. The results obtained in this study point to a relatively high upwelling frequency also off the Vistula Spit. In the remaining areas studied, winter upwelling was less frequent. Particularly rare was upwelling off the southern coast of Skåne, although the effect was very frequent in the summer [6,16,17,29]. It is worth pointing out that upwelling identification in SST images involved the detection of water with a temperature higher than in adjacent areas; therefore, detection was not always possible when the temperature of the upwelled water was similar to that in the vicinity.

The upwelling frequencies computed in this study throughout the southern Baltic Sea in January and February are somewhat different than those calculated using the analysis of upward currents [34]. The upwelling frequencies calculated in this work from both SST scenes and numerical simulations were, off the southern and eastern coast of the southern Baltic Sea, higher than the frequencies of the upward currents of a velocity above 10^{-4} ms^{-1} (indicating strong upwelling). In particular, the frequencies determined from SST images in HP and VS were 38% and 28%, respectively, the PM3D producing frequencies of 43% and 29%, respectively. On the other hand, the frequency of currents > 10^{-4} ms^{-1} for the two areas, calculated by Kowalewski and Ostrowski [34] were 25% and 21%, respectively. The frequencies calculated in this study for areas off the Swedish Baltic coast were somewhat lower than the probability of strong upwelling as given by Kowalewski and Ostrowski [34]. It should be pointed out that the methodological differences show no grounds for expecting tight correspondence between the results reported in this paper and those by Kowalewski and Ostrowski [34]. The reason is, inter alia, that upwelling (a vertical upward current) may precede an SST rise by as long as several days. Moreover, the upwelled warm water may stay at the surface for a long time after the vertical upward current has ceased.

Identifying coastal upwelling on satellite images as areas of warmer water than that of the surrounding area is a difficult task because the presence of warmer water near the shore is not always caused by upwelling. Warmer water may occur as a result of the heating of coastal waters caused by a positive heat exchange balance through the sea surface. This phenomenon is observed during sunny days at the end of winter when the afflux of solar radiation is large. Nevertheless, this phenomenon is particularly visible in shallower areas where coastal upwelling does not occur. Another problem arises from the spread of river waters, which, in some cases, may be warmer than sea water. Although during the interpretation of satellite scenes, the factors associated with bathymetry and warmer fresh waters were accounted for, the identification of the coastal upwelling, especially near the river mouths, e.g., off the Vistula Spit (VS), may be uncertain. The identification of warm upwellings in SST images may also be difficult in case of fog over the southern Baltic Sea, which sometimes results in the increase of SST calculated using the split-window algorithm. However, as distinct from cold events, warm upwellings do not trigger fog. Therefore in winter the shape of fog in the satellite scene hardly resembles the shape of upwelling. Hence, the presence of fog could hardly be interpreted as the occurrence of warm upwelling. The use of the PM3D enables one to minimize such errors, as it is possible to also analyse the temperature and salinity of subsurface waters, as well as the direction of sea currents.

5. Conclusions

the identification of upwellings events in all 12 areas of the southern Baltic Sea, delimited in earlier works as sites of summer upwellings [6,17,29,34].

The statistical descriptors of the model's performance calculated in this study made it possible to demonstrate that assimilation of satellite SST data and high-resolution grid applied in the PM3D resulted in a more realistic representation of the water temperature variability. The correlation coefficients calculated for PM3D simulations and temperature readings from the sea and coastal stations were higher than 0.96. A very good fit was shown by the sea stations, whereas that of the coastal station was somewhat weaker, with bias and RMSE at some stations reaching 0.5 °C and exceeding 1.0 °C, respectively.

The model results proved to be in good agreement with in situ observations and satellite data. The results demonstrate temporal and spatial concordance of upwelling events in the winter months of interest (91.5% agreement). Most upwelling identified in the SST images was reflected by the model simulations. In the two areas featuring the most frequent upwellings, i.e., in HP and R, the agreement amounted to 89.8% and 91.3%, respectively. The agreement for VS was lower (77.8%). To arrive at a better fit between the satellite and computed SST data, the model will be subjected to further tuning.

The comparison of upwelling events recorded in January and February of 2010–2017 observed in satellite images and simulated by the high-resolution model showed a high similarity of spatial SST distributions. During the upwelling events discussed in this work (off the Hel Peninsula in February of 2011 and 2013, off the southern Baltic Sea coast in January 2012, off the Island of Rügen in January 2016, and off the Swedish coast in February 2017), the PM3D generated upwellings with an adequate accuracy and provided a good representation of the spatial SST variability.

This study demonstrates the advantage of combining infrared satellite imagery and high-resolution hydrodynamic modelling for identifying sites and the frequency of warm upwelling in the southern Baltic Sea in winter. The results of the model validation allowed us to regard the high spatial resolution PM3D as a reliable tool with which to predict small-scale phenomena such as upwelling zones, hydrographic fronts, and eddies. In the winter months, when the cloud cover over the Baltic Sea is extensive, the PM3D may be a valuable source of information on physical processes occurring on small temporal and spatial scales. The information obtained by applying the model will make it possible to continue studying warm winter upwellings, their underpinnings and environmental importance, including inter alia effects on primary production and the development of pelagic communities in spring as well as sensitivity to climate change. The daily, updated 72 h forecasts of, inter alia, water temperature, generated by the PM3D with 1 NM resolution for the Baltic and Skagerrak and 0.5 NM resolution for the southern Baltic Sea as well as the archived records dating back to 2010 (with 6 h of temporal resolution) are available free of charge at the SatBałtyk system (http://satbaltyk.iopan.gda.pl).

Author Contributions: Conceptualization, methodology, formal analysis, validation, visualization, writing original draft, writing—review and editing: H.K.-K. and M.K.; software and data curation: M.K.; funding acquisition: H.K.-K. and M.K.

Funding: This work was carried out within the framework of the SatBałtyk project funded by the European Union through the European Regional Development Fund (contract No. POIG.01.01.02-22-011/09) titled 'The Satellite Monitoring of the Baltic Sea Environment'. A part of this work was supported through the funds of the Institute of Marine and Environmental Sciences, University of Szczecin, Poland.

Acknowledgments: The water temperature series were drawn from IMGW-PIB in Poland (Kołobrzeg and Władysławowo), BSH in Germany (Kap Arkona, Arkona and Oder Bank), DMI in Denmark (Tejn), SMHI in Sweden (Kungsholmsfort), and IBW PAN in Poland (Lubiatowo). Water temperature series in Kołobrzeg and Władysławowo were downloaded from the website https://danepubliczne.imgw.pl/ and in Kap Arkona – from the website https://www2.bsh.de/aktdat/bm/KapArkona_wtinfoe.htm. The data from Tejn, Kungsholmsfort, Lubiatowo, Oder Bank and Arkona were intercepted from the website http://www.emodnet-physics.eu/. The satellite SST data and water temperature from the SatBaltic buoy at the Słupsk Furrow were available through the SatBałtyk System (http://satbaltyk.iopan.gda.pl/). Water temperature simulations were also validated with data contained in the ICES database (https://ocean.ices.dk/helcom/). Validation was carried out for the following monitoring stations: BY1 and BY2 in the Arkona Basin; BY5 in the Bornholm Basin; PL-P1 in the Gdańsk Deep; and BCS III-10 in the Gdańsk Basin.

Conflicts of Interest: The authors declare no conflicts of interest.

References

1. Bowden, K.F. *Physical Oceanography of Coastal Water*; Ellis Harwood Ltd.: Chichester, UK, 1983; 302p.
2. Krężel, A.; Ostrowski, M.; Szymelfenig, M. Sea surface temperature distribution during upwelling along the Polish Baltic coast. *Oceanologia* **2005**, *47*, 415–432.
3. Myrberg, K.; Andrejev, O.; Lehmann, A. Dynamical features of successive upwelling in the Baltic Sea. *Oceanologia* **2010**, *52*, 77–99. [CrossRef]

4. Bychkova, I.A.; Viktorov, S.V. Use of satellite data for identification and classification of upwelling in the Baltic Sea. *Oceanology* **1987**, *27*, 158–162.
5. Jankowski, A. Variability of coastal water hydrodynamics in the Southern Baltic hindcast modelling of an upwelling event along the Polish coast. *Oceanologia* **2002**, *44*, 395–418.
6. Lehmann, A.; Myrberg, K.; Höflich, K. A statistical approach to coastal upwelling in the Baltic Sea based on the analysis of satellite data for 1990–2009. *Oceanologia* **2012**, *54*, 369–393. [CrossRef]
7. Gurova, E.; Lehmann, A.; Ivanov, A. Upwelling dynamics in the Baltic Sea studied by a combined SAR/infrared satellite data and circulation model analysis. *Oceanologia* **2013**, *55*, 687–707. [CrossRef]
8. Siegel, M.; Gerth, M.; Neuman, T.; Doerfer, R. Case studies on phytoplankton blooms in coastal and open waters of the Baltic Sea using Coastal Zone Color Scanner data. *Int. J. Remote Sens.* **1999**, *20*, 1249–1264. [CrossRef]
9. Kowalewski, M. The influence of Hel upwelling (Baltic Sea) on nutrient concentrations and primary production—The results of ecohydrodynamic model. *Oceanologia* **2005**, *47*, 567–590.
10. Lehmann, A.; Myrberg, K. Upwelling in the Baltic Sea—A review. *J. Mar. Syst.* **2008**, *74*, 3–12. [CrossRef]
11. Dabuleviciene, T.; Kozlov, I.E.; Vaiciute, D.; Dailidiene, I. Remote sensing of coastal upwelling in the south-eastern Baltic Sea: Statistical properties and implications for the coastal environment. *Remote Sens.* **2018**, *10*, 1752. [CrossRef]
12. Leppäranta, M.; Myrberg, K. *The Physical Oceanography of the Baltic Sea*; Springer: Berlin/Heidelberg, Germany, 2009.
13. Von Storch, H.; Omstedt, A. Introduction and summary. In *Assessment of Climate Change for the Baltic Sea Basin*; Regional Climate Studies; The BACC II Author Team, Ed.; Springer: Cham, Switzerland, 2008; pp. 1–34.
14. Suursaar, U. Waves, currents and sea level variations along the Letipea–Sillamäe coastal section of the southern Gulf of Finland. *Oceanologia* **2010**, *52*, 391–416. [CrossRef]
15. Horstmann, U. Distribution patterns of temperature and water colour in the Baltic Sea as recorded in satellite images: Indicators for phytoplankton growth. *Ber. Inst. Meeresk.* **1983**, *1*, 106.
16. Gidhagen, L. Coastal upwelling in the Baltic Sea–satellite and in situ measurements of sea-surface temperatures indicating coastal upwelling. *Estuar. Coast. Shelf Sci.* **1987**, *24*, 449–462. [CrossRef]
17. Bychkova, I.A.; Viktorov, S.V.; Shumakher, D.A. A relationship between the large-scale atmospheric circulation and the origin of coastal upwelling in the Baltic Sea. *Meteorol. Gidrol.* **1988**, *10*, 91–98. (In Russian)
18. Kahru, M.; Håkansson, B.; Rud, O. Distributions of the sea-surface temperature fronts in the Baltic Sea as derived from satellite imagery. *Cont. Shelf Res.* **1995**, *15*, 663–679. [CrossRef]
19. Uiboupin, R.; Laanemets, J. Upwelling characteristics derived from satellite sea surface temperature data in the Gulf of Finland, Baltic Sea. *Boreal Environ. Res.* **2009**, *14*, 297–304.
20. Kozlov, I.E.; Kudryavtsev, V.N.; Johannessen, J.A.; Chapron, B.; Dailidienė, I.; Myasoedov, A.G. ASAR imaging for coastal upwelling in the Baltic Sea. *Adv. Space Res.* **2012**, *50*, 1125–1137. [CrossRef]
21. Uiboupin, R.; Laanemets, J. Upwelling parameters from bias-corrected composite satellite SST maps in the Gulf of Finland (Baltic Sea). *IEEE Geosci. Remote Sens. Lett.* **2015**, *12*, 529–592. [CrossRef]
22. Esiukova, E.E.; Chubarenko, I.P.; Stont, Z.I. Upwelling or differential cooling? Analysis of satellite SST images of the southeastern Baltic Sea. *Water Resour.* **2017**, *44*, 69–77. [CrossRef]
23. Bednorz, E.; Czernecki, B.; Półrolniczak, M.; Tomczyk, A.M. Atmospheric forcing of upwelling along the southeastern Baltic coast. *Baltica* **2018**, *31*, 73–85. [CrossRef]
24. Krechik, V.; Myslenkov, S.; Kapustina, M. New possibilities in the study of coastal [illegible]

[illegible] on Climate Monitoring: Offenbach am Main, Germany, 2016. [CrossRef]

26. Lass, H.-U.; Schmidt, T.; Seifert, T. On the dynamics of upwelling observed at the Darss Sill. In Proceedings of the 19th Conference of Baltic Oceanographers, Sopot, Poland, 29 August–1 September 1994; pp. 247–260.
27. Fennel, W.; Seifert, T. Kelvin wave controlled upwelling in the western Baltic. *J. Mar. Syst.* **1995**, *6*, 286–300. [CrossRef]
28. Lehmann, A.; Krauss, W.; Hinrichsen, H.-H. Effects of remote and local atmospheric forcing on circulation and upwelling in the Baltic Sea. *Tellus* **2002**, *54*, 299–316. [CrossRef]

29. Myrberg, K.; Andreyev, O. Main upwelling regions in the Baltic Sea—A statistical analysis based on three-dimensional modelling. *Boreal Environ. Res.* **2003**, *8*, 97–112.
30. Golenko, N.N.; Golenko, M.N.; Shchuka, S.A. Observation and modeling of upwelling in the southeastern Baltic. *Oceanology* **2009**, *49*, 15–21. [CrossRef]
31. Fennel, W.; Radtke, H.; Schmidt, M.; Neumann, T. Transient upwelling in the central Baltic Sea. *Cont. Shelf Res.* **2010**, *30*, 2015–2026. [CrossRef]
32. Väli, G.; Zhurbas, V.; Lips, U.; Laanemets, J. Submesoscale structures related to upwelling events in the Gulf of Finland, Baltic Sea (numerical experiments). *J. Mar. Syst.* **2017**, *171*, 31–42. [CrossRef]
33. Zhurbas, V.M.; Stipa, T.; Malkki, P.; Paka, V.T.; Kuzmina, N.P.; Sklyarov, E.V. Mesoscale variability of the upwelling in the southeastern Baltic Sea: IR images and numerical modeling. *Oceanology* **2004**, *44*, 619–628.
34. Kowalewski, M.; Ostrowski, M. Coastal up- and downwelling in the southern Baltic. *Oceanologia* **2005**, *47*, 453–475.
35. Svansson, A. Interaction between the coastal zone and the open sea. *Merentutkimuslait. Julk. Havsforskningsinst. Skr.* **1975**, *24*, 385–404.
36. Pickart, R.S.; Moore, G.W.K.; Torres, D.J.; Fratantoni, P.S.; Goldsmith, R.A.; Yang, J. Upwelling on the continental slope of the Alaskan Beaufort Sea: Storms, ice, and oceanographic response. *J. Geophys. Res.* **2009**, *114*, C00A13. [CrossRef]
37. Lind, S.; Ingvaldsen, R.B. Variability and impacts of Atlantic water entering the Barents Sea from the north. *Deep Sea Res. I* **2012**, *62*, 70–88. [CrossRef]
38. Falk-Petersen, S.; Pavlov, V.; Berge, J.; Cottier, F.; Kovacs, K.M.; Lydersen, C. At the rainbow's end: High productivity fueled by winter upwelling along an Arctic shelf. *Polar Biol.* **2015**, *38*, 5–11. [CrossRef]
39. Randelhoff, A.; Sundfjord, A. Short commentary on marine productivity at Arctic shelf breaks: Upwelling, advection and vertical mixing. *Ocean Sci.* **2018**, *14*, 293–300. [CrossRef]
40. Konik, M.; Kowalewski, M.; Bradtke, K.; Darecki, M. The operational method of filling information gaps in satellite imagery using numerical models. *Int. J. Appl. Earth Obs. Geoinf.* **2019**, *75*, 68–82. [CrossRef]
41. Labrot, T.; Lavanant, L.; Whyte, K.; Atkinson, N.; Brunel, P. AAPP Documentation. Scientific Description, NWP SAF Satellite Application Facility for Numerical Weather Prediction. Document NWPSAF-MF-UD-001, Version 7.0, October 2011. Available online: https://nwpsaf.eu/site/download/documentation/aapp/NWPSAF-MF-UD-001_Science.pdf (accessed on 20 November 2019).
42. Lavanant, L. MAIA AVHRR Cloud Mask and Classification, MAIA v3 Scientific and Validation Document, MeteoFrance, 07/11/2002. Ref: MF/DP/CMS/R&D/MAIA3. Available online: http://www.meteorologie.eu.org/ici/maia/maia3.pdf (accessed on 20 November 2019).
43. Woźniak, B.; Krężel, A.; Darecki, M.; Woźniak, S.B.; Majchrowski, R.; Ostrowska, M.; Kozłowski, Ł.; Ficek, D.; Olszewski, J.; Dera, J. Algorithm for the remote sensing of the Baltic ecosystem (DESAMBEM), Part 1: Mathematical apparatus. *Oceanologia* **2008**, *50*, 451–508.
44. Kowalewski, M.; Krężel, A. System of automatic registration and geometric correction of AVHRR data. *Arch. Fotogram. Kartogr. Teledetek* **2004**, *13*, 397–407. (In Polish)
45. Blumberg, A.F.; Mellor, G.L. A description of a three-dimensional coastal ocean circulation model. In *Three-Dimensional Coastal Ocean Models*; Heaps, N.S., Ed.; American Geophysical Union: Washington, DC, USA, 1987; pp. 1–16.
46. Kowalewski, M. A three-dimensional, hydrodynamic model of the Gulf of Gdańsk. *Oceanol. Stud.* **1997**, *26*, 77–98.
47. Kowalewski, M. An operational hydrodynamic model of the Gulf of Gdańsk. In *Research Works Based on the ICM's UMPL Numerical Weather Prediction System Results*; Wyd. ICM: Warsaw, Poland, 2002; pp. 109–119.
48. Kowalewski, M. The flow of nitrogen into the euphotic zone of the Baltic Proper as a result of the vertical migration of phytoplankton: An analysis of the long-term observations and ecohydrodynamic model simulation. *J. Mar. Syst.* **2015**, *145*, 53–68. [CrossRef]
49. Ołdakowski, B.; Kowalewski, M.; Jędrasik, J.; Szymelfenig, M. Ecohydrodynamic model of the Baltic Sea, Part I: Description of the ProDeMo model. *Oceanologia* **2005**, *47*, 477–516.
50. Herman, A.; Jędrasik, J.; Kowalewski, M. Numerical modelling of thermodynamics and dynamics of sea ice in the Baltic Sea. *Ocean Sci.* **2011**, *7*, 257–276. [CrossRef]

51. Kowalewski, M.; Kowalewska-Kalkowska, H. Performance of operationally calculated hydrodynamic forecasts during storm surges in the Pomeranian Bay and Szczecin Lagoon. *Boreal Environ. Res.* **2011**, *16*, 27–41.
52. Woźniak, B.; Bradtke, K.; Darecki, M.; Dera, J.; Dzierzbicka, L.; Ficek, D.; Furmańczyk, K.; Kowalewski, M.; Krężel, A.; Majchrowski, R.; et al. SatBaltic—A Baltic environmental satellite remote sensing system—An ongoing project in Poland, Part 1: Assumptions, scope and operating range. *Oceanologia* **2011**, *53*, 897–924. [CrossRef]
53. Woźniak, B.; Bradtke, K.; Darecki, M.; Dera, J.; Dudzińska-Nowak, J.; Dzierzbicka, L.; Ficek, D.; Furmańczyk, K.; Kowalewski, M.; Krężel, A.; et al. SatBaltic—A Baltic environmental satellite remote sensing system—An ongoing project in Poland, Part 2: Practica applicability and preliminary results. *Oceanologia* **2011**, *53*, 925–958. [CrossRef]
54. Kowalewski, M.; Kowalewska-Kalkowska, H. Sensitivity of the Baltic Sea level prediction to spatial model resolution. *J. Mar. Syst.* **2017**, *173*, 101–113. [CrossRef]
55. Jędrasik, J.; Kowalewski, M. Mean annual and seasonal circulation patterns and long-term variability of currents in the Baltic Sea. *J. Mar. Syst.* **2019**, *193*, 1–26. [CrossRef]
56. Herman-Iżycki, L.; Jakubiak, B.; Nowiński, K.; Niezgódka, B. UMPL—Numerical weather prediction system for operational applications. In *Research Works Based on the ICM's UMPL Numerical Weather Prediction System Results*; Wyd. ICM: Warsaw, Poland, 2002; pp. 14–27.
57. Krężel, A.; Kozłowski, Ł.; Paszkuta, M. A simple model of light transmission through the atmosphere over the Baltic Sea utilising satellite data. *Oceanologia* **2008**, *50*, 125–146.
58. Zujev, M.; Elken, J. Testing marine data assimilation in the Northeastern Baltic using satellite SST products from the Copernicus marine environment monitoring service. *Proc. Est. Acad. Sci.* **2018**, *67*, 217–230. [CrossRef]
59. Cressman, G.P. An operational objective analysis system. *Mon. Weather Rev.* **1959**, *87*, 367–374. [CrossRef]
60. Gandin, L.S. *Objective Analysis of Meteorological Fields*; Israel Program for Scientific Translation: Jerusalem, Israel, 1965; 242p.
61. Evensen, G. Sequential data assimilation with a nonlinear quasigeastrophic model using Monte Carlo methods to forecast error statistics. *J. Geoph. Res.* **1994**, *99*, 10143–10162. [CrossRef]
62. Evensen, G. The Ensemble Kalman filter: Theoretical formulation and practical implementation. *Ocean Dyn.* **2003**, *53*, 343–367. [CrossRef]
63. Pham, D.T.; Verron, J.; Gourdeau, L. Singular evolutive Kalman filters for data assimilation in oceanography. *Comptes Rendus Acad. Sci. Paris Earth Planet. Sci.* **1998**, *326*, 255–260.
64. Parrish, D.; Derber, J. The National Meteorological Center's spectral statistical interpolation analysis system. *Mon. Weather Rev.* **1992**, *120*, 17471763. [CrossRef]
65. Sokolov, A.; Andrejev, O.; Wulff, F.; Rodriguez Medina, M. The data assimilation system for data analysis in the Baltic Sea. *Syst. Ecol. Contrib.* **1997**, *3*, 1–66.
66. Zhuang, S.Y.; Fu, W.W.; She, J. A pre-operational three dimensional variational data assimilation system in the North/Baltic Sea. *Ocean Sci.* **2011**, *7*, 771–781. [CrossRef]
67. Fu, W.; She, J.; Zhuang, S. Application of an ensemble optimal interpolation in a North/Baltic Sea model: Assimilating temperature and salinity profiles. *Ocean Model.* **2011**, *40*, 227–245. [CrossRef]
68. Losa, S.N.; Danilov, S.; Schröter, J.; Nerger, L.; Maβmann, S.; Janssen, F. Assimilating NOAA SST data into the BSH operational circulation model for the North and Baltic Seas: [illegible]

[illegible] Dzierzbicka-Głowacka, L.; Janecki, M.; Kałas, M. Assimilation of the satellite SST data in the 3D CEMBS model. *Oceanologia* **2015**, *57*, 17–24. [CrossRef]

Article

Quantifying Spatiotemporal Patterns and Major Explanatory Factors of Urban Expansion in Miami Metropolitan Area During 1992–2016

Shaikh Abdullah Al Rifat and Weibo Liu *

Department of Geosciences, Florida Atlantic University, Boca Raton, FL 33431, USA; srifat2017@fau.edu
* Correspondence: liuw@fau.edu; Tel.: +1-561-297-4965

Received: 25 September 2019; Accepted: 23 October 2019; Published: 25 October 2019

Abstract: Urban expansion is one of the most dramatic forms of land transformation in the world and it is one of the greatest challenges in achieving sustainable development in the 21st century. Previous studies analyzed urbanization patterns in areas with rapid urban expansion while urban areas with low to moderate expansion have been overlooked, especially in developed countries. In this study, we examined the spatiotemporal dynamics of urban expansion patterns in South Florida, United States (US) over the last 25 years (1992–2016) using Remote Sensing and GIS techniques. The main goal of this paper was to investigate the degree and spatiotemporal patterns of urban expansion at different administrative level in the study area and how spatiotemporal variance in different explanatory factors influence urban expansion in this region. More specifically, this research quantifies the rates, types, intensity, and landscape metrics of urban expansion in Miami-Fort Lauderdale-Palm Beach, Florida Metropolitan Statistical Area (Miami MSA) which is the 7th largest MSA and 4th largest urbanized area in the US using remote sensing (satellite imageries) data from National Land Cover Datasets (NLCD) and Coastal Change Analysis Program (C-CAP) at 30 m spatial resolution. We further investigated the urban growth patterns at the county and city areas that are located within this MSA to portray the local 'picture' of urban growth in this region. Urban expansion in this region can be divided into two time periods: pre-2001 and post-2001 where the former experienced rapid urban expansion and the later had comparatively slow urban expansion. Results suggest that infilling was the dominant type of urban expansion followed by edge-expansion and outlying. Results from landscape metrics represent that newly developed urban lands became more aggregated and simplified in form as the time progressed in the study region. Also, new urban lands were generated away from the east coast and historic cities which eventually created new urban cores. We also used correlation analysis and multiple linear stepwise regression to address major explanatory factors of spatiotemporal change in urban expansion during the study period. Although the influence of factors on urban expansion varied temporally, Population and Distance to Coast were the strongest variables followed by Distance to Roads and Median Income that influence overall urban expansion in the study area.

1. Introduction

The level of urbanization went up by almost 80 percent in some parts of the world in 2003 including North America, Europe, and Australia and more than half of the world's population lives in urban areas today [1]. This number is expected to increase even more in the coming years. Urbanization is more intense and complex in the coastal areas given that they are more densely populated with higher rate of urban expansion than surrounding areas [2]. According to United Nations (2018) [1], about 66% of total population is projected to be living in urban areas by 2050. As the total urban and coastal population is expected to increase at an alarming rate, urban growth and urbanization have become a crucial issue among the city planners, policymakers, and scientific community.

Urbanization or urban expansion mainly characterized by population change from rural to urban areas and conversion of urban lands from non-urban lands [3] which presents both prospects to the society by enhancing economic development and challenges by bringing different social, cultural, and environmental problems, which may affect the overall living quality of people [3–8]. Also, the physical growth of urban land is considered to be one of the most radical and unalterable forms of land alterations on the planet [9,10]. Most often, the impacts of urban expansion exceed its boundary [6,11] and include landscape change [12,13], loss of agricultural land [9], biodiversity [14,15], air and water pollution [4,16,17], biogeochemical cycles [18,19], and local and regional climate change [4,20,21] at different scales.

The population in Florida (FL) increased from 12 million in 1990 to almost 21 million in 2018. South Florida is not an exception. During the past two decades, South Florida experienced a moderate but noticeable urban expansion, especially in the urban region. The economic development in this region had a boost-up during the start of the nineteenth century by involving the private-sector induced tourism and urban land conversion from natural land was accounted for more than 50% of the total natural land conversion between the period 1973–1995 [22]. The south Florida region consists of one of the largest Metropolitan Statistical Area (MSA) in the United States named 'Miami-Fort Lauderdale-Palm Beach, FL MSA'. This region is our study area of this paper and will be referred to as 'Miami MSA' in the rest of the paper.

Remote sensing is an important source of data when it comes to land use land cover (LULC) change analysis with high spatial and temporal accuracy [23–25]. With the help of Geographical Information Systems (GIS), remote sensing improved the evaluation and examination of the urban growth process [26]. Landscape metrics have been widely used to analyze landscape patterns of urban expansion [24,27–29].

Many studies have been conducted utilizing the remote sensing data and landscape metrics combined with GIS techniques to quantify the spatiotemporal dynamics of urban expansion [3,13,26,30–35]. Among them, most of the studies concentrated on developing countries like China considering their rapid urbanization over the last few decades. Zhao et al. (2015) [31] studied urbanization processes in two moderately developed cities in China by quantifying and comparing dynamics of urban expansion between those two cities and analyzing the trend of landscape metrics and growth types. Shi et al. (2018) [35] examined the dynamics of urban expansion over the 15 years' period in southeastern China by utilizing Nighttime Light Data from National Oceanic and Atmospheric Administration (NOAA). Sun et al. (2014) [26] studied spatiotemporal pattern of urban expansion in northeast China over the last three decades by quantifying the urban expansion rates and their spatial extent, different types of urban expansion, urban expansion intensity, and landscape metrics in their study region. Li et al. (2016) [34] studied the change of urban land areas at regional scale over the last two decades in 15 metropolitan areas in the southeastern United States. Chen et al. (2018) [36] quantified the urban expansion pattern in northeast China over 25 years using the Landsat imageries and analyzed the influence of socioeconomic factors on urban expansion between that time period. Yu and Zhou (2017) [37] analyzed spatiotemporal patterns of urban land expansion at regional and city scale and examined how geographical location, city size, and expansion rate influence overall urban expansion at different scales. However, there is still not enough research done that concentrates on the urban expansion process in the moderately growing urban areas in an already developed region like south Florida. Most of the previous research studied areas where rapid urban expansion occurred. Moreover, many of those studies only quantified spatiotemporal patterns of urban expansion in a single administrative boundary (e.g., city, metropolitan area, province, etc.).

Generally, urban expansion patterns are analyzed by examining the explanatory factors of urban growth. Previous studies have found that urban expansion is driven by socioeconomic factors like population and economy [36,38,39]. In most cases, a qualitative approach was taken to describe those guiding factors due to a lack of data where quantitative analysis would be more effective [36]. Furthermore, it is not only socioeconomic factors that drive urban expansion in an area but physical factors like elevation [40,41] and proximity factors like distance to river and water [42,43] and distance to major roads [44,45] also influence urban expansion. Physical factors influence urban expansion in two ways: they may provide a spatial direction of urban land development (e.g., a mountain) or may serve as limiting factors for urban development such as extreme slope, unsuitable soils for development, etc. However, quantitative analysis of urban expansion patterns that incorporate various explanatory factors (socioeconomic, physical, and proximity factors) has been inadequate. Additionally, the spatial and temporal variation of those factors of urban expansion were hardly studied [46].

In this study, we quantified and analyzed urban expansion and its spatiotemporal patterns in Miami MSA for the last 25 years (1992–2016) at a 5-year temporal scale using the National Land Cover Datasets (NLCD) and NOAA (under the Coastal Change Analysis Program (C-CAP)) combined with landscape metrics. We then further examined the urban expansion process at the county level and compared and analyzed the spatiotemporal pattern of urban growth in the three counties (Miami-Dade, Broward, and Palm Beach) in Miami MSA. Additionally, to get the local representation of urban expansion in this region for the last 25 years, we further explored the urban expansion scenario at the city level in the study area. We also examined the influence of spatiotemporal variations in guiding factors on urban expansion in the study area. The explanatory/guiding factors were chosen based on the previous literature and local knowledge. The objectives of this study were to (1) analyze and compare the extent of urban expansion at metropolitan, county, and city level, (2) illustrate the spatiotemporal patterns of urban expansion at different administrative levels, (3) analyze and compare the landscape metrics of urban expansion and their types at those administrative levels, and (4) quantify the influence of major explanatory factors on urban expansion.

2. Materials and Methods

2.1. Study Area

The study area of this research is the Miami Metropolitan Area which is also known as the Greater Miami Area or South Florida. It is the 73rd largest metropolitan area in the world and the seventh-largest metropolitan area in the United States. It is located in the most southern part of the State of Florida and is the most populous region in the State of Florida.

The Miami Metropolitan Area is defined by the Office of Management and Budget as the Miami-Fort Lauderdale-West Palm Beach, FL MSA, which consists of Miami-Dade, Broward, and Palm Beach Counties. Miami-Dade, Broward, and Palm Beach county are the first, second, and third most populous counties in Florida, respectively. This MSA consists of 15,895 km^2 of land area (US Census Bureau, 2017). The major cities in the MSA include Miami, Fort Lauderdale, Miami Beach, West Palm Beach, Jupiter, and Boca Raton which are also known collectively as 'Gold Coast' (Figure 1). There are [illegible] to the north, Everglades and Atlantic Ocean to the west. The climate in this region is tropical monsoon where most of the rainfall is in summer and winter is typically dry. The average rainfall is about 1500 mm per year in this region. The temperature during the summer ranges between 24–39 °C and during the winter ranges between 15–24 °C. The highest elevation is 16.2 m and the lowest elevation is 0 m in this region. Due to the geographical location (surrounded by Atlantic Ocean), south FL is often hit by deadly hurricanes. Hurricane season runs from June 1st to November 30th while the most dangerous period is from mid-August to end of September. This region is known as one of the most dangerous areas likely to hit by hurricanes. Major hurricanes that hit this region in the recent period are Hurricane Andrew (1992), Irene (1999), Katrina and Wilma (2005), and Irma (2017).

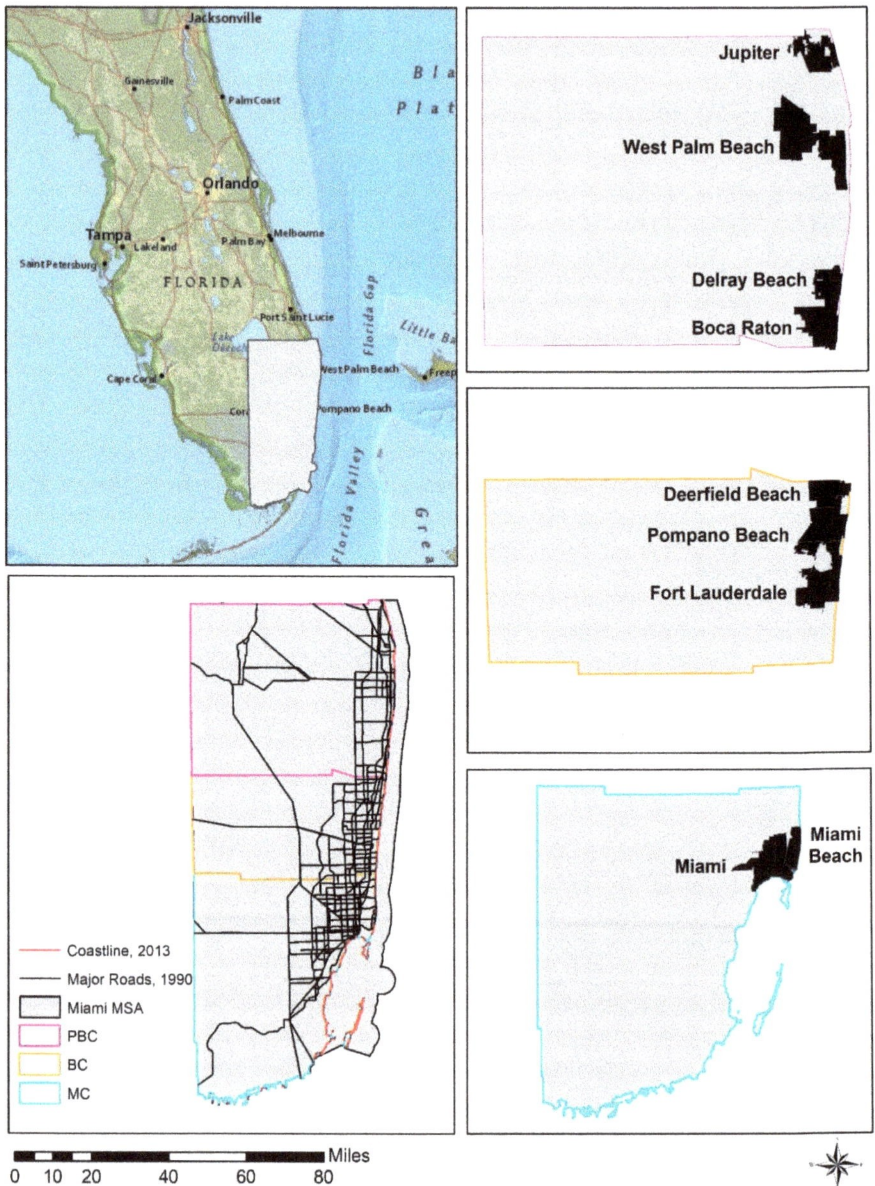

Figure 1. Location of the Florida Metropolitan Statistical Area (Miami MSA), three counties, major cities, along with coastline in 2013 and major roads in 1990 within the study area.

This MSA is about 161 km long and 32.2 km wide which makes it the second-longest urbanized area in an MSA after New York. The urbanized area in Miami-Fort Lauderdale-Palm Beach, FL MSA consists of 2,890 km^2 with a population of 4,919,036 and a population density of 1702.1 per km^2 in 2000. The total population in Miami MSA in 2000 was 5,007,564 and the estimated population in this area in 2018 is 6,198,782. The population increased by about 23.8% from 2000 to 2018. The urbanized area in Miami MSA was the fourth largest urbanized area in the US in 2010 census.

2.2. *Data*

2.2.1. LULC Data

LULC data were obtained from the Multi-Resolution Land Characteristics Consortium (MRLC, https://www.mrlc.gov/) [47,48]. MRLC provides NLCD for the year 1992, 2001, 2006, 2011, and recently released (May 2019) 2016 in ArcGIS grid format [49]. NLCD data products are derived from Landsat imageries, geometrically and radiometrically corrected, and uses unsupervised classification method for the LULC classification. The LULC data provided by NLCD contains land cover and land cover change data for the entire United States with a 30 m spatial resolution. Land covers in NLCD are classified into several classes but the 8 broad categories are water, developed, barren, forest, shrubland, herbaceous, planted/cultivated, and wetlands. The overall accuracy of these datasets is 80% or more [50–54].

To be consistent with the temporal resolution of the dataset and keep it at a roughly 5-year temporal scale, we have also used land cover data sets for the year 1996 provided by the National Oceanic and Atmospheric Administration (NOAA) under the Coastal Change Analysis Program (C-CAP). This dataset covers the coastal portion of the NLCD datasets that contains LULC data with a 30 m spatial resolution and are based on Landsat imageries like NLCD. C-CAP data is also classified into several categories. The overall accuracy of these datasets is 85% or more [55]. Since the focus of this research is to examine the spatiotemporal change of the urban expansion in the study area, the NLCD and C-CAP data were re-classified into two land covers namely urban and non-urban using 'Raster Reclassification' tool in ArcGIS. Every land cover that is not urban was reclassified as non-urban. Table 1 shows a detailed description of different classes in NLCD and C-CAP datasets and the classes that were used in this research to define urban areas.

Table 1. Land cover classes in National Land Cover Datasets (NLCD) and Coastal Change Analysis Program (C-CAP) datasets and land covers used to define urban areas in this research.

Datasets	Classes in the Original Dataset	Classes Used to Define Urban Area in this Research	Overall Accuracy of the Datasets
NLCD (1992, 2001, 2006, 2011, 2016)	Open Water; Perennial Ice/Snow; Developed, Open Space; Developed, Low Density; Developed, Medium Density; Developed, High Density; Barren Land; Deciduous Forest; Evergreen Forest; Mixed Forest; Dwarf Scrub; Shrub/Scrub; Grassland/Herbaceous; Sedge/Herbaceous; Lichens; Moss; Pasture/Hay; Cultivated Crops; Woody Wetlands; Emergent Herbaceous Wetlands	Developed, Open Space; Developed, Low Density; Developed, Medium Density; Developed, High Density	≥80%
C-CAP (1996)	Developed, High Intensity; Developed, Medium Intensity; Developed, Low Intensity; Developed, Open Space; Cultivated Crops; Pasture/Hay; Grassland/Herbaceous; Deciduous Forest; Evergreen Forest; Mixed Forest; Scrub/Shrub; Palustrine Forested Wetland; Palustrine Scrub/Shrub Wetland; Palustrine Emergent Wetland; Estuarine	Developed, High Intensity; Developed, Medium Intensity; Developed, Low Intensity;	≥ 85%

[illegible] Factors of Urban Expansion

In this study, we considered socioeconomic, proximity, and physical factors that affect urban land expansion based on previous studies and local knowledge. Not all these factors act as drivers or facilitators of urban land expansion. Some of the factors act as limiting factors of urban growth as well. For socioeconomic factors, we selected population and median household income as the explanatory factors of urban land expansion. For proximity factors that affect urban land expansion, we considered distance to major roads and distance to coastal boundary, and for physical factors, we selected elevation of the study area. Table 2 shows a detailed description of these variables and their data sources.

Table 2. Selected variables of major explanatory factors of urban expansion.

Variable Category	Description	Variable	Sources
Socioeconomic factors	People (10^3 per km^2)	Population	Population grid datasets from NASA's Socioeconomic Data and Applications Center (SEDAC) website (https://sedac.ciesin.columbia.edu/data/collection/gpw-v4 and https://sedac.ciesin.columbia.edu/data/collection/grump-v1) for the year 1990, 1995, 2000, 2005, 2010, and 2015 as raster surface at 1 km resolution.
	Median Household Income (10^3 per km^2)	Median Income	Median household income data were derived at the block group level from National Historical GIS (NHGIS) website (https://www.nhgis.org/) [56] for the year 1990, 2000, 2010, 2011, and 2016 which were later converted to raster layers at 1 km resolution.
Proximity factors	Distance to Major Roads (km)	Distance to Roads	Major road data for the years 2007, 2011, and 2016 were derived from TIGER/Line Shapefiles website (https://www.census.gov/geographies/mapping-files/time-series/geo/tiger-line-file.html) and for the year 1990, 1993, and 2000 were derived from Florida Geographic Data Library (FGDL) website (https://www.fgdl.org/metadataexplorer/about.html). Nearest distance to major roads for above years was calculated using the Euclidean Distance tool in ArcMap at a 1 km resolution.
	Distance to Coastal Boundary (km)	Distance to Coast	Coastal boundary data were derived from the TIGER/Line Shapefiles website (https://www.census.gov/geographies/mapping-files/time-series/geo/tiger-line-file.html) for the year 2013 and 2016. Nearest distance to coastline was calculated using the Euclidean Distance tool in ArcMap at a 1 km resolution.
Physical factors	Elevation (km)	DEM	Digital Elevation Data at 30 m resolution were derived from USGS National Elevation Dataset (https://catalog.data.gov/dataset/usgs-national-elevation-dataset-ned) and calculated using the Zonal Statistics tool in ArcMap at 1 km resolution.

Based on available data, datasets for selected influential factors of urban expansion were derived for the study period (1992–2016) from different data sources. Population and income are important factors as part of the socioeconomic factors of urban expansion. 1 km gridded population datasets were derived for the years 1990, 1995, 2000, 2005, 2010, and 2015 and used for subsequent study years (1990 data for study year 1992, 1995 data for study year 1996, and so on). Median household income data were derived at the block group level for the years 1990, 2000, 2010, 2011, and 2016. For the study year 1992 we used the 1990 data, 1996 we used the average of 1990 and 2000 data, 2001 we used 2000 data, 2006 we used the average of 2000 and 2010 data. Population and economic development (in this case, median income) are commonly used as a driving factor that facilitates urban expansion [46].

Urban expansion near the major roads is considered one of the most common patterns of urban expansion that guides urban growth [46]. As a result, it is expected that distance to major roads will have a negative effect on urban expansion. Both the primary and secondary roads were considered as major roads in this research. Major road data for the years 1990, 1993, 2000, 2007, 2011, and 2016 were derived. We used 1990 data to represent the year 1992, 1993 data to represent the year 1996, 2000 data to represent the year 2001, 2007 data to represent the year 2006, and 2011 and 2016 data to represent the year 2011 and 2016, respectively.

Since the study area is a coastal area, we considered distance to the coastal boundary as one of the proximity factors to urban expansion. Proximity to water, or in this case, the coast, affect urban expansion in two different ways: while urban expansion could be restricted by the presence of a water body (Atlantic Ocean for this study), it could also advance water resources and waterborne advantages to facilitate urban development at the same time [46]. Generally, people tend to live near the coast due to the high recreational value, which is evident in south Florida. Therefore, it is expected that distance to coastal boundary will have a negative impact on urban expansion in the study area. Based on the available data, coastal boundary in 2013 and 2016 were derived. Coastal boundaries in 2013 were used to represent the study years between 1992 and 2011, and 2016 were used to represent the study year in 2016. Since the factors like coastal boundary and elevation do not change much over time, we kept elevation constant for each year.

2.3. Methods

2.3.1. Annual Urban Expansion Rate

Annual urban expansion rate (AUE_a) was calculated for the study area to examine the annual expansion rate of an urban area over the last 25 years. AUE_a shows the temporal patterns of urban expansion in Miami MSA and the tri-county area distinctly. Along with AUE_a, standardized annual urban growth rate (AUE_s) was also calculated. AUE_s can be useful to compare the urban expansion rate between the three counties as it does not consider the impact of initial sizes of the counties [26]. The indexes that were used to calculate the AUE_a and the AUE_s are given below:

$$AUE_a\ (km^2\ year^{-1}) = UA_{n+i} - UA_n/i \quad (1)$$

$$AUE_s\ (\%) = ((UA_{n+i}/UA_n)^{1/i} - 1) \times 100\% \quad (2)$$

where AUE_a and AUE_s are the annual urban expansion rate ($km^2\ year^{-1}$) and standardized annual urban expansion rate (%) from the year n to n + i, respectively. UA_{n+i} and UA_n are the total area of urban land (km^2) at the year n + i and n, respectively, and i is the difference between year n + i and n (years).

2.3.2. Annual Expansion Type

It is really important to identify the different types of urban expansion in an area to successfully analyze the patterns of urban expansion [26]. According to Xu et al. (2007) [30], there are three types of urban expansion that can be identified including outlying, edge-expansion, and infilling. Outlying urban expansion happens when the newly developed urban patch has no spatial connection with existing urban land, edge-expansion type denotes the new urban land that spreads out from the border of existing urban land, and infilling urban expansion occurs when the non-urban land that is surrounded by existing urban land converts to urban land [32]. However, outlying urban expansion could either follow a scattered or random development or it could be directed or guided by features like roads and canals. This pattern of outlying urban expansion with major roads in the study area was also explored in this research. Following Xu et al. (2007) [30], an E index was created to identify different types of urban expansion in the study area using the following equation:

$$E = L_{com}/P_{new} \quad (3)$$

where E means the type of urban expansion, L_{com} is the length of the common edge between newly developed urban land and existed urban land, and P_{new} is the perimeter of a newly developed urban land. The value of E ranges between 0 to 1. Urban expansion type is outlined as outlying when $E = 0$, edge-expansion when $0 < E \leq 0.5$ and infilling when $E > 0.5$.

2.3.3. Urban Expansion Intensity

Urban expansion intensity was calculated to examine the spatial [illegible]

where $UII_{i,\ t\ to\ t+n}$ is the urban expansion intensity for spatial unit i between the time period t and t+n, $UA_{i,t+n}$ and $UA_{i,t}$ refer to the total urban land area of spatial unit i at the time t+n and t, respectively, and TA_i denotes the total area of spatial unit i. In this research, the spatial unit is a 2 km × 2 km grid. The urban expansion intensities were classified into five groups by a custom standard including standard <10%, 10–20%, 20–40%, 40–70%, and 70–100%, which refer to the urban expansion intensity level of very low, low, moderate, rapid, and highly rapid, respectively.

2.3.4. Landscape Metrics

Four different metrics at class and landscape levels were calculated to see the impacts of urban growth in the study area. FRAGSTATS version 4.2 was used to calculate these metrics. Since the main objective of this research is to examine the spatiotemporal pattern of urban expansion over the last 25 years in the study area and there are a lot of metrics available, we chose four metrics to identify the shape, landscape and distribution of urban patches including Number of Patches (NP), Largest Patch Index (LPI), Landscape Shape Index (LSI), and Area-weighted Mean Shape Index (SHAP_AM) [28]. NP was calculated using the 8-cell neighborhood rule in FRAGSTATS version 4.2. An increase in NP means a more fragmented urban surface while decrease in NP suggests opposite [57]. Similarly, increase in LPI means the increase in urban center, increase in LSI suggests a more complicated and irregular shape of urban patches, increase in SHAP_AM represents increasing complexity of urban land, and vice-versa [26]. NP, LSI, and SHAP_AM were calculated at class level and LSI was calculated at landscape level for urban land. Table 3 below shows the detailed description of these metrics.

Table 3. Landscape metrics based on McGarigal and Marks (1995) [28].

Landscape Metric	Abbreviation	Description	Range
Number of Patches	NP	Total number of urban land cover patches surrounded by non-urban land cover types	$NP \geq 0$
Largest Patch Index	LPI	The proportion of total area occupied by the largest patch of a land cover type	$0 < LPI \leq 100$
Landscape Shape Index	LSI	A modified perimeter-area ratio of the form that measures the shape complexity of the urban land cover type	$LSI > 0$
Area-weighted Mean Shape Index	SHAP_AM	The shape index weighted by relative patch area which measures the average shape complexity of individual patches for the urban land cover type	$SHAP\_AM > 0$

2.3.5. Urban Expansion Direction

Identifying urban expansion directions helps to understand the spatiotemporal pattern of urban development. In this study, we used weighted mean center via ArcMap to depict the change in direction of urban expansion in each study year. In calculating the mean center of urban lands for each year, we used area (km^2) of urban patches as the weight. As a result, urban patches with larger areas get higher weight in calculating the mean center.

2.3.6. Factors Influencing Urban Area

A linear correlation between the urban area and each of the five explanatory/independent variables was calculated. Then a regression model was created for each of the six study years. Urban area was considered as the dependent variable and variables of urban expansion with statistically significant relationship with urban area were considered as explanatory/independent variables. In doing this, 1 km × 1 km grids were created over the urban land areas of each year and the total urban area for each grid in each year was calculated to be consistent with the datasets used. Then the values from each influential factor of urban expansion were extracted for each grid each year using ArcMap.

Correlation Analysis

The Pearson Correlation Coefficient (r) was calculated between the urban land area and the explanatory factors of urban expansion using SPSS (version 26) at a 95% confidence interval. The value of r ranges from -1 to 1. The value 0 means no linear correlation between urban land area and explanatory factors of urban expansion, value above 0 means positive linear correlation and value below 0 means negative linear correlation. The value of r was considered statistically significant when the *p*-value is below 0.05.

Regression Analysis

A multiple linear regression analysis was performed in SPSS (version 26) for each study year where the urban land area was the dependent variable and explanatory factors of urban land expansion were independent variables. Independent variables with statistically significant positive or negative correlation with the urban areas were considered only in the regression analysis. The stepwise method was chosen while running the multiple linear regression analysis since the stepwise regression method is a method of fitting regression models where the independent variables are automatically chosen and the influence of each independent variable to the model can be easily determined. Multicollinearity between the independent variables was also tested in each regression model using the Tolerance and Variance Inflation Factors (VIF) in correlation matrix. Generally, if the tolerance value is high (close to 1 and greater than 0) and VIF value is low (smaller than 4 but greater than 1), it is considered as low degree of multicollinearity [58]. Before running the multiple linear stepwise regression, the dependent and independent variables were normalized using the below equation to remove the effect of factor dimension and magnitude [36,59]:

$$x'_i = (x_i - x_{min})/(x_{max} - x_{min}) \quad (5)$$

where x'_i is the normalized value of the ith cell of variable x, x_i is the value of ith cell of variable x, x_{max} is the maximum value of variable x, and x_{min} is the minimum value of variable x. After normalization, the value of the dependent and independent variables ranges from 0 to 1. The regression model is acceptable when the p-value is below 0.05 and a high Adjusted R^2 value means a better model fit. Figure 2 illustrates the methodology of this study below.

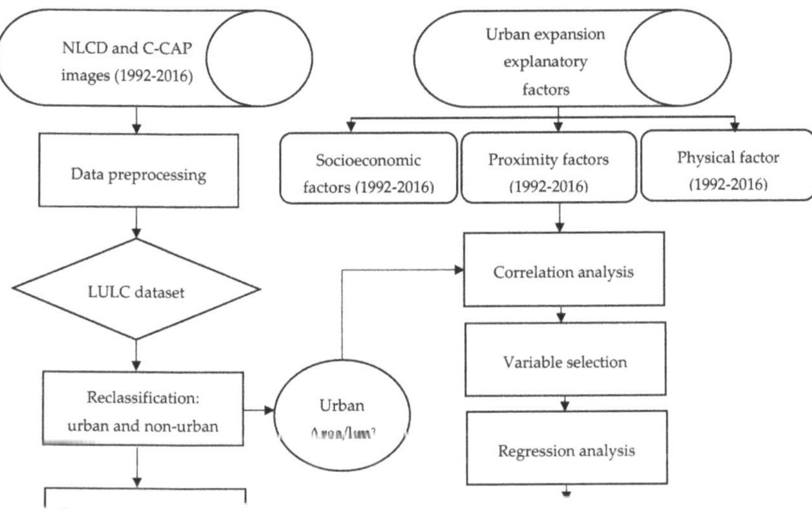

Figure 2. Methodological flowchart of the study.

3. Results

3.1. Urban Expansion Rate

The Miami MSA has gone through a moderate but noticeable urban expansion over the last 25 years (Figure 3 and Appendix A) where the proportion of urban land increased and non-urban land decreased. From the period 1992 to 2016, the urban land increased from 2308.28 km^2 to 3167.78 km^2

which is 859.5 km^2 of newly developed urban land in total in the last 25 years. The AUE$_a$ in this period was 34.38 km^2 year^{-1}. To get the sense of urban expansion in the study area more locally, we then looked into the urban expansion in the last 25 years in the three counties (Palm Beach, Broward, and Miami-Dade) individually. The urban land in Palm Beach County (PBC) increased substantially where the total urban land in 1992 was 780.01 km^2 and it increased to 1183.77 km^2 in 2016 with an average annual growth rate of 16.15 km^2 year^{-1}. Urban land expanded with an average growth rate of just over 9 km^2 year^{-1} in both Broward County (BC) and Miami-Dade County (MC) (Figure 3).

However, these urban expansions were not the same across different time periods. In Miami MSA, the AUE$_a$ was 45.7 km^2 year^{-1} from 1992 to 1996 and 96.57 km^2 year^{-1} between 1996 and 2001. After 2001, there was a sharp decrease in urban land growth in this area. The rate decreased to 18.41 km^2 year^{-1} from 2001 to 2006 and continued to decrease till 2011 when it was found as 8.91 km^2 year^{-1}. It then again increased to 11.45 km^2 year^{-1} between the period 2011–2016 (Table 4).

Similarly, PBC experienced rapid urban expansion between the periods 1992–1996 and 1996–2001 where the AUE$_a$ were 23.64 and 48.39 km^2 year^{-1}. After 2001, it had a steep decrease in AUE$_a$ until 2011 when the rate was 3.56 km^2 year^{-1}. After 2011, it again had a slight increase with 4.02 km^2 year^{-1}. AUE$_a$ in MC between 1992 –1996 and 1996 –2001 were the lowest among three counties when the rate was found as 9.63 and 20.82 km^2 year^{-1}, respectively. After 2001, it experienced a sharp decrease until 2016. However, between the period 2001–2006 and 2011–2016, MC had the highest AUE$_a$ in the tri-county region where the rate was 8.34 and 5.38 km^2 year^{-1}, respectively. In BC, AUE$_a$ was 12.31 km^2 year^{-1} between 1992–1996 and 27.28 km^2 year^{-1} between 1996–2001. Following the same pattern as other counties, it then continued to decrease till the end when the rate was only 1.72 km^2 year^{-1} in the period 2011–2016 (Table 4). PBC had the highest rate of overall (1992–2016) AUE$_a$ (16.15 km^2 year^{-1}) among these three counties and MC and BC had almost same overall AUE$_a$ (9.08 and 9.04 km^2 year^{-1}, respectively).

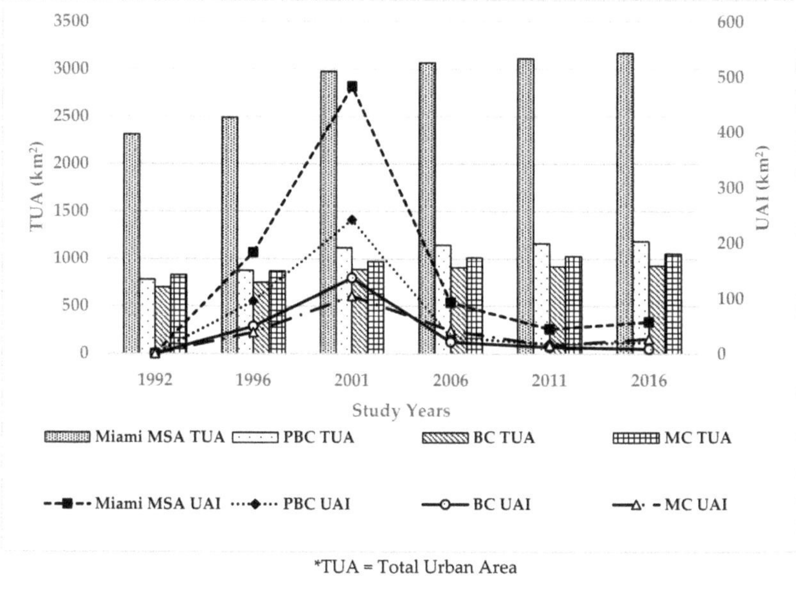

*TUA = Total Urban Area

Figure 3. Urban Area Expansion (in km^2) in the Study Area over Different Time Periods.

Table 4. Annual urban expansion rate (AUE_a) (km^2 $year^{-1}$) and AUE_s (%) in Miami MSA and three counties during 1992–2016.

	1992–1996		**1996–2001**		**2001–2006**		**2006–2011**		**2011–2016**		**1992–2016**	
	AUE_a	**AUE_s**	**AUE_a**	**AUE_s**	**AUE_a**	**AUE_s**	**AUE_a**	**AUE_s**	**AUE_a**	**AUE_s**	**AUE_a**	**AUE_s**
Miami-MSA	45.7	1.92	96.57	3.61	18.41	0.61	8.91	0.29	11.45	0.37	34.38	1.27
PBC	23.64	2.90	48.39	5.01	5.89	0.52	3.56	0.31	4.02	0.34	16.15	1.68
BC	12.31	1.71	27.28	3.40	4.21	0.47	2.35	0.26	1.72	0.19	9.08	1.13
MC	9.63	1.14	20.82	2.29	8.34	0.84	2.96	0.29	5.38	0.52	9.04	0.97

The overall AUE_s in Miami MSA between the period 1992–2016 was found as 1.27%. Again, in the beginning, between 1992–1996, the rate was almost 2%. Then it increased to 3.61% between 1996–2001. After 2001, it experienced a sharp decrease till the end (2011–2016) where the AUE_s were below 1% in this area. The AUE_s in PBC were almost 3% between 1992–1996 which accounts for the highest rate of standardized annual urban expansion between the three counties in this period. It then increased to 5% between 1996–2001. After 2001, it had a sharp decrease till 2016 where the AUE_s were below 1% in all the period making the overall average AUE_s in PBC as 1.68%. In BC and MC, the AUE_s were 1.71% and 1.14% respectively between the period 1992–1996. It then increased to 3.4% and 2.29% respectively in the following period (1996–2001). After 2001, BC experienced a steep decrease in AUE_s as it went below 1% for the rest of the periods. MC also experienced a noticeable decrease in AUE_s as it was 0.84% in the period (2001–2006), 0.29% between 2006–2011, and increased to 0.52% again between 2011–2016. Interestingly, both the BC and MC had almost the similar overall average AUE_s including 1.13% and just below 1%, respectively, over the last 25 years considering the AUE_a in MC was substantially lower than that of BC (Table 4).

3.2. *Urban Expansion Types*

Figure 4 shows the percentages of urban expansion types for the newly developed urban lands in Miami MSA, PBC, BC, and MC area over the five different time periods. Infilling was found as the leading urban expansion type (more than 50%) and outlying as the least dominant expansion type in all the time periods for all the areas. In Miami MSA, the infilling expansion type accounts for 65.06% in 1992–1996. Then it continues to increase till 2006 when the infilling type accounts for over 87% of the total proportion of extension types. During 2006–2011, it decreased to 57.65% and then again increased to 64.42% during 2011–2016 making the overall average of infilling type as 84.4% in Miami MSA. The outlying type in Miami MSA during 1992–1996 was 13.44%. It then started to decrease until 2011 when the proportion of outlying type was below 1%. During 2011–2016, it again increased to 2.14% in this area. The proportion of edge-expansion ranges from 12.53% (2001–2006) to 41.36% (2006–2011) in Miami-MSA.

All three counties (PBC, BC, and MC) more or less follow the same trend as the overall study area (Miami MSA) as infilling was the dominant expansion type over all the study periods. Outlying was the least dominant type of expansion in PBC as it was 15.9% during 1992–1996. Then it started to decrease in the following periods until 2006 when it was almost zero (0.13%). During 2006–2011 [illegible]

again increased to almost 65% during the following period (2011–2016). In BC, the highest proportion of infilling type (93.37%) and the lowest proportion of outlying type (0.06%) was during 2001–2006. Outlying was the highest during 1992–1996 (7.35%) and infilling was the lowest during the period 2006–2011 (70.6%). In MC, infilling was also the highest during 2001–2006 (almost 83%). However, during 2006–2011, edge-expansion was the very dominant type of expansion type as it accounted for almost 47% of the total proportion and infilling was the lowest (51.82%). Edge-expansion was 38.62%

during 2011–2016 in MC and outlying was the highest at the beginning (1992–1996) as it accounted for almost 15% of the total proportion.

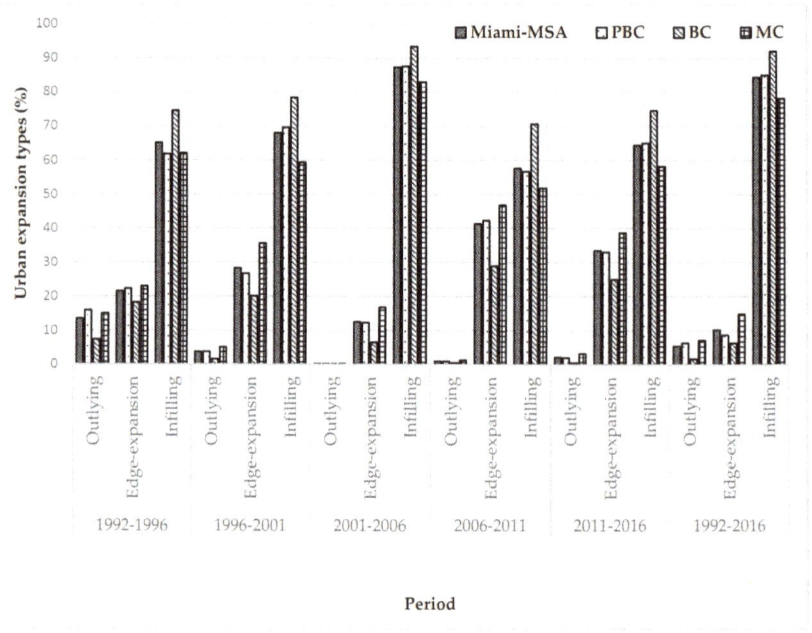

Figure 4. Proportion of Urban Expansion Types in the Study Area over the last 25 Years.

3.3. *Urban Expansion Intensity Index*

Urban expansion intensity has been examined to evaluate the spatial distribution of urban expansion over the study area from 1992 to 2016 in a five-year interval. To understand the pattern of urban expansion intensity more locally, we overlayed the county boundary over the MSA map (Figure 5) and boundaries of most historical and populous cities over the county map (Figure 6). Figure 5 shows that urban expansion mainly occurred on the east coast of the study area. Most of the rapid and highly rapid urban expansion grids are between 1992–2001 in Miami MSA (Figures 5 and 7). More specifically, rapid and highly rapid expansion grids account for over 3% in the periods 1992–1996 and 1996–2001. After 2001, rapid and highly rapid urban expansion grids decreased significantly.

The proportion of rapid and highly rapid grids in the following periods and there were no rapid and highly rapid expansion grids overall (1992–2016) in this region (Figure 7). Very low and low urban expansion grids account for over 85% of the total proportion over all the study periods in this region. There are some 'hotspots' of rapid and highly rapid urban expansion grids during 1992–1996 and 1996–2001 (Figure 5). In PBC, rapid and highly rapid urban expansion grids are clustered together in the south-eastern part of the county which is near the coast during 1992–1996. During 1996–2001, the 'hotspots' of rapid and highly rapid urban expansion grids were moved away from the coast and can be seen in the northern part of the county. In BC, rapid and highly rapid expansion grids are dispersed during the period 1992 to 1996. During 1996–2001, there is clearly a cluster of the grids of such kinds at the southern part of the county which is further away from the east coast. In MC, most of the rapid and highly rapid expansion grids are located in the northern part of the county during 1992–1996 and they are away from the coast. During 1996–2001, that cluster is no longer available, and the grids of rapid and highly rapid expansions are rather dispersed. After 2001, the cluster of rapid and highly rapid urban expansion grids started to diminish in this county like the other two counties. However, during 2001–2006, there were still some rapid and highly rapid dispersed grids in this county which completely disappeared after 2006.

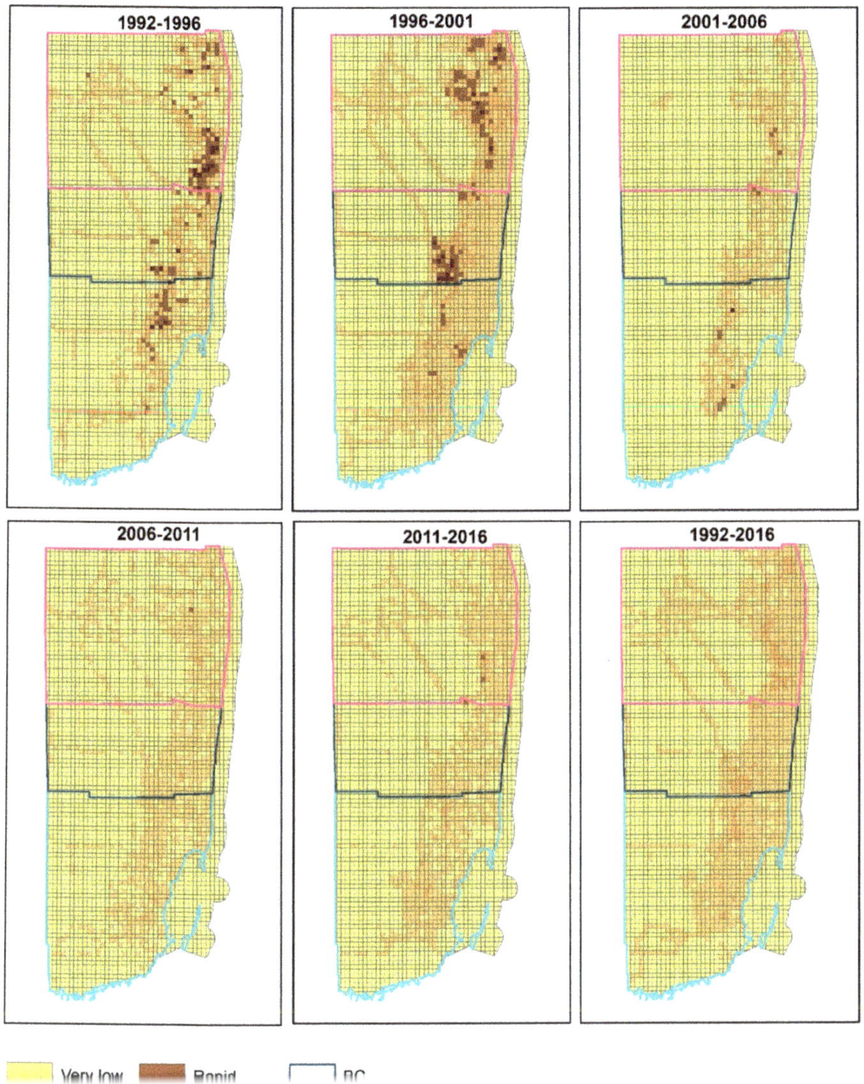

Figure 5. Spatial distribution of UII at MSA and County level at 2 km × 2 km spatial unit.

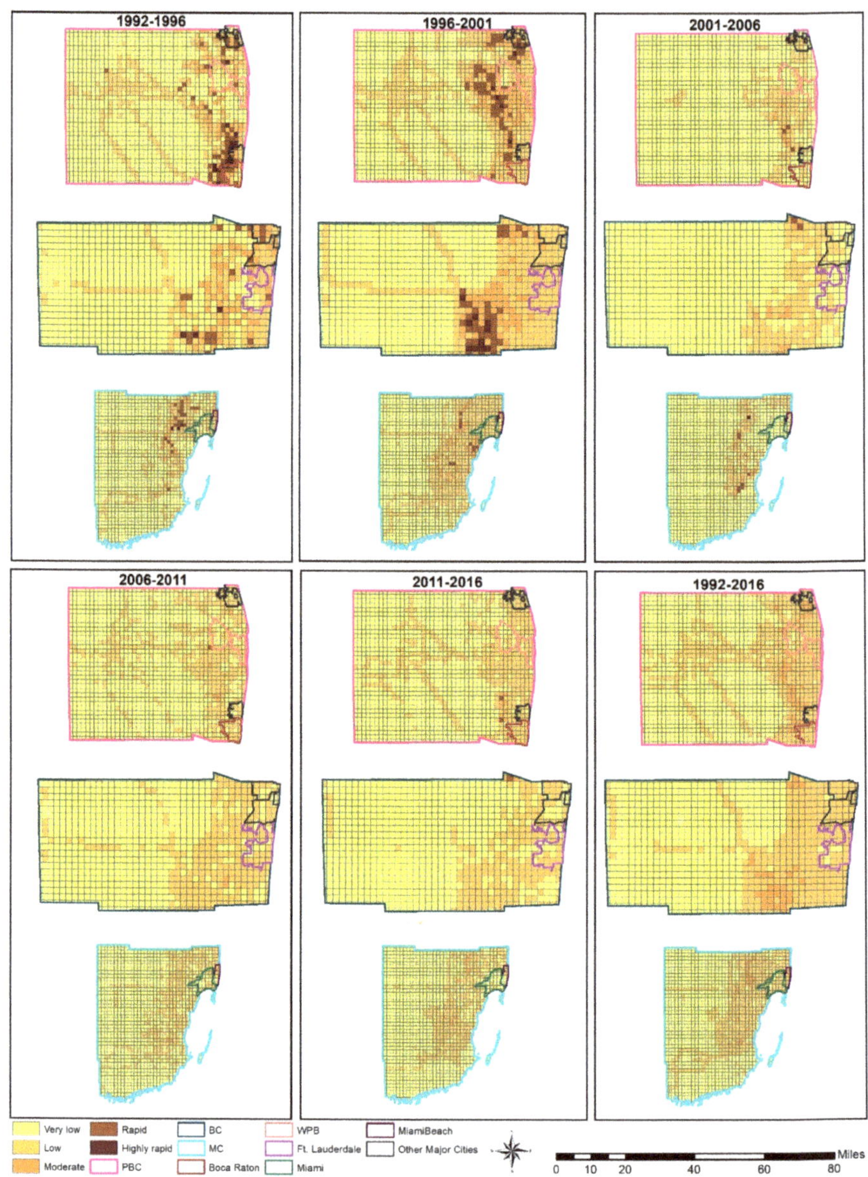

Figure 6. Spatial distribution of UII at County and City level at 2 km × 2 km spatial unit.

Figure 6 shows the spatial pattern of urban expansion intensity at the city level. In PBC, most of the rapid and highly rapid expansion grids are located within or around some major city areas like Boca Raton during 1992–1996. Interestingly, during 1996–2001, most of these grids were shifted away from the major city areas. Since all these major cities are adjacent to the east coast of the study area, it means that most of these grids were shifted away from the beach as well. The same happens in the cities in BC. Although the rapid and highly rapid expansions in this county are dispersed during 1992–1996, it was still within or around the major cities in the county (e.g., Fort Lauderdale). However, during 1996–2001, there was a clustered pattern of these grids which was further away from the major cities and the coast. This dispersed pattern of rapid and highly rapid urban expansion grids suggests that high-intensity urban expansion was occurring outside the major city areas in the study area. Cities in the MC follows the same pattern as well. After 2001, rapid and highly rapid urban expansion grids are no longer visible in this region.

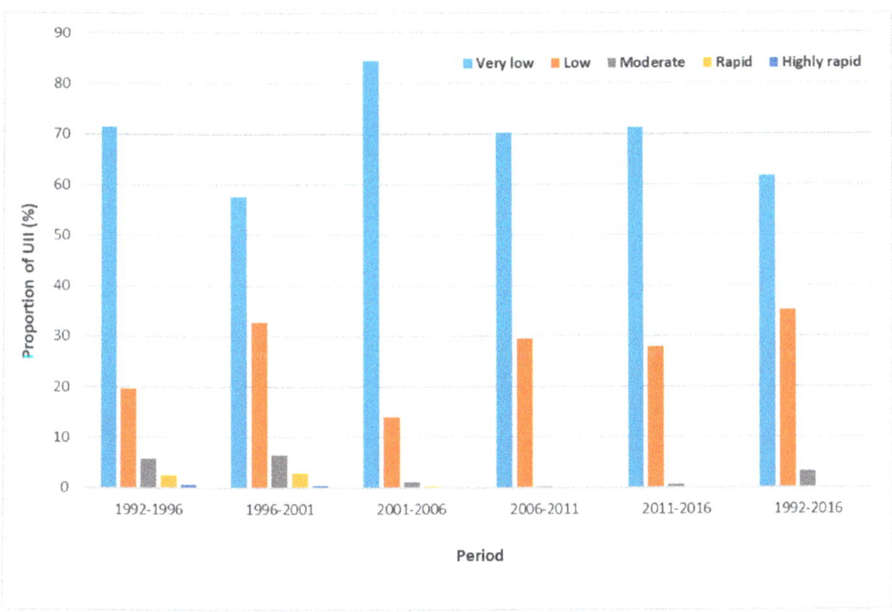

Figure 7. Proportion (%) of Urban Expansion Intensity Grids in Miami MSA.

3.4. Landscape Metrics

To further examine the patterns of urban expansion in the study area, three class-level metrics including NP, LPI, and SHAP_AM and one landscape-level metric (LSI) are calculated at the MSA and County level (Figure 8). To minimize the size effect, NP and LPI were divided by the respective MSA and County area. NP decreased from 1992 to 2001 and then it was stable for the Miami MSA. The three counties (PBC, BC, and MC) follows the same pattern. However, the highest value of NP was different for the three counties. PBC had the highest value of NP among these three counties followed by MC, and then BC. LPI increased rapidly from 1992 until 2001 in Miami MSA which means rapid urban expansion in this period. After 2001, LPI increased very slowly in this area. Although all the counties followed this pattern more or less exactly, BC among the three counties had the highest percentages of LPI in all the time periods. PBC had the lowest percentage of LPI at the beginning (1992). After 1996, it had higher percentages of LPI than MC. SHAP_AM, which measures the mean shape complexity of

last period (2011–2016). In BC, SHAP_AM decreased right after 1992 till 2011. After 2001, it shows almost a flat line, which means that it was stable till 2016. In MC, it was slowly decreasing from 1996 until 2001. From 2001 to 2011, it was stable and had a rapid increase in the last period (2011–2016).

LSI shows a rapid decreasing pattern from 1992 to 2011 in Miami MSA where LSI decreased in almost half (from almost 200 to 100). After 2001, it was stable till the end (2016) which indicates patch shape was becoming complex. PBC and BC follow the same exact pattern as Miami MSA. In MC,

LSI was still decreasing during 1992–2001, but it was rather a slow decrease compared to other two counties. After 2011, it shows a slight increase in LSI also.

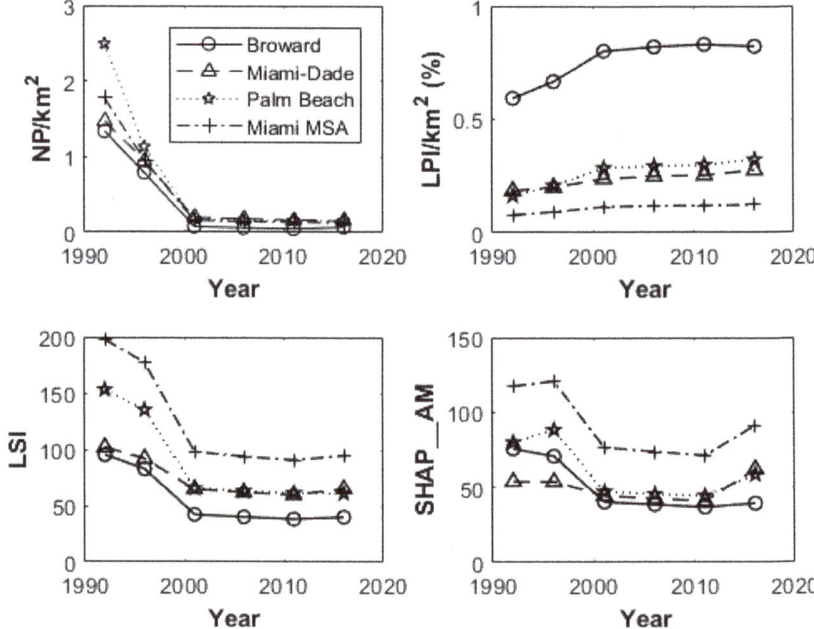

Figure 8. Patterns of Urban Land Expansion Metrics (NP, LPI, LSI, and SHAP_AM).

The decreasing number of NP throughout the study period suggests that newly developed urban lands became more aggregated after 1992 in Miami MSA and in all three counties. After a sharp decrease until 2001, the rate of decrease in NP slowed down till the end (2016). Unlike NP, LPI increased during the entire study period at all administrative levels (MSA and County) which indicates an increase in urban centers in the study area. The decreasing trend in LSI and SHAP_AM leads to simplified form of urban patches and land areas in the study area.

3.5. Urban Expansion Direction

Figure 9 shows the shift in the mean center of the urban areas in Miami MSA from 1992 to 2016. There is a northward shift in the mean center of urban areas from 1992 to 2001. During this period (1992–2001), the mean center first turned towards the east in 1992–1996 and then towards west in 1996–2001. After 2001, the mean center displays a slightly inverse shift from north to south in 2001–2016 while turning towards further west. Although the mean center in the entire study period remained almost stable, there is clearly a westward turn in the mean center in this period. We then further explored the change in mean center of urban areas at the county level. Figure 10 shows the mean center change from 1992–2016 in PBC, BC, and MC. In PBC, the mean center of urban areas shifted towards southeast in 1992–1996. After 1996, there is a significant opposite turn in the mean center in 1996–2001 when it shifted towards a northwest direction. The period 2001–2011 remained steady until 2016 when the mean center in PBC shows a slight turn towards southeast direction. Unlike PBC, BC shows a southwest shift in mean center of urban areas overall (1992–2016). Although after 2001, the mean center remained almost stable in BC while turning towards further west. In MC, urban area means center shifts towards north in 1992–1996. After 1996, it experienced a reverse change towards southwest until 2016.

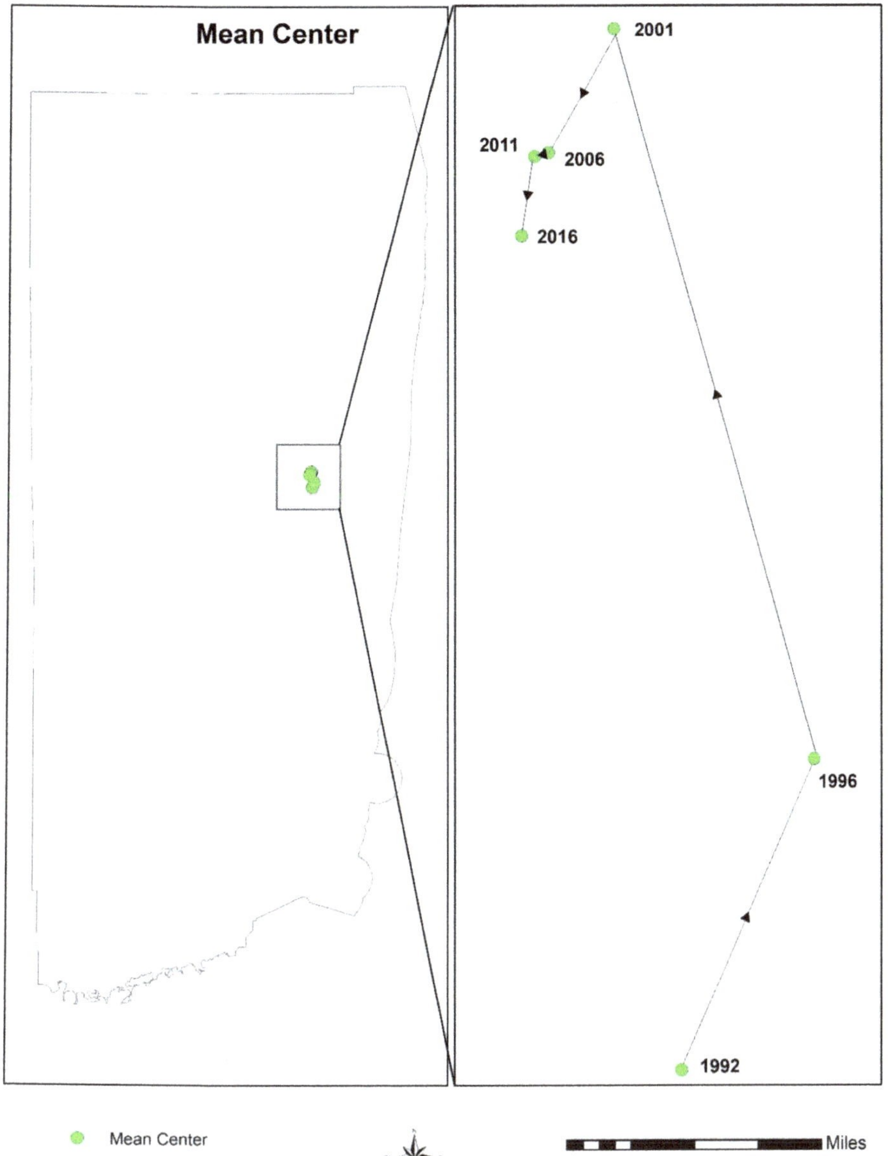

Mean Center
2001
2011
2006
2016
1996
1992
Mean Center
Miles

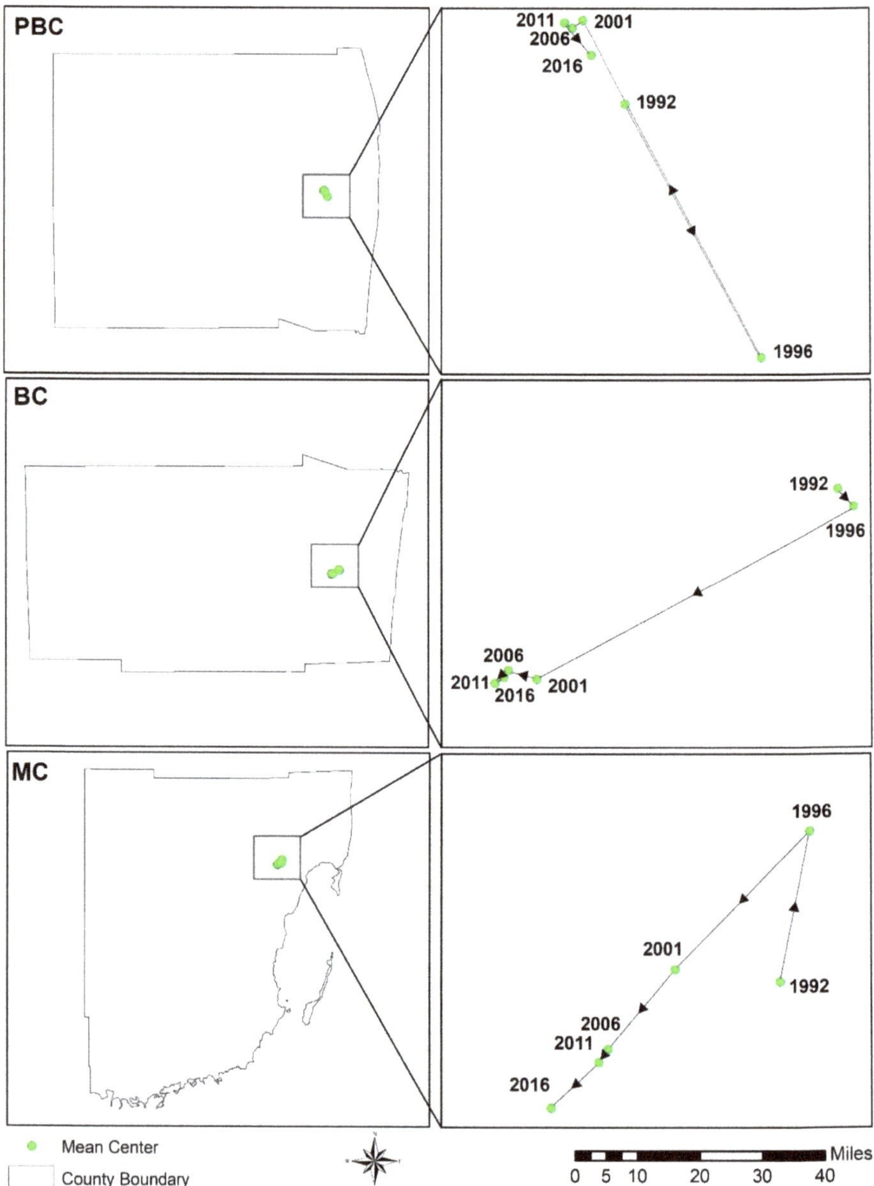

Figure 10. Change of direction in the mean center in the tri-county region.

3.6. Statistical Analyses

The results of the correlation and regression analysis are portrayed in Table 5. Six multiple linear stepwise regression models were built, one for each study year from 1992 to 2016. Each model explained more than half (Adjusted R^2 is above 0.5) of the total variation in urban land. The population was found as the variable that explains most of the variance in urban land area in each year except at the end (year 2016). In 2016, variable Distance to Coast was the most significant variable in explaining the total variance. In 1992 and 1996, Distance to Coast entered in the model as the second explanatory variable followed by Distance to Roads. Median Income was not an entry variable in the first two years' model (1992 and 1996) and last year's model (2016). From 2001 to 2011, Distance to Coast was the second entry variable in each year's model followed by Distance to Roads and Median Income, respectively. In 2016, Population was the second entry variable in the model and Distance to Roads was the last to enter the model. Elevation (DEM) did not have any statistically significant strong correlation with urban land and therefore, elevation was not selected as one of the independent factors in the models. Since the land surface is mostly flat in the south Florida region and elevation does not change very frequently, our findings of elevation not having any strong relationship with urban land area are admissible.

In all models, Population, Distance to Coast, and Distance to Roads are accounted for approximately 90% of the total variation of urban land explained. Median Income was not found influential to the variation of urban land areas explained in 1992, 1996, and 2016. For the rest of the years (2001–2011), Median Income had a weak positive correlation with urban land areas and limited influence in explaining the variance in urban land areas. The population had a strong positive correlation with urban land areas and positive model coefficients in all the study areas indicating that higher population increases urban land areas in each of the study periods. Additionally, Distance to Coast and Distance to Roads had strong negative correlations with urban land areas and negative model coefficients in each year indicating that shorter distances from coast and major roads increase urban land areas regardless of the study year. High tolerance value (close to 1) and low VIF values (1<VIF>3) in the correlation matrix in each year's model suggest that there was a very low degree of multicollinearity among the independent variables.

Table 5. Summary of the multiple linear stepwise regression models. For each of the six models, the dependent variable (urban land area in 1 km^2 grid) was explained by four variables. The sequence of the explanatory variables in each model is the order these variables entered the regression models. S-coefficient means standardized coefficients, which could be used to determine the relative significance of explanatory variables. r means Pearson Correlation Coefficient, which indicates the state (positive/negative) and the level (weak/strong) of correlation of dependent variable and explanatory variables. All the values are statistically significant at 0.001 level (two-tailed).

1992 Model	**Population**	**Distance to Coast**	**Distance to Roads**		**Adjusted R^2**
r	0.691	−0.605	−0.409		0.605
S-coefficient	0.521	−0.360	−0.065		
1996 Model	**Population**	**Distance to Coast**	**Distance to Roads**		
r	0.681	−0.578	−0.319		0.590
S-coefficient	0.522	−0.360	−0.084		
2001 Model	**Population**	**Distance to Coast**	**Distance to Roads**	**Median Income**	**Adjusted R^2**
r	0.606	−0.587	−0.367	0.106	0.538
S-coefficient	0.438	−0.365	−0.132	0.095	
2006 Model	**Population**	**Distance to Coast**	**Distance to Roads**	**Median Income**	**Adjusted R^2**
r	0.611	−0.599	−0.376	0.144	[illegible]
S-coefficient	0.436	−0.376	−0.137	0.124	
2011 Model	**Population**	**Distance to Coast**	**Distance to Roads**	**Median Income**	**Adjusted R^2**
r	0.611	[illegible]	[illegible]	[illegible]	
[illegible]	[illegible]	0.381	−0.181		

4. Discussion

4.1. Spatiotemporal Patterns of Urban Expansion

The results portrayed above show the spatiotemporal changes of urban expansion in Miami MSA and the three counties (PBC, BC, and MC) within this metropolitan area (Figure 11) over the last

25 years (1992–2016). The urban land area expanded around 1.37 times in 2016 than the initial land area in 1992 in Miami MSA (Section 3.1). However, urban expansion at the county level was different. Urban land in PBC expanded almost 1.52 times in 2016 than the initial size in 1992. Urban land expansion in BC and MC were comparatively slower than PBC but almost like each other (1.32 and 1.28 respectively) in this period. Urban land expanded rapidly during the period 1992 to 2001 in the Miami MSA and at the county level, especially between 1996–2001 when urban land expansion almost doubled than the previous study year (1992–1996), but decreased dramatically afterward (2001–2016) in this region (Figure 3). In late 1980s, this region experienced a rapid population growth due to industrial and economic development, which generated lavish job opportunities and attracted migrants from other parts of the country. These intra-national migrants along with booming tourism sector resulted thriving housing sector, and hotel and recreational facilities development. All these factors contributed to rapid urbanization in 1990s in the study area. However, population growth started to decline considerably since the beginning of 2000s. The population growth rate was found 23.5% between 1990 and 2000 in this region. However, between 2000 and 2010, this population growth rate declined to 11.4% in this area. Although the rate of population growth slowed down after 2000, it is estimated that population will still continue to increase in this region and therefore urban land expansion will further increase considerably after 2016 in the study area.

The types of urban expansion varied between the study periods in the study area (Section 3.2). While the urban expansion was isolated at the beginning (1992–1996), it became more aggregated and showed clusters of urban lands in the following periods which might be related to the formation of infilling and edge-expansion types of urban growth [26,30]. This compactness in urban expansion continued to increase with the increase of infilling expansion type in the Miami MSA and the three counties (Figures 12 and 13). Interestingly, most of these urban expansions occurred outside of the most prominent and historic city boundaries in the study area (Figure 13). Additionally, major roads of the ending year in each study period (e.g., 2016 roads for period 2011–2016) were used to see if the outlying expansion types were influenced/guided by roads (Figure 12). Figure 12 shows that the outlying expansion types were rather dispersed/scattered in pattern than following or guided by the road networks in each year in Miami MSA. Similarly, most of the rapid and highly rapid urban expansion intensity grids were found between the years 1992 and 2001 (Section 3.3). After 2001, rapid and highly rapid expansion grids decreased dramatically as urban expansion rate slowed down substantially after 2001. Additionally, most of the moderate-highly rapid expansion grids were outside of the major city boundaries which are in line with above results. Urban expansion in the developed region causes a few problems including extreme land price increase, environmental pollution, and rapid rising of living costs which make people move outside of the city area to the suburban region [35] and eventually construct new urban cities/cores. Results from the landscape metrics are consistent with these findings (Section 3.4). The decreasing trend of NP, LSI, and SHAP_AM indicates that urban lands are becoming more aggregated and less complex than the past with their expansion as evidenced in previous studies [26] while continuous increase in LSI represents an increase in the urban center. This also suggests that many non-urban patches initially located in the urban areas might have been transformed into urban patches which are also evident in previous studies [57]. Additionally, a decrease in NP and increase in LPI is an indication of infilling expansion type along with aggregated development [57]. Since infilling expansion type is dominant during the whole study period, along with a continuous decreasing trend in NP and increasing trend in LPI, our results of a more aggregated form of urban expansion process with simplified form of urban lands are reasonable.

Furthermore, urban land expansions are moving away from the east coast of the study area and directed towards the west (Section 3.5), which is evident in mean center of urban lands from 1992–2016 (Figures 9 and 10). Although very slowly, this westward turn of mean center of Miami MSA and the three counties proves our previous hypothesis that urban lands are expanding towards suburb region (west side) of the study area since the east part of the study area are already developed. Being very near to the coastline which poses high recreational value and better transportation access, demands on land parcels are extremely high on the east side of the study region and eventually the land prices are skyrocketing. All these reasons might be responsible for the urban lands expanding toward west part of the study area and away from the east coast. With the increasing conversion of nonurban lands to urban land, it will be extremely difficult to protect the environment and biodiversity in the coming years [14,15,60].

Figure 11. Urban expansion over different time periods in the tri-county region.

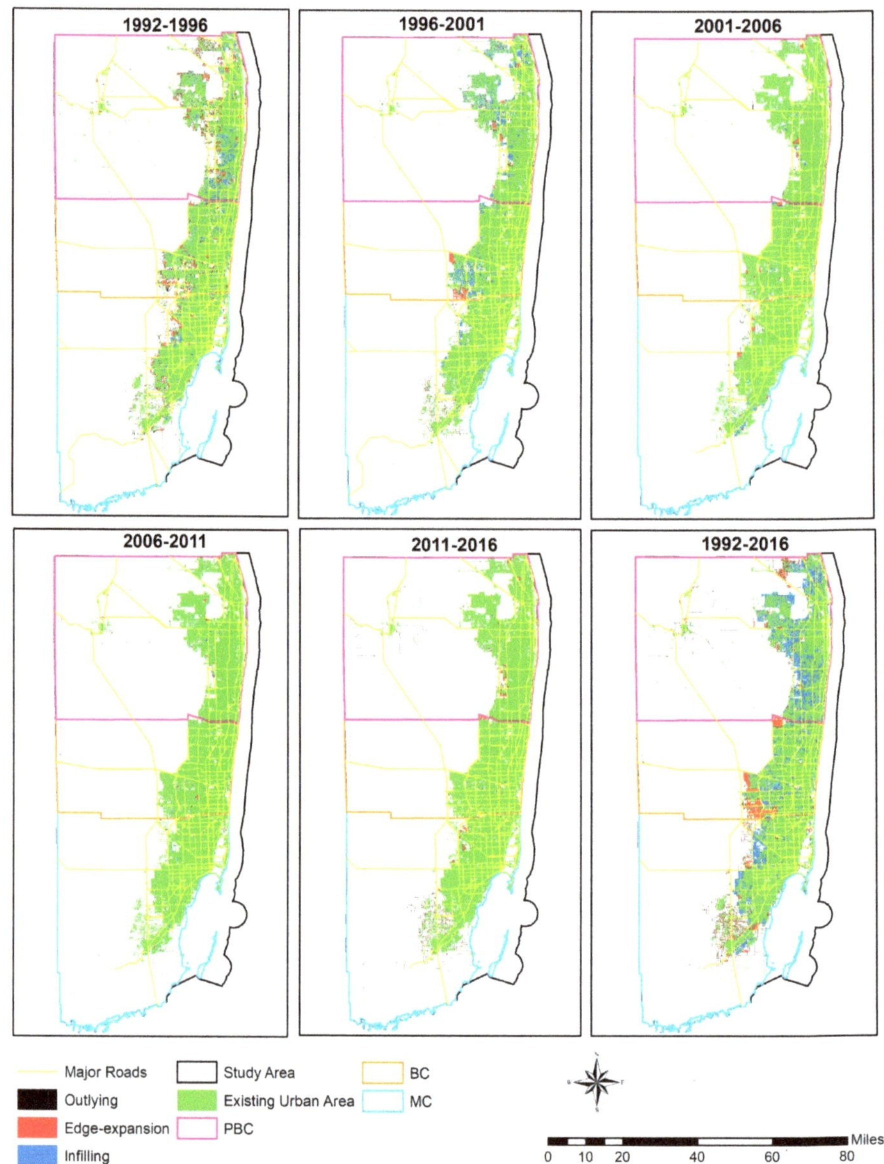

Figure 12. Urban expansion types in different time periods in Miami MSA.

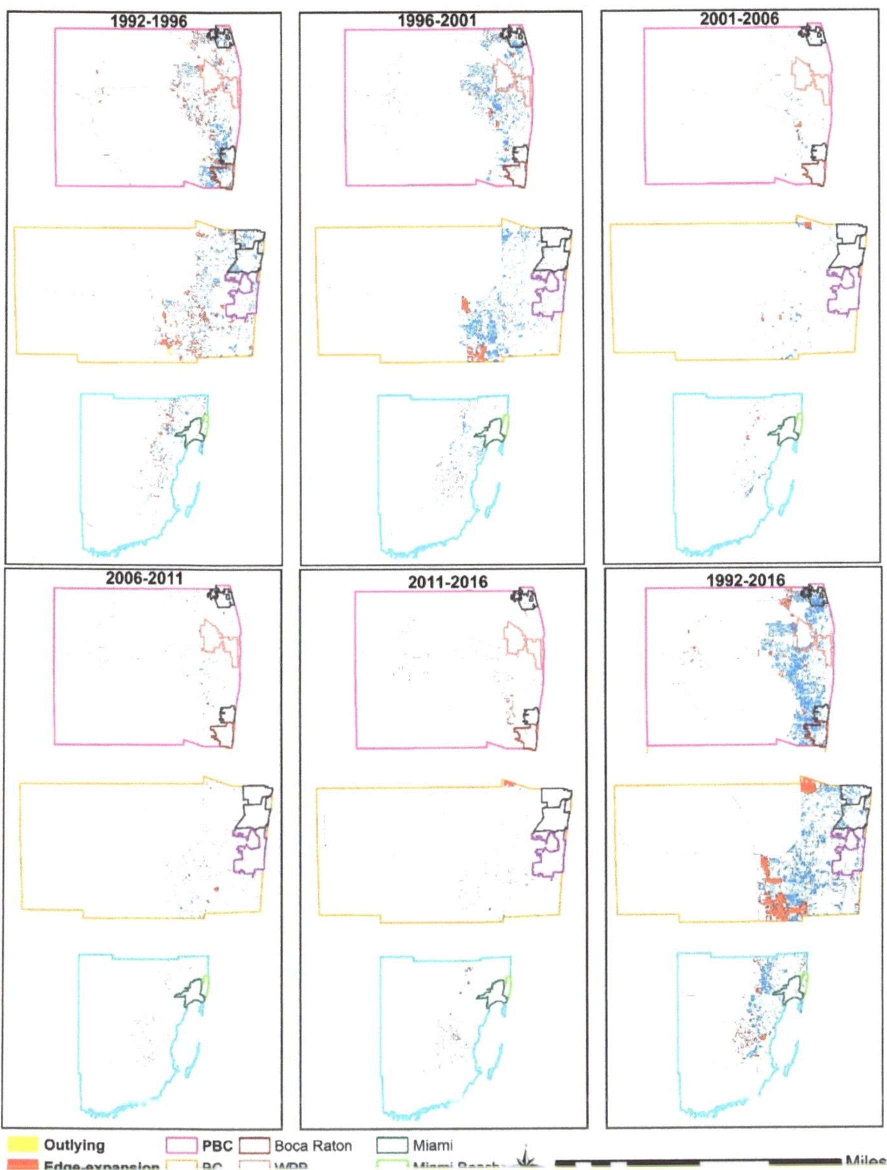

4.2. Major Explanatory Factors of Urban Expansion

Four of the five selected variables of urban expansion had strongly statistically significant relationships with the urban land areas during the study period (Section 3.6). Elevation, which was preliminarily selected as a physical factor of urban land expansion did not have significant relationship with urban land area in any study year (1992–2016). Since the land surface is almost flat throughout the study region and there is not much change in elevation from mean sea level in this region, this result was not surprising. The population had a very strong positive relationship with urban land in all six study years since it is found as one of the most important socio-economic factors driving urban expansion in previous studies [36,46]. Surprisingly, median household income did not have a very strong relationship with urban land expansion except in 2011 and did not even enter the regression models in year 1992, 1996, and 2016. This suggests that median income had very little to do in guiding urban expansion in this region. Unlike developing countries, people here in the US with high median household income tend to live outside of the urban core which might cause a weak statistical relationship with urban lands and median income. However, previous studies found other factors like gross domestic product (GDP), secondary and tertiary industry product, gross output of construction industry as factors of economic development that influence urban expansion [36,46]. However, based on available data and spatial scale of the study area, median household income was selected as one of the socioeconomic factors that influence urban expansion in this study. Distance to Coast and Distance to Roads both had strong negative relationships with urban land in all the years which suggest that urban land area expanded within proximity to major roads and coastal boundary.

Population and Distance to Coast had the strongest relationship with urban land in all the six models followed by Distance to Roads and Median Income except in 1992. In 1992, the population was the least strong variable. This is due to the increase in population in this area. According to US Census Bureau, the total population in Miami MSA was about 4,056,100 in 1990. In 2000, total population increased to 5,007,564 which is about 23.5% increase from 1990. In 2010, total population increased to 5,564,635 which is about 11.1% increase from 2000. In 2018, it is estimated that the total population is 6,198,782 which is 11.4% increase from 2010 in this region.

In previous studies, factors of urban expansion were analyzed either very briefly [61,62] or using only socioeconomic factors [36]. In this research, we quantified the influence of socioeconomic, proximity, and physical factors on urban land expansion. However, these factors explained over 50% of the variation in urban land during the study period (1992–2016) and inclusion of other factors like urban planning and zoning, policies, hydrological factors, neighborhood impacts, etc., might improve the results.

4.3. Limitations of the Study and Future Work

To retrieve information about urban land and urban expansion at a finer scale, high spatial resolution imagery like QuickBird and Lidar could be used in the future. Due to the unavailability of those high spatial resolution imageries for the entire study year (1992–2016) in our study area, we used moderate resolution (30 m) land cover maps in this study. This research is limited to the horizontal expansion of urban lands only. However, previous studies found that urban expansion is not necessarily limited to horizontal dimension and urban areas could also be expanded vertically [63,64]. Investigating the vertical expansion of urban areas in the study area could help understand the urban growth dynamics more profoundly in future. To understand the effects of urban expansion on non-urban lands, analyzing the relationship between urban expansion and the loss of agriculture, protected and reserved areas would be useful for this region. Since the physical factor used in this research does not have any statistically significant relationship with urban land area, other physical factors like soil characteristics, flood risk zones, physical features like schools and entertainment facilities, etc., could be used in future for a comprehensive analysis of influential factors of urban expansion in the study area. Besides, other factors of urban expansion like hydrological factors, policy, and building codes, development control zones, etc. should be used in explanatory factor analysis in future. Additionally, since the study area region has always been a target of hurricanes over the past few decades, analyzing the impact of hurricane damages on urban land expansion in this area could be advantageous.

5. Conclusions

Urban expansion is one of the most irretrievable land transformation processes in the world and poses major impacts on the environment and challenges for the entire world at global, regional, and local scale. There is still a lack of research that focuses on urban expansion process at local scale in the US. Being a developed country, less attention has been paid on the dynamics of urban expansion in this region than necessary and timely monitoring and evaluation of urban land expansion is extremely important.

In this study, we used NLCD data and C-CAP data combined with urban growth type and landscape metrics analysis to quantify the dynamics of the spatiotemporal pattern of urban expansion in the moderately developed Miami MSA. We then further extended our analysis to examine the spatiotemporal patterns of urban growth at the county level and compared the results between the three counties (PBC, BC, and MC) that are located within the MSA area. More specifically, we analyzed and compared the spatial and temporal distribution of urban expansion rate, pattern, expansion types, urban expansion intensity, and landscape metrics. Results suggest that this region experienced a moderate but noticeable urban expansion over the last 25 years. Urban expansion in this region can be divided into two phases: pre-2001 phase when urban expansion rate was very rapid in this MSA and post-2001 phase when urban expansion rate was moderate/slow. Urban expansion at the county level was almost the same as the MSA where the overall annual urban expansion rate in PBC was the highest (16.15 km^2 $year^{-1}$) followed by BC and MC (9.08 and 9.04 km^2 $year^{-1}$, respectively). The standardized annual urban growth rate for PBC, BC, and MC are 1.68%, 1.13%, and 0.97%, respectively, which suggests a moderate urban expansion overall in these counties.

Similarly, infilling was the dominant expansion type in all the study periods in the MSA and in all three counties. Additionally, they all followed the same pattern of decreasing amount of NP and LSI from the beginning of the study period (1992) which suggests that as the urban land expanded and time progressed in this region, they became more aggregated and simplified. Another interesting finding from the analysis was that the newly developed urban lands were generated away from the east coast in this region over the entire study period which was further corroborated by mean center analysis.

As per the cities were concerned, it was evident that the newly developed urban lands were moving away from the major historical and prominent cities in this region. As urban growth creates pressure on the existing population and increases the cost of living and environmental pollution, people might have forced to move away from the city areas to sub-urban areas and eventually created new 'urban cores'. Statistical analysis of the major explanatory factors of urban expansion reveals that Population and Distance to Coast had strongest relationships with urban lands where the former had a positive relationship and latter had a negative relationship. Distance to Roads was another variable that had a strong relationship with urban lands followed by Median Household Income in this region. With continuous population growth and economic development, dealing with urban expansion and environment and biodiversity protection in this region would be perplexing in the coming years. Future research should emphasize the improvement of the spatial and spectral resolution of remote sensing data in order to more accurately portray the urban expansion patterns and should include additional factors of urban expansion including that are used in this research to explain the urban expansion characteristics comprehensively. Finally, the methods and techniques utilized in this research could be applied in future studies focusing on the urban expansion patterns in other metropolitan areas.

Author Contributions: [illegible]

[illegible] comments and suggestions which helped improve the manuscript greatly.

Conflicts of Interest: The authors declare no conflict of interest. The funders had no role in the design of the study; in the collection, analyses, or interpretation of data; in the writing of the manuscript, or in the decision to publish the results.

Appendix A

Table A1. Changes of Urban Land Covers in last 25 years (in km^2) in Miami MSA and three counties.

	Urban Area						**Non-Urban Area**					
	1992	**1996**	**2001**	**2006**	**2011**	**2016**	**1992**	**1996**	**2001**	**2006**	**2011**	**2016**
Miami-MSA	2308.28	2491.09	2973.93	3065.99	3110.55	3167.78	11848.94	11666.13	11183.29	11091.23	11046.67	10989.97
PBC	780.01	874.58	1116.52	1145.85	1163.67	1183.77	4982.09	4887.52	4645.58	4616.25	4598.43	4578.33
BC	699.60	748.83	885.22	906.26	917.99	926.6	2470.23	2421.0	2284.61	2263.57	2251.84	2243.23
MC	828.67	867.19	971.3	1013.0	1027.8	1054.69	4396.62	4358.1	4253.99	4212.29	4197.49	4170.6

References

1. United Nations. *World Urbanization Prospects, the 2018 Revision*; United Nations: New York, NY, USA, 2018; Available online: http://esa.un.org/unpd/wup/index.htm2018 (accessed on 13 July 2019).
2. Neumann, B.; Vafeidis, A.T.; Zimmermann, J.; Nicholls, R.J. Future Coastal Population Growth and Exposure to Sea-Level Rise and Coastal Flooding—A Global Assessment. *PLoS ONE* **2015**, *10*, e0118571. [CrossRef] [PubMed]
3. Zhao, S.; Zhou, D.; Zhu, C.; Qu, W.; Zhao, J.; Sun, Y.; Huang, D.; Wu, W.; Liu, S. Rates and patterns of urban expansion in China's 32 major cities over the past three decades. *Landsc. Ecol.* **2015**, *30*, 1541–1559. [CrossRef]
4. Zhao, S.; Da, L.; Tang, Z.; Fang, H.; Song, K.; Fang, J. Ecological consequences of rapid urban expansion: Shanghai, China. *Front. Ecol. Environ.* **2006**, *4*, 341–346. [CrossRef]
5. Shi, Y.; Li, S. Research on Rational Urban Growth and Land-Use Issues. *J. Urban Plan. Dev.* **2007**, *133*, 91–94. [CrossRef]
6. Grimm, N.B.; Faeth, S.H.; Golubiewski, N.E.; Redman, C.L.; Wu, J.; Bai, X.; Briggs, J.M.; Grimm, N. Global Change and the Ecology of Cities. *Science* **2008**, *319*, 756–760. [CrossRef]
7. Jenerette, G.D.; Potere, D. Global analysis and simulation of land-use change associated with urbanization. *Landsc. Ecol.* **2010**, *25*, 657–670. [CrossRef]
8. Zheng, Z.; Bohong, Z. Study on Spatial Structure of Yangtze River Delta Urban Agglomeration and Its Effects on Urban and Rural Regions. *J. Urban Plan. Dev.* **2012**, *138*, 78–89. [CrossRef]
9. Tan, M.; Li, X.; Xie, H.; Lu, C. Urban land expansion and arable land loss in China—A case study of Beijing–Tianjin–Hebei region. *Land Use Policy* **2005**, *22*, 187–196. [CrossRef]
10. Radeloff, V.C.; Stewart, S.I.; Hawbaker, T.J.; Todd, J.; Gimmi, U.; Pidgeon, A.M.; Flather, C.H.; Hammer, R.B.; Helmers, D.P. Housing growth in and near United States protected areas limits their conservation value. *Proc. Natl. Acad. Sci. USA* **2010**, *107*, 940–945. [CrossRef]
11. Folke, C.; Jansson, A.; Larsson, J.; Costanza, R. Ecosystem appropriation by cities. *Ambio* **1997**, *26*, 167–172.
12. Jenerette, G.D.; Wu, J. Analysis and simulation of land-use change in the central Arizona—Phoenix region, USA. *Landsc. Ecol.* **2001**, *16*, 611–626. [CrossRef]
13. Wu, J.; Jenerette, G.D.; Buyantuyev, A.; Redman, C.L. Quantifying spatiotemporal patterns of urbanization: The case of the two fastest growing metropolitan regions in the United States. *Ecol. Complex.* **2011**, *8*, 1–8. [CrossRef]
14. Seto, K.C.; Gu¨neralp, B.; Hutyra, L.R. Global forecasts of urban expansion to 2030 and direct impacts on biodiversity and carbon pools. *Proc. Natl. Acad. Sci. USA* **2012**, *109*, 16083–16088. [CrossRef] [PubMed]
15. He, C.; Liu, Z.; Tian, J.; Ma, Q. Urban expansion dynamics and natural habitat loss in China: A multiscale landscape perspective. *Glob. Chang. Boil.* **2014**, *20*, 2886–2902. [CrossRef] [PubMed]
16. Shao, M.; Tang, X.; Zhang, Y.; Li, W. City clusters in China: Air and surface water pollution. *Front. Ecol. Environ.* **2006**, *4*, 353–361. [CrossRef]
17. Chen, Y.; Zhang, Z.; Du, S.Q.; Shi, P.J.; Tao, F.L.; Doyle, M. Water quality changes in the world's first special economic zone, Shenzhen, China. [illegible] 2011, 47 [CrossRef]
18. Baker, L.A.; Hope, D.; Xu, Y.; Edmonds, J.; Lauver, L. Nitrogen Balance for the Central Arizona-Phoenix (CAP) Ecosystem. *Ecosystems* **2001**, *4*, 582–602. [CrossRef]

19. Kaye, J.P.; Groffman, P.M.; Grimm, N.B.; Baker, L.A.; Pouyat, R.V. A distinct urban biogeochemistry? *Trends Ecol. Evol.* **2006**, *21*, 192–199. [CrossRef]
20. Trusilova, K.; Jung, M.; Churkina, G.; Karstens, U.; Heimann, M.; Claussen, M. Urbanization Impacts on the Climate in Europe: Numerical Experiments by the PSU–NCAR Mesoscale Model (MM5). *J. Appl. Meteorol. Clim.* **2008**, *47*, 1442–1455. [CrossRef]
21. Yang, X.; Hou, Y.; Chen, B. Observed surface warming induced by urbanization in east China. *J. Geophys. Res. Atoms.* **2011**, *116*, 116. [CrossRef]
22. Walker, R. Urban sprawl and natural areas encroachment: Linking land cover change and economic development in the Florida Everglades. *Ecol. Econ.* **2001**, *37*, 357–369. [CrossRef]
23. Jensen, J.R.; Cowen, D.C. Remote sensing of urban/suburban infrastructure and socio-economic attributes. *Photogramm. Eng. Remote Sens.* **1999**, *65*, 611–622.
24. Herold, M.; Goldstein, N.C.; Clarke, K.C. The spatiotemporal form of urban growth: Measurement, analysis and modeling. *Remote Sens. Environ.* **2003**, *86*, 286–302. [CrossRef]
25. Michishita, R.; Jiang, Z.; Xu, B. Monitoring two decades of urbanization in the Poyang Lake area, China through spectral unmixing. *Remote Sens. Environ.* **2012**, *117*, 3–18. [CrossRef]
26. Sun, Y.; Zhao, S.; Qu, W. Quantifying spatiotemporal patterns of urban expansion in three capital cities in Northeast China over the past three decades using satellite data sets. *Environ. Earth Sci.* **2015**, *73*, 7221–7235. [CrossRef]
27. Turner, M.G.; Gardner, R.H. *Quantitative Methods in Landscape Ecology*; Springer: New York, NY, USA, 1991.
28. McGarigal, K.; Marks, B.J. *FRAGSTATS: Spatial Pattern Analysis Program for Quantifying Landscape Structure*; USDA Forest Service: Washington, DC, USA, 1995; Volume 351.
29. Sudhira, H.; Ramachandra, T.; Jagadish, K. Urban sprawl: Metrics, dynamics and modelling using GIS. *Int. J. Appl. Earth Obs. Geoinf.* **2004**, *5*, 29–39. [CrossRef]
30. Xu, C.; Liu, M.; Zhang, C.; An, S.; Yu, W.; Chen, J.M. The spatiotemporal dynamics of rapid urban growth in the Nanjing metropolitan region of China. *Landsc. Ecol.* **2007**, *22*, 925–937. [CrossRef]
31. Zhao, J.; Zhu, C.; Zhao, S. Comparing the Spatiotemporal Dynamics of Urbanization in Moderately Developed Chinese Cities over the Past Three Decades: Case of Nanjing and Xi'an. *J. Urban Plan. Dev.* **2015**, *141*, 5014029. [CrossRef]
32. Zhao, S.; Zhou, D.; Zhu, C.; Sun, Y.; Wu, W.; Liu, S. Spatial and Temporal Dimensions of Urban Expansion in China. *Environ. Sci. Technol.* **2015**, *49*, 9600–9609. [CrossRef]
33. Ramachandra, T.V.; Aithal, B.H.; Sanna, D.D. Insights to urban dynamics through landscape spatial pattern analysis. *Int. J. Appl. Earth Obs. Geoinf.* **2012**, *18*, 329–343.
34. Li, Q.; Lu, L.; Weng, Q.; Xie, Y.; Guo, H. Monitoring Urban Dynamics in the Southeast U.S.A. Using Time-Series DMSP/OLS Nightlight Imagery. *Remote Sens.* **2016**, *8*, 578. [CrossRef]
35. Shi, G.; Jiang, N.; Li, Y.; He, B. Analysis of the Dynamic Urban Expansion Based on Multi-Sourced Data from 1998 to 2013: A Case Study of Jiangsu Province. *Sustainability* **2018**, *10*, 3467. [CrossRef]
36. Chen, L.; Ren, C.; Zhang, B.; Wang, Z.; Liu, M. Quantifying Urban Land Sprawl and its Driving Forces in Northeast China from 1990 to 2015. *Sustainability* **2018**, *10*, 188. [CrossRef]
37. Yu, W.; Zhou, W. The Spatiotemporal Pattern of Urban Expansion in China: A Comparison Study of Three Urban Megaregions. *Remote Sens.* **2017**, *9*, 45. [CrossRef]
38. Hu, Y.; Jia, G.; Pohl, C.; Feng, Q.; He, Y.; Gao, H.; Xu, R.; Van Genderen, J.; Feng, J. Improved monitoring of urbanization processes in China for regional climate impact assessment. *Environ. Earth Sci.* **2015**, *73*,

[CrossRef]

40. Li, X.M.; Zhou, W.Q.; Ouyang, Z.Y. Forty years of urban expansion in Beijing: What is the relative importance of physical, socioeconomic, and neighborhood factors? *Appl. Geogr.* **2013**, *38*, 1–10. [CrossRef]
41. Tian, G.; Qiao, Z.; Zhang, Y. The investigation of relationship between rural settlement density, size, spatial distribution and its geophysical parameters of China using Landsat TM images. *Ecol. Model.* **2012**, *231*, 25–36. [CrossRef]

42. Luo, J.; Wei, Y.D. Modeling spatial variations of urban growth patterns in Chinese cities: The case of Nanjing. *Landsc. Urban Plan.* **2009**, *91*, 51–64. [CrossRef]
43. Aspinall, R. Modelling land use change with generalized linear models—A multi-model analysis of change between 1860 and 2000 in Gallatin Valley, Montana. *J. Environ. Manag.* **2004**, *72*, 91–103. [CrossRef]
44. Ye, Y.; Zhang, H.; Liu, K.; Wu, Q. Research on the influence of site factors on the expansion of construction land in the Pearl River Delta, China: By using GIS and remote sensing. *Int. J. Appl. Earth Obs. Geoinf.* **2013**, *21*, 366–373. [CrossRef]
45. Vermeiren, K.; Van Rompaey, A.; Loopmans, M.; Serwajja, E.; Mukwaya, P. Urban growth of Kampala, Uganda: Pattern analysis and scenario development. *Landsc. Urban Plan.* **2012**, *106*, 199–206. [CrossRef]
46. Li, G.; Sun, S.; Fang, C. The varying driving forces of urban expansion in China: Insights from a spatial-temporal analysis. *Landsc. Urban Plan.* **2018**, *174*, 63–77. [CrossRef]
47. Vogelmann, J.E.; Howard, S.M.; Yang, L.; Larson, C.R.; Wylie, B.K.; Van Driel, N. Completion of the 1990s National Land Cover Data Set for the Conterminous United States from Landsat Thematic Mapper data and ancillary data sources. *Photogramm. Eng. Remote Sens.* **2001**, *67*, 650–662.
48. Homer, C.; Dewitz, J.; Fry, J.; Coan, M.; Hossain, N.; Larson, C.; Herold, N.; McKerrow, A.; Van Driel, J.N.; Wickham, J.D. Completion of the 2001 National Land Cover Database for the conterminous United States. *Photogramm. Eng. Remote Sens.* **2007**, *73*, 337–341.
49. Mitsova, D.; Shuster, W.; Wang, X. A cellular automata model of land cover change to integrate urban growth with open space conservation. *Landsc. Urban Plan.* **2011**, *99*, 141–153. [CrossRef]
50. Yang, L.; Jin, S.; Danielson, P.; Homer, C.; Gass, L.; Bender, S.M.; Case, A.; Costello, C.; Dewitz, J.; Fry, J.; et al. A new generation of the United States National Land Cover Database: Requirements, research priorities, design, and implementation strategies. *ISPRS J. Photogramm. Remote Sens.* **2018**, *146*, 108–123. [CrossRef]
51. Wickham, J.; Stehman, S.V.; Gass, L.; Dewitz, J.A.; Sorenson, D.G.; Granneman, B.J.; Poss, R.V.; Baer, L.A. Thematic accuracy assessment of the 2011 National Land Cover Database (NLCD). *Remote Sens. Environ.* **2017**, *191*, 328–341. [CrossRef]
52. Wickham, J.; Stehman, S.V.; Gass, L.; Dewitz, J.; Fry, J.A.; Wade, T.G. Accuracy assessment of NLCD 2006 land cover and impervious surface. *Remote Sens. Environ.* **2013**, *130*, 294–304. [CrossRef]
53. Wickham, J.; Stehman, S.; Fry, J.; Smith, J.; Homer, C. Thematic accuracy of the NLCD 2001 land cover for the conterminous United States. *Remote Sens. Environ.* **2010**, *114*, 1286–1296. [CrossRef]
54. Stehman, S.V.; Wickham, J.D.; Smith, J.H.; Yang, L. Thematic accuracy of the 1992 National Land Cover Data (NLCD) for the eastern United States: Statistical methodology and regional results. *Remote Sens. Environ.* **2003**, *86*, 500–516. [CrossRef]
55. Coastal Change Analysis Program (C-CAP) Regional Land Cover. Available online: https://coast.noaa.gov/data/digitalcoast/pdf/ccap-faq-regional.pdf (accessed on 13 October 2019).
56. Manson, S.; Schroeder, J.; Van Riper, D.; Ruggles, S. *IPUMS National Historical Geographic Information System: Version 14.0 [Database]*; IPUMS: Minneapolis, MN, USA, 2019. [CrossRef]
57. Fang, C.; Li, G.; Wang, S. Changing and Differentiated Urban Landscape in China: Spatiotemporal Patterns and Driving Forces. *Environ. Sci. Technol.* **2016**, *50*, 2217–2227. [CrossRef] [PubMed]
58. Hair, J.F., Jr.; Black, W.C.; Babin, B.J.; Anderson, R.E. *Multivariate Data Analysis*, 7th ed.; Pearson: New York, NY, USA, 2010; pp. 151–230.
59. Hossain, M.K.; Meng, Q. A thematic mapping method to assess and analyze potential urban hazards and risks caused by flooding. *Comput. Environ. Urban Syst.* **2020**, *79*, 101417. [CrossRef]
60. Güneralp, B.; Seto, K.C. Futures of global urban expansion: Uncertainties and implications for biodiversity conservation. *Environ. Res. Lett.* **2013**, *8*, 014025. [CrossRef]
61. Li, L.; Lu, D.; Kuang, W. Examining Urban Impervious Surface Distribution and Its Dynamic Change in Hangzhou Metropolis. *Remote Sens.* **2016**, *8*, 265. [CrossRef]
62. Dou, Y.; Liu, Z.; He, C.; Yue, H. Urban Land Extraction Using VIIRS Nighttime Light Data: An Evaluation of Three Popular Methods. *Remote Sens.* **2017**, *9*, 175. [CrossRef]

63. Zhang, W.; Li, W.; Zhang, C.; Ouimet, W.B. Detecting horizontal and vertical urban growth from medium resolution imagery and its relationships with major socioeconomic factors. *Int. J. Remote Sens.* **2017**, *38*, 3704–3734. [CrossRef]
64. Handayani, H.H.; Murayama, Y.; Ranagalage, M.; Liu, F.; Dissanayake, D. Geospatial Analysis of Horizontal and Vertical Urban Expansion Using Multi-Spatial Resolution Data: A Case Study of Surabaya, Indonesia. *Remote Sens.* **2018**, *10*, 1599. [CrossRef]

MDPI
St. Alban-Anlage 66
4052 Basel
Switzerland
Tel. +41 61 683 77 34

Remote Sensing Editorial Office
E-mail: remotesensing@mdpi.com
www.mdpi.com/journal/remotesensing

www.ingramcontent.com/pod-product-compliance
Lightning Source LLC
LaVergne TN
LVHW070202170726
843515LV00007B/1978

* 9 7 8 3 0 3 6 5 2 6 1 3 3 *